D0930536

WASTE MANAGEMENT
and
RESOURCE RECOVERY

Charles R. Rhyner
Professor, Natural and Applied Sciences (Physics)
University of Wisconsin-Green Bay

Leander J. Schwartz
Professor, Natural and Applied Sciences (Biology)
University of Wisconsin-Green Bay

Robert B. Wenger
Professor, Natural and Applied Sciences (Mathematics)
University of Wisconsin-Green Bay

Mary G. Kohrell
Recycling Markets Specialist
University of Wisconsin-Extension

LEWIS PUBLISHERS

Boca Raton   New York   London   Tokyo

Library of Congress Cataloging-in-Publication Data

Waste management and resource recovery / Charles R. Rhyner . . . [et al.].
   p.   cm.
Includes bibliographical references and index.
ISBN 0-87371-572-1
   1. Refuse and refuse disposal.   2. Recycling (waste, etc.).   I. Rhyner, Charles R.
TD791.W23   1995
628.4—dc 20   95-10964

© 1995 by CRC Press, Inc.
Lewis Publishers is an imprint of CRC Press

No claim to original U.S. Government works
International Standard Book Number 0-87371-572-1
Library of Congress Card Number 95-10964
Printed in the United States of America 1 2 3 4 5 6 7 8 9 0
Printed on acid-free paper

We dedicate this book to our spouses—
Lenora, Helen, Lena, and Mark—
who have been most understanding and have
provided much encouragement throughout
this writing project.

Acknowledgments

Material for this textbook had its beginning in a class project in the Practicum in Environmental Sciences course at the University of Wisconsin-Green Bay in 1990. Because of the absence of textbooks on this subject, the students in this course prepared a study guide on waste management. The students who contributed to the study guide were Jacqueline Felder, Kurt Foth, Thomas Novak, Gregory Schmidbauer, and Jodi Trzebiatowski. We thank them for their efforts and contributions.

We give special thanks to John Reindl, Dane County Public Works Department, Madison, WI; Steven M. Grenier, Robert E. Lee and Associates, Green Bay, WI; Harold J. Day, Emeritus Professor, University of Wisconsin-Green Bay; Arthur E. Peterson, Emeritus Professor, University of Wisconsin-Madison; Ronald D. Stieglitz, Professor, University of Wisconsin-Green Bay for reviewing all or portions of the manuscript and providing valuable suggestions for its improvement.

Illustrations for the book were prepared by Michael J. Christopherson.

Charles R. Rhyner is Professor of Natural and Applied Sciences (Physics) at the University of Wisconsin-Green Bay. Dr. Rhyner earned his B.S., M.S., and Ph.D. degrees from the Department of Physics at the University of Wisconsin-Madison. He was Assistant Professor at the University of Wisconsin-Kenosha Center in 1967, then joined the University of Wisconsin-Green Bay faculty in 1968. Dr. Rhyner served as Director of Graduate Studies from 1978 to 1982 and Chair of Physics from 1987 to 1993.

Dr. Rhyner's involvement in solid waste management began in 1970 as a member of the Brown County Solid Waste Committee. He was subsequently appointed to the newly formed Brown County Solid Waste Board in 1973 where he served as vice-chair from 1974 to 1991, and chair from 1992 to present. He was a research scientist at Oak Ridge National Laboratory working on a multicounty solid waste project in the summer of 1972, and was a member of the Wisconsin Legislative Council Committee on Solid Waste and Hazardous Waste Management from 1976 to 1978. He has taught waste management courses with Drs. Schwartz and Wenger since the early 1970s.

Dr. Rhyner is a member of the American Physical Society, the American Association of Physics Teachers, the Wisconsin Association of Physics Teachers, the honorary society, Sigma Xi, and the Wisconsin Counties Solid Waste Management Association. He served as vice-president of the Wisconsin Association of Physics Teachers in 1981 and president in 1982.

Dr. Rhyner is the author of over 20 papers. He has also written computer software for weighing vehicles at disposal facilities and contributed to the development of the Wisconsin Recycling Markets Database for use by University of Wisconsin Extension field agents.

Leander J. Schwartz is Professor of Natural and Applied Sciences (Microbiology) at the University of Wisconsin-Green Bay. Dr. Schwartz received his B.S. degree from Wisconsin State University, Platteville, and his M.S. and Ph.D. degrees from the Department of Botany, University of Wisconsin, Madison. He served as an Assistant Professor from 1963 to 1969 and Associate Professor and Dean from 1969 to 1972 at the University of Wisconsin Fox Valley Center (Menasha, WI). Dr. Schwartz joined the faculty of the University of Wisconsin-Green Bay in 1972 and served as Chair of Biology from 1972 to 1975, Associate Vice Chancellor for Academic Affairs from 1982 to 1986, and Dean of Natural Sciences and Mathematics from 1986 to 1988.

Dr. Schwartz is the author of 18 papers and reports about laboratory investigations on biological waste treatment. Additionally, he has made numerous presentations at scientific meetings. He has received grants from industry and wastewater treatment districts to support his work in biological waste treatment. His continuing major research interest focuses on anaerobic treatment of industrial waste streams and bioremediation of soils contaminated with petroleum fuel.

Dr. Schwartz is a member of the American Association for the Advancement of Science, the American Society for Microbiology and the honorary society, Sigma Xi.

Robert B. Wenger is a member of the Department of Natural and Applied Sciences at the University of Wisconsin-Green Bay. All of Dr. Wenger's academic degrees are in mathematics: the B.S. degree from Eastern Mennonite College in 1958, the M.A. degree from Pennsylvania State University in 1962, and the Ph.D. degree from the University of Pittsburgh in 1969. Before joining the faculty at the University of Wisconsin-Green Bay in 1969, he was an Instructor and Assistant Professor of Mathematics at the McKeesport Campus of the Pennsylvania State University from 1961 to 1969. At the University of Wisconsin-Green Bay, Dr. Wenger has served two terms as chair of the Natural and Applied Sciences Department, a department which offers a major in environmental science and has had a strong tradition of interdisciplinary research in this field. Dr. Wenger has held visiting professor appointments at the University of Aalborg in Denmark and at Beijing Normal University in the People's Republic of China.

Dr. Wenger's major research interest is the application of mathematical models to environmental problems, including problems in the solid waste management field. He has worked on solid waste management problems in the United States, Denmark, and the People's Republic of China. He has published several papers on topics in solid waste management, most of which are joint publications with colleagues from the United States and abroad, with a variety of disciplinary backgrounds.

Dr. Wenger is a member of the Mathematical Association of America, the Institute for Operations Research and the Management Sciences, the International Association for Mathematical and Computer Modelling, the American Association for the Advancement of Science, the honorary society Sigma Xi, and the Society for Values in Higher Education.

Mary G. Kohrell is a recycling specialist with the University of Wisconsin Extension, and a lecturer in the Department of Natural and Applied Sciences at the University of Wisconsin-Green Bay.

Ms. Kohrell received her B.S. degree in geography from Carroll College, Waukesa, Wisconsin and her M.A. degree in geography and environmental planning from the University of Nebraska, Lincoln. She worked for nonprofit recycling organizations in Nebraska and New Hampshire, and served as a recycling consultant in the eastern U.S.before assuming her present position in 1991.

Ms. Kohrell is a member of the National Recycling Coalition and the Associated Recyclers of Wisconsin. She has been the recipient of research grants from the University of Wisconsin Solid Waste Research Council, the U.S. Environmental Protection Agency, and the U.S. Department of Agriculture. She has provided education and technical assistance on recycling issues to over 10,000 communities, businesses, and industries in Wisconsin and across the United States. Her current research interests include methods to maximize recycling efficiencies and local recycling market development.

Preface

In the latter part of the twentieth century waste management issues are a major concern at the local, national, and international levels. The "out-of-sight, out-of-mind" approach to waste problems which was predominant at the beginning of the century has been replaced in the second half by a number of new initiatives designed to enhance environmental protection. Governmental regulations have been put in place to provide protection to our air and water resources. The discharge of untreated wastewater and the indiscriminant dumping or burning of wastes, common practices in many communities and industries in the past, have been replaced with new wastewater treatment facilities and sanitary landfills. Despite the improvements brought about by stricter regulations and improved technologies, many challenges remain as we approach the end of the century. Landfills are reaching capacity, treatment and disposal costs are rising, and effective recycling programs need to be implemented. Finding sites for new processing and disposal facilities in the face of local citizen opposition is also a continuing challenge.

The aim of this book is to provide a thorough understanding of the waste management problems and issues faced by modern society. Scientific, technical, and environmental principles are emphasized to aid in understanding how waste management processes work and the nature of impacts resulting from waste dispersal and disposal in the environment. The establishment of effective waste management policies and efficient waste management systems requires insights from the social sciences as well. Therefore economic, social, and regulatory aspects of waste management are also addressed in the volume.

The topics contained in this book are those which have been included in the Waste Management and Resource Recovery course taught at the University of Wisconsin-Green Bay since 1977. One of the challenges in teaching this course has been a dearth of suitable textbooks. Most of the available books on waste management are research monographs, handbooks consisting of chapters written by different authors, or nontechnical books written for the general public. In addition, wastewater treatment and biological methods are generally omitted from solid waste management books, even though these treatment methods generate sludges that are handled as solid waste.

This book is designed to address the deficiences described above and to provide a suitable textbook for students in upper level undergraduate and beginning graduate courses in waste management. It includes examples and problem sets. The level of exposition assumes that students have completed introductory courses in the physical and biological sciences. We have tried to provide accurate, authoritative, and current information that will benefit students and be useful to practitioners in the waste management field.

Standard International (SI) units are used throughout the book. For the convenience of practitioners in the United States who customarily use English units, the equivalent quantities are given in parentheses, or conversion factors are provided. A table of conversion factors is included in Appendix A. In the text, we refer to kilograms as a "weight" in displaying data derived by weighing materials on a scale. Proper use of SI units would dictate that we express weight, a force, in units of Newtons and use kilograms as a mass.

Table of Contents

CHAPTER 5. PROCESSING SOLID WASTES AND RECYCLABLE MATERIALS

CHAPTER 7. BIOLOGICAL TREATMENT OF WASTE SOLIDS

CHAPTER 12. HAZARDOUS WASTE

CHAPTER 13. COSTS AND MANAGEMENT OF WASTE FACILITIES AND SYSTEMS

CHAPTER 1

Waste Management and Resource Conservation

1.1 INTRODUCTION

Waste is material perceived to have little or no value by society's producers or consumers. Nearly all human activities produce waste. Solid wastes are generated during the acquisition of raw materials, during refining and manufacturing processes, and when products are used by consumers. Large quantities result from agricultural and mining operations, and in the form of residues from wastewater treatment and electrical power generation. In some cases wastes are hazardous and require special handling or treatment in order to prevent serious harm to humans or ecosystems. In addition to solid waste, human activities generate liquid and gaseous by-products which often exceed the assimilative capacity of the natural environment. No matter what type of waste is being considered, individuals, organizations, and governmental bodies have a responsibility to find ways to minimize waste generation, control harmful waste emissions, recover material and energy resources from the waste stream, and dispose of waste in a manner which protects human health and minimizes environmental degradation.

This textbook focuses on a portion of the overall problem of managing society's wastes by providing an integrated approach to the subjects of solid waste management and wastewater treatment. A unified approach which includes a study of all forms of waste—solid, liquid, and gaseous —is theoretically appealing but beyond the scope of this writing project. A practical justification for giving priority to solid waste management and wastewater treatment is that many other pollution problems tend to be reduced to one or both of these domains. Thus, for example, an electrostatic precipitator may be very successful in reducing the dispersal of airborne pollutants but this very success adds to the problem of solid waste management because of the accumulation of fly ash, a solid material which is often landfilled.

There is also a strong interplay between the fields of solid waste management and wastewater treatment. Wastewater treatment processes generate large quantities of sludge, a semisolid or solid material which is sometimes handled within solid waste management systems. Sanitary landfills often generate leachate which must be treated in wastewater treatment facilities. It is also to be noted that there are processes—composting and incineration are examples—which are utilized in both solid waste systems and wastewater treatment operations.

The study of solid waste management and wastewater treatment processes incorporates scientific, technical, and environmental principles. All are emphasized in this book along with economic, social, political, and legal aspects. Scientific concepts are needed in order to understand how waste management and treatment processes work and the nature of impacts resulting from waste disposal and dispersal in the environment. However, scientific and technical insights by themselves are insufficient because the establishment of effective and efficient waste management policies requires knowledge from the social sciences, including economics, political science, and law.

Waste management problems are high on the list of environmental concerns of the general public, particularly for people in the developed countries. In the United States there have been a number of incidents in recent years which have created a heightened public awareness of waste disposal problems and issues. One of the most famous is the Love Canal episode in Niagara Falls, New York, where several decades ago the Hooker Electrochemical Corporation dumped 19,800 tonnes of hazardous waste in a 6.5 hectare site. A school and a housing subdivision were later built on and adjacent to the site. In the late 1970s when hazardous wastes began seeping into basements of buildings constructed in the subdivision, over 900 families were evacuated from the area. Federal agencies incurred over $140 million in cleanup costs and the lives of many individuals were disrupted and threatened with real and potential health problems.

Other famous cases where industrial wastes have badly polluted groundwater or surface water (Blackman, 1993) include the Hardeman County, Tennessee, landfill where wells of 40 families were polluted with the chemicals endrin, dieldrin, aldrin, and heptachor; the LaBounty Dump near Charles City, Iowa, where arsenic and organic chemicals leached from the dump into the groundwater and the Cedar River; the Life Sciences Products operation in Hopewell, Virginia, which contaminated the sediments in the James River and Chesapeake Bay with kepone; and the Reilly Tar and Chemical Corporation discharges of contaminated wastewaters into wetlands. Industrial wastes are not the only problem, however. In Madison, Wisconsin, seepage of methane from a municipal land-fill into basements in an adjoining residential area caused an explosion and destruction of a house in 1983 (O'Leary et al., 1988). Two tenants were burned in the explosion.

Most current discussions, however, do not focus on localized dra-matic episodes of the type listed above, but upon the burgeoning amounts of waste arising from what is often referred to as the "throw-away" soci-eties existing in the developed countries. Televised images of barges or railroad cars laden with garbage traveling circuitous routes over a period of many days in search of a disposal site dramatize this issue. Problems faced by cities and local communities in finding alternatives to landfills which have reached capacity, or are close to it, have forced citizens to confront waste management issues on a daily basis. Where once it was possible to throw all household wastes into a single container and carry it to curbside for pickup, now, in the interest of saving landfill space, resi-dents are often required to segregate wastes into two or more containers for recycling. In addition, some materials—yard waste is an example—are banned from disposal and are not collected from households in many states.

Furthermore, costs of waste disposal have risen greatly in recent years. For example, local governments which only a few years ago may have paid $10 or $15 per tonne for the disposal of solid waste at a land-fill, may now be paying as much as $75 or $100 per tonne. Other commu-nities are making major financial commitments to cover the cost of new solid waste processing facilities. Collection costs have also increased due to the more complex and labor intensive systems required to collect an array of materials which are segregated from each other. Despite inconve-niences and higher costs, citizens often willingly support recycling activi-ties in the interest of conserving resources and improving environmental quality. This gives rise to the additional problem of finding and creating markets for products manufactured from recycled materials.

State governments and the federal government have responded to cit-izen concerns about waste management problems by assuming a regula-tory role and providing financial assistance for the implementation of im-proved waste management and wastewater treatment systems. For example, since the beginning of the "Environmental Decade" in the 1970s the United States Government has enacted the Clean Water Act (CWA); the Clean Air Act (CAA); the Toxic Substances Control Act (TSCA); the Safe Drinking Water Act (SDWA); the Resource Conservation and Re-covery Act (RCRA); and the Comprehensive Environmental Response, Compensation, and Liability Act (CERCLA), also known as "Superfund." All have resulted in new regulations or, through financial incentives, stimulated improvements in waste management or wastewater treatment practices with the goal of achieving a cleaner environment.

Despite the fact that some gains have occurred in protecting or improving environmental quality, many challenges remain in the waste management field. Individuals, businesses, industries, and governmental bodies will be required to make strategic and difficult choices in order to maintain or, in many cases, restore ecosystem health. With a rapidly growing global population which many people predict will reach 8 to 10 billion by the year 2025, decisions leading to sound waste management practices and wise use of resources will be crucial.

1.2 HISTORICAL SKETCH OF WASTE MANAGEMENT

Problems of waste management have existed ever since humans made the transition from hunting and gathering societies to settled communities. In early references to problems associated with waste generated by humans, the primary concern seems to have been with the nuisance factor and its potential impact on health. Wastes close at hand were unsightly, filthy, and foul-smelling, thereby bringing discomfort and inconvenience. Technological innovations in the construction of houses designed to remove garbage and human wastes from the immediate presence of the household appeared to have occurred in India, Egypt, and China as early as the third and second centuries B.C. According to Melosi (1981), the first municipal dumps were established somewhat later in Athens, when about 500 B.C. "the Council of Athens began requiring scavengers to dispose of wastes no less than a mile from the city walls." Throughout much of the medieval period an "out of sight, out of mind" approach to the problems of human waste were very much in evidence. For example, in about 1400 A.D. a regulation was enacted in Paris which stated that those who brought a cart of sand or gravel into the city were to leave with a load of mud or refuse (Melosi, 1981).

The Industrial Revolution, which brought with it the crowding of large numbers of people into cities, first in Europe and then somewhat later in the United States, gave rise to pollution problems on an unprecedented scale. Smoke from factory chimneys, the discharge of industrial wastes into rivers and streams, and the piling up of garbage and other types of solid waste in vacant lots, alleys, and street corners created an unsightly and filthy mess. But a deeper level of concern arose as people gradually became aware of the connection between communicable diseases and the wastes which were so prevalent in their environment.

One of the most significant advances in understanding the connection between wastes and diseases occurred as a result of a classic study conducted by John Snow, a London physician (Rosenberg, 1962). Dr. Snow's theory was that the cause of the cholera epidemics which occurred in London in the mid-nineteenth century was a poison found in the excreta and vomitus of cholera patients. When a new epidemic arose in 1854 he was prepared to test his theory based on the sources of Londoners' water supply. Water was supplied to the residents of London by two companies, the Lambeth and the Southwark and Vauxhall. The Southwark and Vauxhall company drew its water supply from the lower Thames after it had been contaminated with London sewage, while Lambeth's came from the upper Thames. Snow was able to show that cholera occurred far more frequently among the users of the water supplied by the Southwark and Vauxhall company. Snow came to his remarkable conclusion before the germ theory of disease had been fully developed, and approximately 30 years before the bacterial organism which causes cholera was isolated and shown to be the specific cause of this food and waterborne disease. As the germ theory of disease came to be more fully understood it stimulated measures to improve public health.

Even the affluent were unable to escape the ravages of the underside of the Industrial Revolution. Since individuals could no longer take actions on their own to escape nuisances and disease, collective approaches were required and pressures mounted for the development of sanitation services. The late nineteenth and early twentieth centuries witnessed the appearance of several types of municipal sanitation services, including garbage collection, street cleaning, and wastewater collection and treatment.

However, even after the institution of garbage collection services, waste disposal methods remained rudimentary and consisted mainly of indiscriminate discharging of wastes in town dumps, on open land, and sometimes into watercourses. In a few cases solid waste was used as fill in roadbeds, or the organic portion was turned into fertilizer or animal feed. Similarly, the first sewerage systems were introduced as a means of collecting and transporting wastewater, but they lacked any form of treatment and the contents were discharged directly into the nearest watercourse. Vesilind and Peirce (1982) report that the first wastewater treatment process consisted of screening for the removal of floatables which were buried or incinerated. The first complete treatment systems were operational by the turn of the century and included land spraying of effluent.

As the inadequacies of waste disposal began to be recognized, interest developed in new forms of waste handling. Foremost was the attention given to the "cremation" or burning of waste. The first systematic "cremators" or incinerators of solid waste were developed in England in the 1870s (Melosi, 1981). Following the English initiative a large number of incineration devices were constructed throughout Europe. A short time later incinerators were constructed in the United States, but, according to Melosi (1981), much of the European technology was adopted impulsively and many of the early furnaces were abandoned as failures after a few years.

During the twentieth century, the economic development that has occurred in Western countries—and in a number of others—has brought
about significant changes in the quantity and composition of solid waste.
Ashes from coal-fired furnaces constituted a major portion of municipal
solid waste in the first part of the century. With the arrival of furnaces
using natural gas or oil as fuels, ashes now comprise only a small fraction
of the municipal solid waste stream. When comparing the portion of solid
waste remaining when ashes are excluded, food discards comprise a much
smaller portion of municipal solid waste than formerly, and paper, plastics, metals, and other materials of modern society have become much
more prevalent in the waste stream. This century has also witnessed a
changing perspective in waste management, brought about primarily as a
result of the advent of the environmental movement in the 1970s. This
new viewpoint holds that it is not sufficient to manage wastes out of concern for the nuisances they create or even their potentially deleterious impacts on human health; consideration must also be given to the broader
concerns of ecological integrity. In short, we are witnessing a movement
away from narrow concerns with human health to the larger context of
ecosystem health. Interwoven with these ecological issues are concerns
arising from the depletion of natural resources and the need for resource
conservation. Hence, resource recovery is now firmly entrenched as a
component of waste management and planning.

Changes in the nature of wastes, concern for environmental protection, and the desire to recover resources from the waste stream have stimulated the development of new waste management technologies and
processes. Carefully engineered sanitary landfills have replaced open
dumps, incinerator technology has been greatly improved over the initial
designs of a century ago and now includes the possibility of heat recovery, and large scale systems have been developed to recover a variety of
materials from municipal solid waste. Wastewater treatment systems now
include secondary and tertiary treatment, in addition to primary treatment.
As will be discussed in more detail later, current approaches to waste
management also include waste minimization, a concept aimed at reducing the amount of material entering the waste stream through improved
product designs and manufacturing processes.

The scientific method and professional expertise have also become a
part of the waste management scene in the twentieth century. It is to be
expected that techniques which have been instrumental in the achieve-
ments attained in the industrial and business realms would eventually be
applied to problems in the public sector. Persons with engineering and
scientific skills are found in public works departments in municipal governments and in regulatory agencies at the state and federal levels. University faculty and staff members, employing a variety of advanced skills,
conduct research in the waste management field. Mathematical modeling

techniques from the fields of management science and operations research have been applied to management and planning issues in solid waste and wastewater treatment systems. The social sciences, as discussed in the next section, are utilized to help gain insights into the nature of the waste stream and the manner in which the complex behaviors of individuals influence it.

1.3 THE ARCHAEOLOGY OF REFUSE

A meaningful characterization of society can be made by examining what it throws away. Many artifacts that archaeologists have used to study past civilizations (for example, tools, weapons, cooking pots, and utensils) have been extracted from old or ancient dumps. Based on the evidence provided by the recovered tools and materials, insights are gained into societal activities such as hunting, fishing, and farming, as well as the manner in which people once went about their daily lives.

Since 1973, Dr. William Rathje (Rathje and Murphy, 1992) and others associated with the University of Arizona Garbage Project have been applying archaeological methods in analyzing the materials contained in municipal solid waste and in old landfills at several sites throughout the United States to obtain information about purchasing, consumption, and disposal habits of persons in contemporary society. In the initial phase of the project, collected trash from households was sorted into 150 specifically coded categories. Later, landfilled waste was sampled and sorted. Since the project's inception, over 15,000 household refuse samples have been collected, sorted, and analyzed. The information from these studies has been stored in a computer database. Included in the database is information about householders' purchasing habits, the percentages of newspapers, bottles, and cans that are recycled, amounts of hazardous waste included in household discards, and the quantity of edible food that is thrown away.

Findings from these studies include the following:

- American families waste between 10% and 15% of the food they buy, with fresh produce accounting for 35% to 40% of the edible food discards. From their study of food wastes, Rathje and his fellow researchers have formulated the First Principle of Food Waste: "The more repetitive your diet—the more you eat the same things day after day—the less food you waste." Support for this principle is derived from several informational sources. A curious piece of evidence is the fact that an increase in the rate of food waste often occurs during food shortages, such as the beef shortages in 1973 and 1974 and the sugar shortage in

1975. In the case of the former, food waste from beef tripled during the shortage. The researchers hypothesized that the increase was caused by crisis buying of less familiar cuts of beef, and a lack of knowledge about storage requirements of meat for long time periods. Collaborative evidence for this principle was obtained from the city of Tucson, Arizona, during the sugar shortage when it was noted that many people stocked up on less-refined Mexican sugar, only to find that it turned hard and had to be thrown out. There is other evidence supporting this principle as well. For example, it was observed that only small amounts of regular bread, a staple of daily diets, are wasted, while specialty breads and buns that are purchased less regularly are discarded at rates of 30% to 60%.

- Lower-income families consistently buy small-sized packages, whereas more affluent families buy larger economy sizes. Thus, the ratio of packaging material to product is larger for those in the low income brackets.

- Although the proportion of hazardous waste is about the same in the trash generated from households in different socioeconomic categories, the content is quite different. The hazardous waste from low income households is primarily related to car care (for example, motor oil and gasoline additives), while that from middle-income households tends to come from home and lawn care products. In 1986, the Garbage Project researchers investigated the impact of a special collection day for household hazardous waste upon the contents of the waste picked up during the regular collection service. To their surprise they found that more than twice as much household hazardous waste appeared in the trash after the special collection day than before. An explanation of this phenomenon, which has also been confirmed in other studies, is that a heightened public awareness resulting from a media campaign stimulates the gathering up of household hazardous wastes. However, many households miss the special collection day and then proceed to dispose of the materials along with other household wastes during regular collection day pickup. Thus, instead of meeting the goal of reducing the quantity of hazardous waste entering a landfill, just the opposite tends to occur. If this explanation is correct, it provides an argument for collecting household hazardous waste in a regular ongoing basis.

- Self-reported personal behavior and perceptions about personal household trash generally do not correspond to what people actually do. The tendency is to report in a manner which enhances one's personal image. For example, consumption levels of healthy foods, such as fresh fruits and vegetables and low-calorie products, tend to be overreported, while those for un-healthy foods, such as snack items, sugar, and fatty meats, is often underreported. In addition, the level of household recy-cling activity tends to be overstated.

Excavations of landfills by the Garbage Project researchers have pro-vided information concerning the changes occurring in waste composition over time. The results of this research have also challenged some com-monly held beliefs about decomposition rates and the fate of solid waste once it is placed in a landfill. Observations from the portion of the study dealing with landfill excavations include:

- While the number of plastic items has increased since the 1970s, the proportion of landfill volume occupied by plastics has not changed appreciably. The explanation appears to be the trend to-ward manufacturing items with less raw material. An example is the reduction in mass of the high density polyethylene (HDPE) one gallon milk container. Twenty-five years ago its mass was 120 g compared to 65 g at present, a reduction of approximately 46%. Once compacted in a landfill, the current container takes up less space than did its counterpart of twenty-five years ago. Overall, in the landfills examined by Garbage Project re-searchers, the volume of all plastics from the 1980s occupied approximately 16% of the total volume. It should be noted that this is less than the estimate of 21% derived from the 1990 data reported by the U.S. Environmental Protection Agency (USEPA) (USEPA, 1992) and shown in Table 2.14 in Chapter 2.
- The degree and rates of biodegradation of organic materials in a landfill are often assumed to be higher than they actually are. Many people are surprised to learn that even after two decades in a landfill about one-third to one-half of the nonplastic organic waste remains recognizable. The modern landfill constructed ac-cording to USEPA standards minimizes moisture, thereby dis-couraging biodegradation. Therefore, arguments in support of a given waste management alternative which are based on the as-sumption of rapid degradation of organic materials in a landfill may lack credence. However, decomposition which does occur, slow as it may be, can give rise to very real practical problems arising from gas generation, odor, settlement, and leachate.
- The public often perceives certain items to be major contributors to landfill volumes when in reality they may account for only a small fraction of the total space. For example, disposable dia-pers account for less than 2% of the space taken up in a typical municipal landfill. Also, polystyrene foam containers, such as

fast-food packaging, meat trays, and egg cartons, are responsible for less than 1% of landfill volume. Paper is the largest occupier of landfill space, accounting for over 40% of the volume in a typical municipal landfill.

Studies such as those by Rathje and his co-workers not only provide insights into the nature and characteristics of a given society but also generate information which is important in making solid waste management decisions. They contribute to our understanding of how much waste is produced and its composition. They also illustrate the complexities of human behavior involved in making individual purchasing decisions and in deciding how to handle household wastes. These types of understandings are necessary in order to design effective waste management policies within a holistic context that take into account resource conservation issues and ecosystem integrity.

1.4 RESOURCE CONSERVATION

Worldwide population growth, urbanization, technological development, and growth in economic activity generate large quantities of waste, but they also place great pressures on the finite material and energy resources of the earth. At issue is our stewardship of the earth's resources and our obligation to consider resource needs of future generations and the sustainability of the earth's ecosystem. Since society's waste streams contain material and energy resources, waste management decision-making must be tied inextricably to resource conservation and utilization issues.

Although proponents of recycling often focus on the importance of conserving materials, the conservation of energy is also important. In some cases consumers who wish to make wise decisions on material and energy conservation have clear choices, but in others the appropriate decision can be baffling and complex. However, careful analyses can often provide insights into the issues involved.

An example where conventional wisdom is supported by the results of detailed analysis is the issue of multiple-use returnable bottles versus single-use throwaway bottles. A returnable bottle can be filled many times. Hannon (1972) estimates that eight refillings are possible; others have estimated that as many as 50 refillings may be realistic for some containers. Whatever the number, the use of refillable containers represents a savings in both raw materials and energy because if a returnable container can be refilled N times it is equivalent to N throwaway containers of the same size. Hannon (1972) tabulated the total energy required to deliver beverages in returnable bottles and other nonreturnable containers. The total energy included the energy required to extract the necessary raw materials, manufacture the containers, fill the containers with beverages, transport the filled containers to market, return refillable containers for cleaning and refilling, and landfill the throwaway containers and discarded returnable containers when their useful lifetime had expired. His detailed analysis showed that approximately three times as much energy is required to deliver beverages in throwaway glass bottles compared to returnable glass bottles when the returnable bottles are filled eight times. It should be noted that air and water pollution considerations were not included in Hannon's analysis. In Problem 8 at the end of this chapter, data are given which enable the reader to calculate the comparative energy requirements for the single- and multiple-use alternatives of beverage bottles and verify Hannon's conclusions.

A more recent evaluation of the energy and environmental impacts for the delivery of soft drinks in selected containers was prepared by Sellers and Sellers (1989). Table 1.1 shows the energy consumption, atmospheric emissions, waterborne wastes, and the amount of solid waste resulting from the delivery of one liter of soft drinks in various types of soft drink containers at 1987 recycling rates (i.e., aluminum cans, 50%; polyethylene terephthalate (PET) bottles, 20%; and glass bottles, 10%). When interpreting these numbers, it is important to recognize that many of them would likely be even smaller if the analysis were based on the situation today rather than 1987. One reason is the manufacturing trend toward decreasing the masses (i.e., "lightweighting") of both returnable and nonreturnable containers. A second reason is the significant increase in the recycling rates for aluminum, plastic, and glass. The data show that refillable glass bottles are preferable from an environmental standpoint— lowest energy consumption, lowest air and water emissions, and a solid waste disposal volume only slightly more than aluminum cans and PET containers. However, the trend over the last three decades has been toward the use of the more convenient nonrefillable cans and bottles.

A second example illustrates the complexity of issues related to waste management and resource conservation. In the late 1970s, fast food restaurants shifted from waxed paper and cardboard food wrapping and packaging materials to polystyrene foam clamshells and cups because they keep food warmer and provide resistance to greases contained in food items. Environmental groups protested the use of polystyrene foam material because, among other reasons, it was produced using chlorofluorocarbons (CFCs) and it is not biodegradable. When the impact of CFCs on the ozone layer became known, other foaming agents which are less harmful to the atmospheric environment were substituted in their place in the manufacture of polystyrene foam. Still, concerns about the impacts of the polystyrene materials in landfills and the high cost of recycling them, as well as broader resource conservation issues, remained.

Table 1.1 Summary of the Environmental Impacts for the Delivery of 1 Liter of Soft Drinks in Containers Manufactured at the 1987 Recycling Rates

Type of Container	Energy (kJ/L)	Air Emissions (g/L)	Waterborne Wastes (g/L)	Solid Waste Mass (kg)	Solid Waste Volume (x 10^{-4} m^3)
0.47 L (16 oz) refillable glass bottle (8 trips)	4290	6.5	1.0	0.18	2.2
0.47 L (16 oz) nonrefillable glass bottle	9700	18.2	2.0	0.52	6.7
1 L nonrefillable glass bottle	10,200	19.8	2.1	0.59	7.5
0.29 L (10 oz) nonrefillable glass bottle	11,600	22.0	2.4	0.63	8.1
0.47 L (16 oz) PET bottle	8810	11.1	1.9	0.10	3.4
1 L PET bottle	7110	8.9	1.6	0.07	2.6
2 L PET bottle	5270	6.7	1.2	0.05	1.8
3 L PET bottle	5180	6.5	1.2	0.05	1.7
0.35 L (12 oz) aluminum can	9170	11.0	3.2	0.63	1.6

Conversions: kJ/L × 3.558 = BTU/gal
g/L × 0.008344 = lb/gal
kg × 2.205 = lb
m^3 × 35.31 = ft^3

Source: Adapted from Franklin Associates, Ltd. *Comparative Energy and Environmental Impacts for Soft Drink Delivery Systems*, a report prepared for The National Association for Plastic Container Recovery (NAPCOR), Charlotte, NC, March 1989.

Some insights into these issues can be gained from a recent analysis conducted by Hocking (Nov/Dec 1991) on the relative merits of polystyrene foam cups and uncoated paper cups. In his life-cycle analysis, Hocking compared initial resource requirements, raw material needs for fabrication, utility usage and emission profiles, use attributes, and recycle and disposal options for these two types of beverage containers. From the results of this analysis, it is not clear that paper is the most desirable material. In fact, Hocking argues that plastics should be given a more even-handed consideration relative to paper in this beverage container debate, as well as in other packaging applications.

Tables 1.2 and 1.3 display some of the data upon which Hocking's analysis is based. Among the conclusions derived from the study were the following: the amount of raw wood required to manufacture a paper cup is about 2.5 times its finished weight; the hydrocarbon fueling requirements are about the same for the paper cup and the polystyrene cup; and about six times as much steam, 13 times as much electric power, and twice as much cooling water are consumed to process the raw materials needed to produce a paper cup as compared to the polystyrene foam cup. In addition, emission rates to air are similar and to water are generally higher for the paper cup.

Once used, both cup types can be recycled. Landfill disposal of the two items under dry conditions will occupy similar landfill volumes after compaction, and decomposition rates for both will be slow to nonexisting. Under wet and anaerobic conditions polystyrene foam will not degrade; paper will biodegrade to produce methane gas and biochemical oxygen demand to leachate. Finally, both materials can be incinerated cleanly in a municipal waste stream with the option of energy recovery to yield an ash content of 2% to 5% of the incoming waste volume.

Detailed as Hocking's analysis may be, others have raised concerns about issues which may not be fully addressed. In response to an article which appeared in *Science* (Hocking, 1 February 1991), several persons wrote responding letters in a later issue (*Science*, 7 June 1991). Among the issues raised were: petroleum is a fuel of economic convenience in papermaking but other fuels may be substituted in its place, including waste biomass; wood is a renewable resource and polystyrene foam cups must be debited to resource depletion; the worst case scenario—large spills—in the use of petroleum is more severe than is the case in the use of wood; litter management is more difficult with plastics than with paper; and trees required to support paper production also play an important positive role in fixing carbon dioxide.

Thus, it is clear that resource conservation issues are indeed complex. It seems equally clear, however, that life-cycle analyses of the type conducted by Hocking can contribute much to our understanding and help us to move beyond decision-making based solely on expediency. In particular, on the issue of disposal of paper versus polystyrene foam cups in landfills which are constructed and operated under dry conditions, the analysis reveals that there is little basis for choosing one over the other. The paper and polystyrene materials from an equivalent number of cups occupy approximately the same landfill volume and both are relatively inert. When moisture is present in a landfill and biodegradation of paper occurs, questions associated with methane and biochemical oxygen demand come to the fore. A well-managed landfill with good leachate control and one which captures methane—a more potent greenhouse gas than carbon dioxide—can mitigate potentially harmful effects resulting

Table 1.2 Raw Material Requirements, Recycling Potential, and Ultimate Disposal Options of Uncoated Paper Versus Polystyrene Foam Cup

Item or Property	Paper Cup	Polyfoam Cup
Raw materials, per cup		
Wood, quality (g)	20 (19-21)	nil
Waste + bark (g)	1.4 (1.2-1.5)	nil
Total (mean) (g)	21 (20-23)	
Petroleum fractions		
Feedstock (g)	nil	~2.4
Process energy (g)	1.8 (1.2-2.4)	1.9
Total (mean) (g)	1.8	4.3
Other chemicals (g)	1.2 (0.9-1.4)	0.08 (0.06-0.10)
Finished weight (g/cup)	8.3 (6.1-10.2)	1.9 (1.4-2.4)
Recycle potential after use	Acceptable	Good
Proper incineration	Clean	Clean
Heat recovery (MJ/kg)	20	40
To landfill, in g	8.3	1.9
Density (g/cm^3)	0.475	0.107
Volume (cm^3)	17.5	17.8
Biodegradability	Yes, BOD to leachate, methane to air	No, inert
Litter	Gradual disintegration	Breakdown slow

Source: Hocking, M.B., Relative merits of polystyrene foam and paper in hot drink cups: Implications for packaging, *Environmental Management*, 15(6), 731, Nov./Dec. 1991. Reprinted with permission.

from the biodegradation of paper. On the other hand, in a poorly managed landfill, the inert polystyrene material would be an advantage. Issues related to the recycling or incineration of the paper and polystyrene foam alternatives are also clarified based on a life-cycle analysis.

1.5 RISKS ASSOCIATED WITH WASTE MANAGEMENT

There is relatively little public opposition to the siting and construction of municipal wastewater treatment plants. The benefits of discharging less polluted water into rivers and other bodies of water are readily recognized and a wastewater treatment plant is often viewed as another

Table 1.3 Utility Requirements and Emission Rates for Fully Bleached Kraft Paper and Polystyrene Resin for Use in Cups

	Per Tonne of Raw Material		Per 10,000 cups	
			8 g	2 g
	Paper	Polyfoam	Paper	Polyfoam
Utilities				
Steam (kg)	9,000–12,000	5500–7000	840	130
Electricity (kWh)	960–1000	260–300	78	6
Cooling water (m^3)	50	130–140	4	3
Water effluent				
Volume (m^3)	50–190	1–4	10	0.5
Suspended solids (kg)	4–16	0.4–0.6	0.8	0.01
BOD (kg)	2–20	0.20	0.9	0.004
Organochlorines (kg)	2–4	nil	0.2	—
Fiber (kg)	0.5–2	nil	0.1	—
Inorganic salts (kg)	40–80	10–20	5	0.3
Air emissions				
Chlorine (kg)	0.2	nil	0.02	—
Chlorine dioxide (kg)	0.2	nil	0.02	—
Reduced sulfides (kg)	1–2	nil	0.1	—
Particulates (kg)	2–15	0.3–05	0.2	0.008
Chlorofluorocarbons (kg)	nil	nil	—	—
Pentane (kg)	nil	30–50	—	0.8
Ethylene styrene (kg)	nil	0.3–5.0	—	0.05
Carbon monoxide (kg)	3.6	0.08	0.3	0.002
Nitogen oxides (kg)	6.0	0.4	0.5	0.008
Sulfur dioxide (kg)	10–16	3–4	1	0.07

Source: Hocking, M.B., Relative merits of polystyrene foam and paper in hot drink cups: Implications for packaging, *Environmental Management*, 15(6), 731–747, Nov./Dec. 1991. Reprinted with permission.

important industrial-type facility. On the other hand, it is rare not to have public opposition to the siting of municipal and industrial solid waste processing and disposal facilities. This opposition is often related to perceptions and concerns about the potential health risks, adverse environmental effects posed by these facilities, lowered property values, and the view that communities with such facilities are second-class and being "dumped on."

The most obvious risks to human health and the environment are associated with the air emissions and ash from incinerators; the leachate and gaseous emissions of landfills; the sludges from wastewater processing; the air emissions, effluents, and residues from composting; and the residues from recycling. Of particular concern are the myriad of organic compounds and heavy metals (e.g., lead, mercury, cadmium) that enter the waste stream from households, businesses, and industries as solvents, paints, plastics, batteries, vehicle fluids, inks, cleaners, fluorescent light bulbs, and insecticides. The disposal of large quantities of these items is controlled by regulations which have been enacted specifically for hazardous materials. Despite these regulations, hazardous materials enter the waste stream from small quantity generators, such as households, farms, and businesses, if for no other reason than a lack of legal alternative disposal methods.

It is common for various public interest and private industry groups to tout one solid waste management method over another. Certain groups may hold positions which conflict with those of other groups. Such conflicting positions may be derived from different economic or environmental philosophies, or, perhaps, they are based on different informational sources. A given group may weigh some factors heavily while minimizing others or ignoring them completely. For example, a waste-to-energy plant which incinerates municipal solid waste and recovers the energy from it may be opposed primarily because of the air emissions emerging from its stacks. Yet, if these wastes are landfilled and a corresponding amount of coal—that is, a quantity of coal with an equivalent heating value to that of the waste—is mined and transported a large distance, and then burned to produce energy, it is not clear that the environmental risks are smaller. The total gases arising from the decomposition of the wastes in the landfill and the air emissions from the mining, transporting, and burning operations may well exceed the emissions from the waste-to-energy facility. Thus, the overall comparative impacts on human and ecosystem health are not obvious.

Visalli (1989–1990) asserts that all methods of managing solid waste—recycling, composting, landfilling, and incineration—have similar environmental impacts. However, others have stated that a comparison of impacts is difficult because the alternative systems generate different pollutants having various toxicities, effects, and risks; produce pollutants in different modes; and affect different populations and ecosystems. Jones (1991) points out that although anti-incineration advocates promote composting as a superior alternative to incineration, composting has not been scrutinized to the same extent as incineration. She conducted an analysis of the comparative health risks resulting from dioxins and trace metals in compost and incineration processes. The results indicated that because human exposure to these substances may be greater through food chain pathways than through inhalation pathways, the risks from dioxins and trace metals may be greater from composting than incineration. Although this is not a definitive analysis, it does indicate the need for risk analysis and careful scrutiny of alternatives.

That the choices we need to make to significantly reduce environment and human health impacts are complex and often unclear, is also emphasized in a 1989 report issued by the U.S. Congress Office of Technology Assessment (USCOTA):

> Quantitative estimation and comparison of the relative risks associated with different management methods is difficult, in part because of problems inherent in risk assessment methodologies and in part because of data deficiencies. For example, it is clear that some potential environmental risks are associated with all MSW management methods because all processing, treatment, or disposal methods result in some type of waste by-product. Many proponents of recycling contend that it poses fewer risks than alternative MSW management methods. However, given current data, it is not possible to quantitatively determine whether recycling produces more or less pollutants, or poses greater or fewer risks, per ton of material processed than do incineration or landfilling. To compare the overall potential risks quantitatively, an in-depth analysis would have to assess the location of all facilities, all waste products from manufacturing and management facilities, exposure pathways and dosages, and potentially affected populations.

The writers of the USCOTA report note further that although comparisons of relative risks within a single management method, such as those for alternative incinerator designs, are possible, risk assessment methodologies have not yet progressed to the point where comparisons among management alternatives are readily possible.

1.6 INTEGRATED WASTE MANAGEMENT

A current perspective in waste management derived from the broad-based environmental and resource conservation concerns described in previous sections is the view that single choice decision-making is unsatisfactory and inadequate. **Integrated waste management**, the term given to this overall concept, regards the following set of elements as comprising a hierarchy of choices:

Reducing the quantity and toxicity of waste
Reusing materials
Recycling
Composting
Incineration with energy recovery
Landfilling
Incineration without energy recovery

Applying this approach in a linear manner means that efforts should first be expended to reduce the materials entering the waste stream. Once materials enter the waste stream first priority should be given to reusing them or to recycling them for the purpose of manufacturing new products. Disposal without any material or energy recovery is to be regarded as a matter of last resort.

Until recently, industries had few incentives to manufacture more durable products, reduce the amount of material used in product design, design products which can be easily repaired, use minimal packaging or use packaging materials which are more amenable to recycling, or purchase post-consumer wastes as raw materials for manufacturing processes. These types of issues are now under the purview of integrated waste management. Most states in the United States have responded by aggressively encouraging recycling and have established goals for the fractions of various components of the waste stream to be recycled. Many businesses and industries have responded with an eye toward reducing wastes arising from manufacturing processes. While doing so, some have discovered that reducing the amount of hazardous and nonhazardous materials used in product manufacture can help save costs.

Many states in the United States have also responded to the guidelines contained in the integrated waste management approach by providing financial incentives for recycling. European countries have long led the United States in recycling, composting, and the recovery of heat or other forms of energy from incineration. A recent example of leadership designed to prevent materials from entering the solid waste steam is the "Green Dot" program in Germany (Redd, 1992). In 1991, the German Parliament passed legislation setting targets for the recycling of 80% to 90% of packaging materials. To comply, domestic and international companies in Germany created the Duales System Deutschland (DSD), a licensing cartel, to license every company selling packaged products within the country and to grant permission to use a green dot indicating compliance with the recycling targets. A small charge, based on the type of packaging but averaging about 1.2 cents per item, is added to each individual package to fund the cost of collecting and recycling the packaging material. The effect is to internalize recycling and disposal costs as part of the total product cost. In a similar manner, France has developed legislation which establishes specific recycling targets. It differs from the German approach in that it includes waste-to-energy systems within its definition of recycling. In 1991 the European Union also prepared a proposal to achieve waste packaging recovery targets of 60% in five years, 90% in ten years, and exclude any packaging that is not recyclable or reusable by the year 2000.

These aggressive approaches using mandated programs, legislated recycling targets, and financial incentives are not without problems. When such programs are initiated, the supply of diverted materials often grows but without a concomitant demand. Consequently, it may be necessary to stockpile or dump the collected materials until such time as new markets are created through incentives, mandates, or the market system itself.

Some people believe it is a mistake to regard the components of the integrated waste management approach as a hierarchy of choices. Alter (1991), for example, avers that linear thinking in the planning of waste management systems raises conflicts in strategic choices among the elements. He asks, "Is it better or worse to choose a lightweight package that cannot be recycled, or a heavier weight one that might be recycled?" He then goes on to suggest that instead of determining this choice on the basis of solid waste management policy and practice, other factors in the marketplace might be more relevant. Matsuto and Ham (1990) subscribe to the integrated waste management approach in a manner that goes beyond a strictly hierarchical or linear perspective when they state that "modern waste management must be flexible, incorporating: (1) waste reduction/minimization; (2) recycling; (3) processing (incineration, composting); and (4) disposal (landfill), in an 'optimal' manner to minimize the cost and negative environmental impacts." This is an excellent principle to follow in waste management planning.

1.7 WASTE MANAGEMENT TERMINOLOGY

In this introductory chapter, the terms solid waste, garbage, refuse, trash, rubbish, and sewage sludge have been used informally and without regard to precise meaning. For the remainder of this book it is important that these terms be defined with some precision in accordance with current technical usage. The definitions which will apply throughout the remaining chapters are the following:

Solid waste—Unwanted or discarded material with insufficient liquid content to be free flowing. The term **refuse** is synonymous with solid waste.

Garbage—Residential or commercial food waste.

Rubbish—Residential and commercial solid waste exclusive of garbage. The term **trash** is synonymous with rubbish.

Sludge—The solid, semisolid, or liquid waste generated from a municipal or industrial wastewater treatment plant or air pollution control facility, excluding the treated effluent.

Another important definition is:

Municipal solid waste (MSW)—Those durable goods, nondurable goods, containers and packaging materials, food wastes and yard trimmings, and miscellaneous organic wastes arising from residential, commercial, institutional, and industrial sources. Other types of waste not typically included in municipal solid waste are industrial waste produced by manufacturing and processing operations, construction and demolition waste, agricultural waste, oil and gas waste, and mining waste resulting from the extracting and processing of minerals.

Other technical terms will arise throughout the book. A glossary providing definitions of all important terms contained in this section and in later chapters is provided at the end of the text.

DISCUSSION TOPICS AND PROBLEMS

1. Which individuals and what companies, municipal departments, state agencies or other entities are responsible for collecting and disposing of the solid waste in your community? For providing wastewater treatment?
2. Which of the elements of an integrated solid waste management system have been implemented in your community? Are there initiatives underway to implement other elements? To what extent is the private sector involved in waste management?
3. Collect newspaper and magazine articles related to waste management. Be prepared to discuss the circumstances and issues addressed by the articles.
4. Provide examples of where a waste problem in one form—solid, liquid, gaseous—can be transformed into a waste problem in another.
5. Go to a library and see if you can discover newspaper accounts from 40 or 50 years ago which discuss issues concerned with solid waste management or wastewater treatment in your local community. In what ways are these issues similar to present-day issues? In what ways are they different?
6. Determine the amount and composition of solid waste produced by your household during one week. This requires segregating paper, glass, plastic, and garbage. You may wish to separate the paper into subcategories of newspaper, junk-mail, and mixed paper. (Suggestion: plastic waste bags are convenient sanitary storage containers). At the end of the week, weigh the waste containers using a kitchen or bathroom scale.
7. One suggestion for reducing waste is to add a product disposal tax to the cost of packaged consumer goods, much like the approach

taken by Germany. Is such an approach fair? Does it disproportionately affect any income level?

8. As noted in the text, the total energy cost of beverage containers is
the sum of the energy required to extract the raw materials, transport the materials to the manufacturer, manufacture the containers,
transport the containers to the bottler, fill the containers with beverage, transport the filled containers to market, return refillable
containers for cleaning and refilling, and landfill the throwaway
containers and discarded returnable containers after their useful
lifetimes.

   (a) Calculate and compare the energy required to provide soft
       drinks to consumers in returnable glass bottles with that re-
       quired to provide soft drinks in nonreturnable glass bottles
       using the data adapted from Hannon (1972). These data, dis-
       played in the table below, are based on a lower mass for the
       modern throwaway glass bottle.

   (b) Calculate the energy for each type of bottle if 30% of the
       waste glass from the throwaway bottles or discarded bottles is
       recycled back to the bottle manufacturer.

	Returnable	Nonreturnable
Volume, in liters	0.47	0.47
Mass of bottle, in kg	0.298	0.184
Mass of soft drink, in kg	0.47	0.47
Average number of times filled	15	1
Extraction of raw material, kJ/kg		2,300
Transport, extraction to manufacturer, kJ/kg		290
Manufacture of bottle, kJ/kg		17,990
Manufacture of metal cap, kJ/cap		255
Manufacture of plastic cap, kJ/cap		250
Transport, manufacturer to bottler, kJ/kg		720
Filling bottles, kJ/liter		1,700
Transport, bottler to retailer, kJ/kg		815
Transport, retailer to consumer (negligible)		0
Transport, return empty bottles (negligible)		0
Waste collection and disposal, in kJ/kg		210
Separation for recycling, kJ/kg		2,560
(assuming 30% recovery)		

9. In spite of the material and energy savings realized by purchasing
beverages in returnable bottles, most consumers purchase their
beverages in nonreturnable bottles or cans. Identify the reasons

that some consumers, retailers, and bottlers favor nonreturnable containers.

10. Hannon (1972) observed that when purchasing beverages, "A beverage in returnable bottles is about 30 percent less expensive than in cans or throwaways." Test the validity of this statement today by gathering price information for a particular beverage that is sold in both returnable and throwaway containers at a local store or supermarket. (Suggestion: Avoid products with special promotional pricing.)

11. Some grocery stores provide both paper and plastic bags to their customers for carrying their purchases. Assuming that both the paper and plastic bags are produced with comparable raw material and energy requirements and effluents as for the paper and polystyrene cups (Tables 1.2 and 1.3), prepare an analysis of the paper versus plastic bag alternatives. The mass of a large paper grocery bag is approximately 66 g and the mass of a plastic bag is approximately 7 g.

REFERENCES

Alter, H., The future course of solid waste management in the U.S., *Waste Management and Research*, 9, 3–20, 1991.

Blackman, W.C., *Basic Hazardous Waste Management*, Lewis Publishers, Boca Raton, FL, 1993.

Franklin Associates, Ltd., Comparative Energy and Environmental Impacts for Soft Drink Delivery Systems, a report prepared for The National Association for Plastic Container Recovery (NAPCOR), Charlotte, NC, March 1989.

Hannon, B.M., Bottles, cans, and energy, *Environment*, 14(2), 11–21, March 1972.

Hocking, M.B., Paper versus polystyrene: A complex choice, *Science*, 251, 504–505, 1 February 1991.

Hocking, M.B., Relative merits of polystyrene foam and paper in hot drink cups: Implications for packaging, *Environmental Management*, 15(6), 731–747, Nov/Dec 1991.

Jones, K.H., Risk assessment: Comparing compost and incineration alternatives, *MSW Management*, 29-32, 36-39, May/June 1991.

Letters re: M.B. Hockings article in *Science*, Feb. 1991; *Science*, 252, 1361–1363, 7 June, 1991.

Matsuto, T. and Ham, R.K., Residential solid waste generation and recycling in the U.S.A. and Japan, *Waste Management and Research*, 8, 229–242, 1990.

Melosi, M.V., *Garbage in the Cities*, Texas A&M University Press, College Station, TX, 1981.

O'Leary, P.R., Walsh, P.W. and Ham, R.K., Managing solid waste, *Scientific American*, 256(12), 36–42, Dec. 1988.

Rathje, W. and Murphy, C., *Rubbish! The Archaeology of Garbage*, Harper Collins Publishers, New York, 1992.

Redd, A., Germany recycles packaging material, *Waste Age*, 23(1), 50–58, Jan. 1992.

Rosenberg, C.E., *The Cholera Years*, The University of Chicago Press, Chicago and London, 1962.

Sellers, V.R. and Sellers, J.D., Comparative Energy and Environmental Impacts for Soft Drink Delivery Systems, prepared for the National Association for Plastic Container Recovery by Franklin Associates, Ltd., Prairie Village, KS, 1989.

U.S. Congress Office of Technology Assessment (USCOTA), Facing America's Trash: What Next for Municipal Solid Waste?, OTA-O-424, U.S. Printing Office, Washington, DC, 1989.

U.S. Environmental Protection Agency, Characterization of Municipal Waste in the U.S.: 1992 Update, EPA/530-R-92-019, U.S. Printing Office, Washington, DC, 1992.

Vesilind, P.A. and Peirce, J.J., *Environmental Engineering*, Butterworth Publishers, Boston, 1982.

Visalli, J.R., The similarity of environmental impacts from all methods of managing solid wastes, *J. Env. Systems*, 19(2), 155–170, 1989–1990.

CHAPTER 2

Waste Generation

2.1 OVERVIEW

Accurate projections of the quantities and composition of waste are essential for the planning of efficient and economical waste transport, processing, and disposal systems. These estimates are used by engineers and planners to determine the type, size, design, and location of facilities, the transportation of the wastes from the sources to the facilities, personnel needs, equipment requirements, the potential for recycling components of the waste stream, and for understanding the environmental impacts of processing and disposal alternatives.

The U.S. Environmental Protection Agency (USEPA) has estimated the quantity of municipal solid waste (MSW) generated in the United States in 1990 to be 177.6 million tonnes (195.7 million tons), of which 147.3 million tonnes (162.3 million tons) are disposed in landfills or incinerators and the remainder recovered or composted (USEPA, 1990). Industrial wastes, wastewater sludges, construction and demolition waste, junked automobiles, agricultural wastes, and mining wastes are not included in the definition of municipal solid waste (MSW), nor are the quantities reported for these categories. Data on the amounts of these materials are not as readily available and, in general, are less accurately known. The estimated annual generation rates for these wastes in the United States are shown in Table 2.1. In addition to solid waste, it is estimated that municipalities and industries discharge over 60 billion cubic meters of wastewater annually. The treatment of these wastewaters removes pollutants, but produces solid and semisolid sludges that become part of the solid waste management system.

Estimating waste quantities and composition is typically the first task in any local waste management study. In communities or regions where in the past wastes have been weighed or metered at processing or disposal

Table 2.1 Estimates of Annual Quantities of Solid Waste Generated in the United States

Type of Work	Annual Quantities (millions of tonnes)
Agricultural	unknown
Construction/demolition	28.1
Household (hazardous)	0.27
Industrial (nonhazardous, dry basis)	390
Industrial hazardous	178
Mining	1270
Municipal solid waste	177.6
Municipal sludge (dry basis)	7.7
Municipal combustion ash	2.1

Conversion: tonnes x 1.102 = tons

Sources: U.S. Environmental Protection Agency, U.S. Environmental Protection Agency Subtitle D Study—Phase I Report, EPA/530-SW 86-054, 1986.

U.S. Congress, Office of Technology Assessment, Facing America's Trash: What Next for Municipal Solid Waste?, OTA-O-424, October 1989.

U.S. Environmental Protection Agency, National Biennial RCRA Hazardous Waste Report, EPA/530-R-92-027, 1993.

U.S. Environmental Protection Agency, Characterization of Municipal Waste in the U.S.: 1992 Update, EPA/530-R-92-019, 1992.

U.S. Federal Register, 54(23), 5745–5902, Feb. 6, 1989.

facilities, such as incinerators, treatment plants, or landfills, the planning for the new facilities or programs can proceed with confidence in the waste quantity data. If no actual weight data are available, then estimating the expected quantities is a more challenging task. In this chapter, waste quantities, composition, and sampling methods are discussed.

2.2 MUNICIPAL SOLID WASTE

As was noted in Chapter 1, municipal solid waste (MSW) includes wastes such as durable and nondurable goods, containers and packaging, food scraps, yard trimmings, and miscellaneous organic and inorganic wastes from residential, commercial, institutional, and industrial sources. Some examples of the types of MSW that come from these sources are listed in Table 2.2. Residential waste is produced by households and other types of dwelling units. Commercial waste results from retail, wholesale,

Table 2.2 Sources of Municpal Solid Waste

Source	Types of Solid Waste
Residential	Appliances, newspapers, clothing, disposable tableware, food packaging, cans, bottles, food scraps, yard trimmings.
Commercial	Corrugated boxes, food wastes, office papers, disposable tableware, yard trimmings.
Institutional	Office papers, cafeteria and restroom wastes, classroom wastes, yard trimmings.
Industrial	Corrugated boxes, lunchroom wastes, office papers, wood pallets.

and service activities in a community. Institutional waste is produced by schools, hospitals, and in buildings housing governmental functions. The industrial waste that is included is primarily from the office and support operations and does not include waste produced by processing and manufacturing operations. The categories of materials typically used to characterize MSW are described in Table 2.3.

Table 2.3 Description of Components of Municipal Solid Waste

Category	Description
Paper and Paperboard	
High grade paper	Office paper and computer paper
Mixed paper	Mixed colored papers, magazines, glossy paper, and other paper not fitting the categories of high grade paper, newsprint, and corrugated
Newsprint	Newspaper
Corrugated	Corrugated boxes, corrugated and brown (kraft) paper
Yard waste	Branches, twigs, leaves, grass, and other plant material
Food waste	All food waste excluding bones
Glass	Clear and colored glass
Plastics	All types of plastics, codes 1-7
Ferrous metals	Iron, steel, tin cans, and bi-metal cans
Nonferrous metals	Primarily aluminum, aluminum cans, copper, brass, and lead
Wood	Lumber, wood products, pallets, and furniture
Rubber	Tires, footwear, wire cords, gaskets
Textiles	Clothing, furniture, footwear
Leather	Clothing, furniture, footwear
Miscellaneous	Other organic and inorganic materials, including rock, sand, dirt, ceramics, plaster, bones, ashes, etc.

The USEPA (USEPA, 1992) estimates the municipal solid waste generation rate in the United States to be 2.0 kg per person per day (4.4 lb per person per day) for 1990. The USEPA values are computed and updated periodically by Franklin Associates Ltd., using a mathematical model. Residential waste accounts for 55% to 65% of the total MSW generation, while commercial waste ranges between 35% and 45% of MSW.

The USEPA report (USEPA, 1992) distinguishes between generation and discards. Generation (called gross discards in earlier USEPA reports) refers to the quantities (weights) of materials as they enter the waste stream. Discards refer to the MSW remaining after materials have been recovered for recycling or composting. The materials in the discard category are presumably landfilled or incinerated, although some are littered, disposed on-site or burned on-site, particularly in rural areas. The USEPA does not have an estimate for these other disposal practices, but the amounts are believed to be small. For 1990, the daily per capita discarded quantity has been estimated to be 1.6 kg (3.6 lb), reflecting a 20% diversion rate.

The USEPA estimates of MSW generation and discards over the past two decades are shown in Table 2.4. The generation of MSW increased by 2.8% per year between 1980 and 1990. The USEPA projects the generation to increase less rapidly, by 1.3% per year, from 1990 to 2000. The

Table 2.4 Historical Changes in the Municipal Waste Stream in the U.S. as Generated and Discarded by Weight

	Millions of tonnes					
	1970		1980		1990	
Component	Generated	Discarded	Generated	Discarded	Generated	Discarded
Paper and paperboard	40.1	33.4	49.6	38.8	66.5	47.5
Yard waste	21.0	21.1	24.9	25.0	31.7	27.9
Food	11.6	11.6	12.0	12.0	12.0	12.0
Glass	11.5	11.3	13.6	12.9	12.0	9.6
Ferrous metals	11.4	11.3	10.5	10.2	11.2	9.4
Nonferrous metals	1.4	0.6	2.6	1.4	3.5	1.8
Plastics	2.8	2.8	7.1	7.1	14.7	14.4
Wood	3.6	3.6	6.1	6.1	11.2	10.8
Rubber and leather	2.9	2.6	3.9	3.8	4.2	4.0
Textiles	1.8	1.8	2.4	2.4	5.1	4.8
Miscellaneous	2.4	2.1	4.6	4.2	5.5	4.8
Total	110.5	102.2	137.3	123.9	177.6	147.0

Conversion: tonnes × 1.102 = tons

Source: U.S. Environmental Protection Agency, Characterization of Municipal Waste in the U.S.: 1992 Update, EPA/530-R-92-019, 1992.

increases in the quantities of paper, plastics, wood, and nonferrous metal are particularly noteworthy. The increase in generation is generally attributed to both an increase in the population and an increase in the per capita generation rate. Neal and Schubel (1987) attributed two-thirds of the increase in the generation rate to population growth and one-third to the increase in the per capita generation rate.

The composition of both generated MSW and discards has changed over time as product design, packaging materials, and buying habits have changed. The composition of the waste stream, and the percentage contributed by each component is shown in Table 2.5. Although the amount of paper and paperboard in the waste stream has increased, its percentage of the total waste stream has remained almost constant. In contrast, the large increase in plastics has resulted in an increase in a percentage of the total as well. The reader should be aware that the USEPA excludes construction and demolition waste quantities and junked automobiles from its MSW data; consequently, the percentages reported in Table 2.5 are larger than would be the case if these wastes were included.

It is widely believed that modern society's demand for convenience packaging and single-use products, attributable in large part to fast-paced lifestyles and increased affluence, is a major contributor to the increase in per capita generation. Although it is true that the quantity of containers

Table 2.5 Historical Changes in the Composition of the Municipal Waste Stream in the U.S. as Generated and Discarded by Weight

	Composition					
	1970		1980		1990	
Component	Generated (%)	Discarded (%)	Generated (%)	Discarded (%)	Generated (%)	Discarded (%)
Paper and paperboard	36.3	32.5	36.1	31.3	37.5	32.3
Yard waste	19.0	20.5	18.2	20.1	17.9	19.0
Food	10.5	11.3	8.7	9.6	6.7	8.1
Glass	10.4	11.0	10.0	10.4	6.7	6.5
Ferrous metals	10.3	11.0	7.7	8.2	6.3	6.4
Nonferrous metals	1.3	1.1	1.9	1.5	2.0	1.2
Plastics	2.5	2.7	5.2	5.7	8.3	9.8
Wood	3.3	3.5	4.4	4.9	6.3	7.3
Rubber and leather	2.6	2.6	2.8	3.1	2.4	2.7
Textiles	1.6	1.8	1.7	1.9	2.9	3.3
Miscellaneous	2.2	2.0	3.3	3.3	3.0	3.4
Total	100.0	100.0	100.0	100.0	100.0	100.0

Conversion: tonnes × 1.102 = tons

Source: U.S. Environmental Protection Agency, Characterization of Municipal Waste in the U.S.: 1992 Update, EPA/530-R-92-019, 1992.

and packaging materials has increased dramatically over the past few decades—from a generation rate of 39.5 million tonnes (43.5 million tons) in 1970 to 58.4 million tonnes (63.9 million tons) in 1990—comparable increases are observed in the nonpackaging materials as well so that, as a percentage by weight, packaging has actually decreased from 35.7% of the generated MSW in 1970 to 32.9% in 1990.

RESIDENTIAL SOLID WASTE

The national daily per-capita generation rate for residential solid waste is about 1.2 kg (2.6 lb). Local variations from this value may be considerable as a result of several factors. Among these are type of dwelling (e.g., single family, apartment), household size, type of housing, urban versus rural households, socioeconomic factors, and season of the year. Household size in particular can be a significant factor. This is illustrated by the results of one study which provided the relationship between per capita generation rate and household size shown in Figure 2.1. As household size increases, so does the total amount of waste produced, but the per capita generation decreases. Little attention has been given to the effect of household size on the per capita generation rate when comparing data among communities and over time as family size has decreased in many communities.

Rural households tend to generate about 25% less waste per capita than their urban counterparts (Rhyner et al., 1976). Household size and

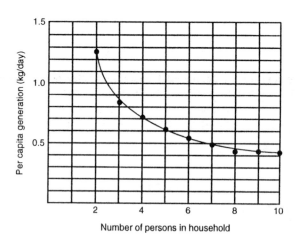

Figure 2.1 The per capita waste generation rate for different household sizes (Rhyner et al., *Waste Age*, 7, April 1976). The study included both urban and rural households. Copyright © 1976 by the National Solid Waste Management Association. Reprinted with permission.

lifestyle may account for much of the difference. These same factors account for some of the differences in generation rates found in different types of housing in a community (e.g., apartments, single-family, multi-family), with a tendency toward higher per capita generation rates in apartments and multifamily dwellings.

COMMERCIAL SOLID WASTE

Local estimates of commercial waste quantities tend to be more difficult to develop than those for residential waste quantities. This results from the greater range in the types and magnitudes of the activities associated with commercial buildings, hotels, restaurants, hospitals, schools, and other private and public service facilities than is typically the case for dwellings in residential areas.

Detailed estimates of the waste generation by commercial activities are sometimes based on measures such as the number of employees in a building, the floor area of a store, the number of meals served at a restaurant, and the number of beds in a hospital. The implicit assumption is that the relationship between the parameter indicating the magnitude of the activity and the quantity of waste produced is linear. A planner may refine these data with information from a specific locality by surveying businesses and industries. Table 2.6 summarizes the typical weekly waste generation multipliers and composition data for various commercial activities based on the floor area utilized by the activity. Typical waste generation rates for hospitals, schools, hotels, and restaurants are summarized in Table 2.7.

Table 2.6 Typical Weekly Commercial Waste Generation Multipliers and Solid Waste Composition Based on Floor Space and Waste Components

Category	Quantity (kg/m²/week)	Paper (%)	OCC[a] (%)	Plastics (%)	Metals (%)	Other (%)
Office	0.2	65	15	6	2	12
Retail	1.1	35	40	8	1	16
Wholesale, warehouse, & distribution	0.3	25	32	25	7	11
Conversions:	kg/m² × 0.2048 = lb/ft²					

Sources: Illinois Dept. of Energy and Natural Resources, 1991. Data provided by Tellus Institute, Waste Plan Default Data Report, Boston, MA, 1990.

[a] OCC = old corrugated containers

Table 2.7 Typical Waste Multipliers for Hospitals, Schools, Hotels, and Restaurants

Category	Waste Generation Rate
Hospital	4.5 to 8.0 kg per occupied bed per day
Schools	0.2 to 0.5 kg per student per day
Hotels	0.5 kg per occupied room per day for noncheckout day (Double this quantity for a suite or for checkout days)
Restaurants	90 kg per thousand dollars of sales

	Recyclables generation/thousand meals/day
	    27 kg of corrugated containers (CC)
	    11 kg of steel (tin) cans
	    1.6 kg of aluminum
	    1.6 kg of glass

Conversion: kg x 2.205 = lb

Sources: Illinois Dept. of Energy and Natural Resources, 1991. Data provided by Camp, Dresser & McKee Inc., Chicago, IL, 1991. Rates for hotels and restaurants are from Westerman, M., Restaurants recycle, *Resource Recycling*, January 1991, pp. 78–83.

In the United States, commercial and industrial activities are identified by a Standard Industrial Classification Code (SIC) (Office of Management and Budget, 1987). The activities are categorized according to specific purpose and identified as single-purpose units called establishments. Industrial establishments are identified by the SIC Major Group Categories in the range 20 through 39 and commercial establishments are identified by categories in the range 40 through 93. The primary use of the SIC classification scheme is to collect economic and employment data, but the SIC codes are also useful in identifying types of waste generating activities. Table 2.8 is a summary of the waste generation rates for commercial activities identified by SIC codes. Representative data showing the composition of commercial waste from various sources are given in Table 2.9.

MUNICIPAL SOLID WASTE IN OTHER COUNTRIES

The generation of MSW is higher in the United States than in other developed countries, but a definitive comparison of MSW generation in other countries to that of the United States is difficult because of differing definitions of MSW and differences in data collection and reporting prac-

Table 2.8 Daily Commercial Waste Generation Rates Based on SIC Codes and Number of Employees

Commercial Activity	SIC Code	Waste Generation (kg/employee/day)
Transportation, communications, utility, and wholesale	40 through 51	1.6–4.0
Retail trade	52 through 59	3.1–5.0
Finance, insurance, real estate, and services	60 through 67 70 through 79 80 through 81 83 through 89	0.9–2.3
Government, education, and institutional	82, and 92 through 93	1.4–2.3

Note: Rates are based on a 7-day week. Multiply by 365 to get annual generation.

Conversion: kg x 2.205 = lb

Sources: Illinois Dept. of Energy and Natural Resources, 1991. Data provided by Camp, Dresser & McKee Inc., Chicago, IL, 1991. Compiled from four locations.

tices. In the United States, for example, the estimate for the generation of MSW includes post-consumer materials that are recycled, whereas in many other countries, these materials are excluded. A further complication is that waste data are often reported for different years. While it is recognized that the accuracy of the data is very limited for the reasons cited, it is nevertheless instructive to compare the MSW of various countries. Comparative data on waste generation rates in various countries are shown in Table 2.10. This table also includes an interesting comparison of waste generation rates with the gross national product (GNP) of these countries. The comparison of waste generation rates with GNP is based on an analysis by Denison and Ruston (1990), but the values in the table are updated. In their analysis they concluded that "generation of MSW is not well correlated with economic output," and that "Many nations have managed to maintain a per capita gross national product close to that of the United States, yet produce far less waste per capita."

The composition of the MSW varies widely in different countries and under different climatic conditions. Waste composition data for various countries are found in Chapter 8, Table 8.5.

Table 2.9 Representative Composition of Waste from Various Commercial Sources Expressed as Percentages by Weight

Waste Component	Retail Trade	Restaurant	Office	School	Government
Paper	**41.5%**	**36.6%**	**64.2%**	**47.8%**	**53.8%**
Newspaper	2.9	2.5	3.6	3.3	6.7
Corrugated	22.0	15.6	11.5	11.6	8.4
Highgrade white	1.4	0.0	10.6	6.3	7.2
Mixed recyclable	10.3	4.4	29.0	21.6	25.0
Nonrecyclable	4.9	14.1	9.5	5.0	6.5
Plastics	**12.0**	**13.7**	**4.3**	**5.1**	**3.5**
PET (1)	0.1	0.0	0.1	0.1	0.1
HDPE (2)	0.0	0.1	0.0	0.0	0.0
Other	11.9	13.6	4.2	5.0	3.4
Glass	**2.5**	**5.9**	**3.9**	**3.2**	**2.7**
Container	2.3	5.8	2.9	1.0	2.4
Nonrecyclable glass	0.2	0.1	1.0	2.2	0.3
Metals	**20.5**	**4.9**	**2.9**	**5.8**	**9.8**
Aluminum cans	0.2	0.5	0.5	0.8	0.5
Tin/steel cans	0.2	3.8	0.2	0.2	0.4
Other ferrous	19.5	0.4	2.2	3.7	8.6
Other nonferrous	0.6	0.2	0.0	1.1	0.3
Organics	**18.8**	**36.6**	**10.8**	**35.0**	**23.2**
Food waste	8.1	36.0	3.0	14.0	3.2
Yard debris & wood	10.7	0.6	7.8	21.0	20.0
Other	**4.7**	**2.3**	**13.9**	**3.1**	**7.0**
Totals	**100.0%**	**100.0%**	**100.0%**	**100.0%**	**100.0%**

Source: Washington State Department of Ecology, *Best Management Practices for Solid Waste: Recycling and Waste Stream Survey*, 1987.

2.3 INDUSTRIAL PROCESS WASTE

Manufacturing wastes include a wide range of process residues—sludges, combustion ashes, slags, kiln dust, foundry sand, food processing residues, and organic and inorganic residues. These wastes are sometimes called Subtitle C wastes (referring to Subtitle C of the Resource Conservation and Recovery Act of 1976) if they are to be managed as hazardous wastes or Subtitle D wastes if they are nonhazardous or exempt from Subtitle C regulation. Only nonhazardous wastes are considered here.

Table 2.10 Municipal Solid Waste Generation per Capita in Selected
Countries, Including a Comparison with Per Capita Economic
Output (GNP)

Country	Year of Waste Estimate	Annual MSW (kg)	Annual GNP Value (US $)	MSW/GNP (kg/$1000 US)
		Per Capita Values		
Austria	1988	355	17,360	20.5
Belgium	1989	349	16,360	21.3
Canada	1989	625	19,020	32.9
Denmark	1985	469	20,510	22.9
England	1989	357	14,570	24.5
Finland	1989	504	22,060	22.8
France	1989	303	17,830	17.0
Germany (FRG)	1987	318	20,750	15.3
Greece	1989	259	5,340	48.5
Italy	1989	301	15,150	19.9
Japan	1988	394	23,730	16.6
Luxembourg	1990	466	24,860	18.7
Netherlands	1988	465	16,010	29.0
New Zealand	1982	670	11,800	56.8
Norway	1989	473	21,850	21.6
Spain	1988	322	9,150	35.2
Sweden	1985	317	21,710	14.6
Switzerland	1989	424	30,270	14.0
U.S.A.	1986	864	21,100	40.9

Conversion: kg x 2.205 = lb

Source: World Resources Institute, *World Resources 1992–93*, Oxford University Press, New York, NY, 1992.

(Hazardous waste is the topic of Chapter 12.) The USEPA estimated that the manufacturing sector produced and handled approximately 5.9 billion tonnes of Subtitle D wastes in on-site land-based units (e.g., landfills, surface impoundments, land application areas, and waste piles) in 1985 (USEPA, 1988). However, because the vast majority of these wastes were initially placed in surface impoundments, it is likely that most were wastewaters with small amounts of solids.

Of the wastes placed in on-site land-based units, the pulp and paper industry accounted for 35% of the total, the largest of any industry. The pulp and paper industry was followed by the primary iron and steel industries and the inorganic chemicals industries, which accounted for 20% and 14%, respectively (USCOTA, 1992).

The annual quantity of coal combustion waste was estimated to be 85 million tonnes in 1984, of which 59 million tonnes were ash and 14.5 million tonnes were flue gas desulfurization sludge (USCOTA, 1992).

Approximately 10% of the weight of coal burned remains in the form of ash. Coal combustion utilities account for 90% of the coal combustion waste.

As one can readily see by comparing the estimate of 5.9 billion tonnes of Subtitle D wastes to the 390 million tonnes estimated on a dry basis (Table 2.1), the difference resulting from including or excluding wastewater is extremely large. Thus, the amount of moisture remaining in the solid waste greatly affects the reported weights.

In the United States, industrial waste quantities (dry basis) are about 2.2 times as great as MSW. However, obtaining an accurate estimate for industrial waste quantities generated in a particular region is difficult. The reasons are the unique mix of industries in a community, the differences in the technologies used by the companies, and the degree to which moisture content affects the reported weights. Gathering information directly from an industrial facility provides the most accurate picture, assuming that the waste generated there is actually weighed or truckloads are accurately counted.

As noted previously, manufacturing establishments are identified by Standard Industrial Code (SIC) Major Group categories in the range 20 through 39. From time to time, industrial waste generation factors are included in waste generation reports. These factors are related to production level, and are commonly expressed in terms of product output or the number of employees engaged in the production activity. Production output is commonly acknowledged to be a better indicator than the number of employees, but production information is often proprietary and confidential, whereas the number of employees is readily available from state agencies.

Several studies in the 1970s reported SIC-based generation rates in terms of tonnes per employee for the major SIC two-digit groups (e.g., Golueke and McGauhey, 1970; Niessen and Alsobrook, 1972; Wisconsin DNR, 1974). The published values of these factors for many SIC groups exhibited significant differences. These differences are attributed to the accuracy of the reported estimates of quantities, the use of small samples in gathering data, variations in manufacturing processes, economies of large-scale production, and regional differences in the mix of industries included in a category. At the 3-digit Group Number or 4-digit Industry Number levels of aggregation, the waste generation rates for industries within the category become more homogeneous. Steiker (1973) developed generation coefficients for 342 manufacturing industries at the four-digit SIC Industry Number level of aggregation. He investigated the optimal level of disaggregation of the SIC categories for accurate predictions and concluded that the accuracy increased as the factors were reported at the 4-digit level. There have been few efforts to refine or update the industrial generation factors based on SIC classications, but a new computer-based waste prediction model INVENT '94 (1994) based on SIC codes is now being marketed.

2.4 MINING WASTE

Mining waste includes the materials that are moved to gain access to minerals and the tailings, slags, and residues that result from the processing of the materials. The low concentrations of many minerals in ore deposits result in large quantities of earth being disturbed in mining operations. Mine waste is the soil or rock that is generated during the process of gaining access to the ore or mineral body. Tailings are the wastes generated from the physical and chemical enrichment (beneficiation) processes used to separate the valuable metal or minerals from the ore. In addition, dump or heap wastes are produced by copper, silver, or gold mining companies which spray low-grade ore, waste rock, and tailings with acid or cyanide to leach out metals.

Traditionally, the mine waste and tailings have been disposed as expeditiously and economically as possible. Ideally, the waste materials should be returned to the excavated area, and the disturbed area restored. This is not always practical or possible, particularly while a mine is still active.

Quantitative data regarding mining wastes are limited, but estimates typically range from 1 to 2 billion tonnes annually in the United States, with about one-half being mine waste and the rest tailings and dump/heap waste. The amount of waste attributed to any particular type of mineral extraction operation can vary greatly. Table 2.11 presents estimates of the ratio of waste to marketable production for several metallic ores, phosphate rock, and oil shale. The waste data were obtained from Rampacek (1980) and the USEPA (1985), while production data are taken from the Statistical Abstract of the U.S. (U.S. Department of Commerce, 1992).

2.5 AGRICULTURAL WASTE

Agricultural waste consists of animal waste and crop residues. These are typically homogeneous organic materials. The quantity of agricultural wastes in the United States is estimated to be several times that of municipal wastes. Much of this waste is recycled to farmland. Crop residues are plowed into the soil, as are the animal wastes from many farms. In these instances, the materials may more appropriately be considered a valued nutrient source rather than a waste. The materials do become a waste problem where there are large concentrations of animals, such as in feedlots and poultry operations, or food processing operations. In such instances the large volumes of wastes generated may exceed the capacity of

Table 2.11 Estimate of Ratio of Mining Waste
to Marketable Production

Industry	Ratio of Mining Waste to Marketable Production
Aluminum[a]	6
Coal[b]	0.2-0.3
Copper	420
Gold	350,000
Iron and steel	6
Lead	19
Oil shale	0.9
Phosphate rock[b]	8
Silver	7500
Uranium	6900
Zinc	27

Sources: U.S. Environmental Protection Agency
Report to Congress, Wastes from the
Extraction and Beneficiation of Metallic
Ores, Phosphate Rock, Asbestos, Over-
burden from Uranium Mining and Oil
Shale, U.S. Environmental Protection
Agency, USEPA/530-SW-85-033, 1985.

[a]Estimated, industry data unavailable.

[b]Rampacek, C., Mining and mineral waste, Accomplishments in Waste Utilization: A Summary
of the Seventh Mineral Waste Utilization Symposium, U.S. Bureau of Mines Circular: 8884, U.S.
Government Printing Office, 1980.

nearby lands to absorb them properly. Therefore, these waste generators
may find it necessary to employ waste processing technologies to reduce
the volume of wastes and/or accept greater transportation costs to more
distant land application sites.

Generation multipliers for some animal wastes are listed in Table
2.12. The annual quantity of agricultural wastes from field and row crops

Table 2.12 Wastes Generated by Farm
Animals (Dry Weight)

Animal	Daily Wastes (kg/animal)
Cattle	4.65 (4.3–5.2)
Hogs	0.54 (0.4–0.7)
Sheep	0.25
Poultry	0.03 (0.02–0.05)

Source: Anderson, L.A. and Tillman,
D.A., Fuels from Waste, Academic Press, New York, 1977,
p.6.

is typically 3.4 to 6.7 t/ha (1.5 to 3.0 ton/acre) (Golueke and McGauhey, 1970). Some common crops that exceed this range are corn, cauliflower, broccoli, and lettuce at 9.0 to 10.1 t/ha (4.0 to 4.5 ton/acre).

2.6 WASTE DENSITIES AND VOLUMES

Although weights of wastes received at solid waste facilities are most easily measured, it is the volume of the wastes that is often of greatest concern when they are disposed of in a landfill. Density relates the mass or weight of a material to the volume it occupies. The mass density is defined as

$$\text{Density (kg/m}^3) = \frac{\text{mass of material in kg}}{\text{volume occupied in m}^3}$$

In this text, we depart from the standard use of the SI system of measurement by referring to kilograms as a weight in displaying data derived by weighing materials on a scale. Although scientists recognize that weights are reported in units of newtons, many engineers more readily identify with pounds and kilograms and convert from one unit to the other.

In English units, the weight density, in terms of pounds per cubic foot or pounds per cubic yard, is most often used. Thus,

$$\text{Density (lb/ft}^3) = \frac{\text{weight of material in lb}}{\text{volume occupied in ft}^3}$$

The uncompacted densities of various components of municipal waste are shown in Table 2.13. Using the composition of municipal waste by weight and the densities, the composition by volume can be calculated. The calculation is shown in Example 2.1.

Example 2.1: Assuming the composition of uncompacted discarded municipal waste in the U.S. by weight in 1990 to be that given in Table 2.5, along with the typical landfilled densities listed in Table 2.13, find the composition by volume when the waste is landfilled.

Solution: Begin by assuming a convenient amount of waste, such as 100 kg; in this case, the weight of each of the components will be numerically the same as its percentage. The volume of each component is found by dividing the weight by the density of the component. The total volume is found by summing the volumes of the individual components. Finally, the

Table 2.13 Typical Densities of Components of Uncompacted Municipal Solid Waste

Component	Uncompacted Density[a] (kg/m^3) Range			Landfilled Density[b] (kg/m^3)
	Low	High	Typical	Typical
Paper and paperboard	32	128	82	465
Yard waste	64	224	104	890
Food	130	480	288	1186
Glass	160	480	194	1345
Ferrous metals	130	1120	320	332
Tin (steel) cans	48	160	88	332
Nonferrous metal	64	240	160	217
Plastics	32	128	64	213
Wood	128	320	240	498
Rubber and leather	96	256	128	205
Textiles	32	96	64	237
Miscellaneous	320	960	480	1186

Conversions: kg/m^3 × 0.06243 = lb/ft^3
kg/m^3 × 1.686 = lb/yd^3

[a]Tchobanoglous, G., Theisen, H., and Eliassen, R., *Solid Wastes: Engineering Principles and Management Issues*, McGraw-Hill Book Company, New York, NY, 1977.

[b]U.S. Environmental Protection Agency, Characterization of Municipal Waste in the U.S.: 1992 Update, EPA/530-R-92-019, 1992.

percentage that each component contributes to the total volume can be calculated.

A computer spreadsheet is extremely useful for doing calculations such as those described above. Construct a table, such as Table 2.14, with column 1 listing the components and column 2, the weight of each component. Enter the densities of each of the components when landfilled in column 3. The volumes of the components (column 4) are found by dividing the mass (column 2 entry) by the corresponding landfilled density (column 3 entry). The total volume is found by summing the values in column 4. The percentage of the total volume contributed by the volume of each component is calculated and displayed in column 5. It is instructive to add another column—the ratio of the volume percent to weight percent. A ratio of 1.0, which is the ratio for paper, means that the material occupies the same proportion of a landfill by volume as it does by weight. Plastics, rubber and leather, textiles, and aluminum have ratios of

Table 2.14 Conversion and Comparison of Composition by Weight to Composition by Volume

Component	Mass (kg)	Landfilled Density (kg/m3)	Volume (m3)	% by Volume	Ratio Vol%/Mass%
Paper & paperboard	32.3	465	0.069	32.0	1.0
Yard waste	19.0	890	0.021	9.8	0.5
Food	8.1	1186	0.007	3.1	0.4
Glass	6.5	1345	0.005	2.2	0.3
Ferrous metals	6.4	332	0.019	8.9	1.4
Nonferrous metals	1.0	217	0.005	2.1	2.1
Plastics	9.8	213	0.046	21.2	2.2
Wood	7.3	498	0.015	6.8	0.9
Rubber and leather	2.7	205	0.013	6.1	2.2
Textiles	3.3	237	0.014	6.4	1.9
Miscellaneous	3.5	1186	0.003	1.4	0.4
Totals	100.0		0.217	100.0	

2.0 or more, indicating that these materials occupy more volume in a landfill than indicated by their percentage by weight. Yard trimmings, food, and glass have ratios less than 1.0 and occupy proportionately less space than their percentages by weight indicate.

In an actual landfill, the total volume of a heterogeneous mixture tends to be less than the value found by summing the components because of intermingling of different materials of different sizes so as to fill voids (USEPA, 1992). Consequently, the actual density may be greater than the density calculated here.

2.7 METHODS OF DETERMINING QUANTITY AND COMPOSITION

Municipal wastes are heterogeneous and have been characterized in many studies conducted since the 1960s. Industrial wastes, although diverse when viewed as a whole, are likely to be homogeneous from a single source. For industrial waste, there is the additional concern about the chemical composition of the material and whether it must be treated as hazardous or not.

There are several methods for developing estimates of waste quantities and composition. These include surveying waste generators, gathering results from published studies, applying a materials flow or materials balance approach (F.L. Smith, 1975), and direct sampling, sorting, and weighing.

SURVEYING INDUSTRIAL WASTE GENERATORS

Surveying industrial waste generators directly often provides the most accurate data, particularly if the wastes are weighed at a disposal site. Industries may be reluctant to disclose information about waste quantities to the general public, for fear of a negative reaction. However, they are often required to provide information to governmental regulatory agencies. They also provide information to their trade organizations. These agencies and organizations may then provide information to the public in summary form.

USING LITERATURE SOURCES

In numerous studies, data on waste quantities and composition have been collected from municipalities and industries. The data are published in USEPA reports, trade magazines, professional journals, and engineering reports. There are often limitations on the usefulness of these data because a standard protocol for acquiring the data is lacking and information about the widely-varying local conditions—climate, implementation of waste reduction strategies and recycling programs, family size, housing type, etc.—often are not reported.

MATERIALS FLOW METHODOLOGY

The materials flow methodology uses the concept of conservation of mass to track quantities of material as it moves through a defined system or region. The methodology is based upon defining a system, then tabulating the inputs to and outputs from the system to arrive at a materials balance. If the United States is considered to be the system, production data provide the input quantity, and after adjusting for imports, exports, and product lifetimes, they become the output to the waste stream. The diagram in Figure 2.2 tracing the flow of plastics in the United States illustrates the materials flow approach.

The limitation of the materials flow approach became evident when, in the early 1970s, a municipal daily waste generation rate of 2.41 kg (5.32 lb) per person in the U.S. was widely quoted in the literature based on a preliminary result of the 1968 National Solid Waste survey (Black et al., 1968). Using this number, along with those for the composition of the waste stream, indicated a serious discrepancy; namely, more paper was being disposed each year than was being manufactured and imported. Based on materials flow calculations, F.A. Smith (1975) estimated the 1971 daily per capita MSW generation rate to be 1.50 kg (3.31 lb). This number consisted of a residential component of 1.09 kg (2.39 lb) and a commercial-institutional component of 0.42 kg (0.92 lb).

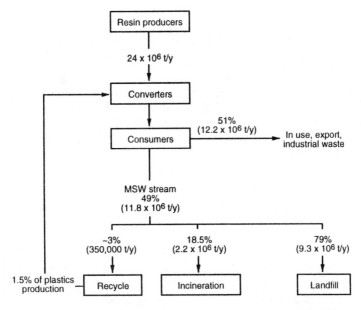

Figure 2.2 Flow of plastics in the U.S. (Eller, Solid waste management in Japan, *Waste Age*, 23(5), 53, May 1992). Copyright © 1992 by the National Solid Waste Management Association. Reprinted with permission.

DIRECT SAMPLING

Direct sampling, sorting, and weighing is particularly useful for obtaining information about local municipal solid waste quantities and composition. Modern automated scale systems—scale equipment interfaced to a computer with software such as the MI-WEIGH scale software (see References at end of chapter)—facilitate the gathering of data by origin, material type, or waste characteristics readily recognizable by a scale operator. These data are usually gathered for the purpose of billing users of the facility but they also provide the waste manager with additional information about the sources of the waste, the types of wastes, and variations in the waste quantities over time.

If information about the components of the waste stream, such as the fractions comprised by aluminum, glass, ferrous metal, paper, or plastic, is desired for planning a recycling program, then sampling and sorting may be necessary. The cost and effort needed to sort the waste increase as the quantity of waste to be sorted and the number of component categories increases. In planning for a waste-to-energy project, the classification scheme might be simply combustible materials and noncombustible materials, whereas planning for a recycling operation requires more detailed information about each component that is to be recycled.

Before beginning a sampling and sorting project, three important questions must be addressed: (1) How will representative samples of the waste be obtained? (2) How large should each sample be? and (3) How many samples should be taken to achieve a desired level of accuracy in the results? The second and third questions are particularly important because sorting waste is labor-intensive and, therefore, expensive. A study by LeRoy et al. (1992) provides an estimate of the labor requirements for the sorting of municipal solid waste. In that project, nine persons sorted 1600 kg/day of solid waste into 12 categories.

One approach is to sample at the source by collecting the waste directly from randomly-selected residences, multifamily dwellings, and commercial businesses. A second approach is to sample the waste after it has been taken to a disposal facility. The percentage composition determined in this manner may differ from that determined by source sampling because of moisture transfer. When wastes are mixed in a collection vehicle, high moisture wastes such as food and yard waste lose moisture, but absorbent materials, primarily paper, gain moisture.

The American Society for Testing and Materials (ASTM) has issued a standard test method for the determination of the composition of unprocessed municipal waste at a disposal facility by manual sorting (ASTM Method D5231–92, 1992). The prescribed sample size is 91 to 136 kg (200 to 300 lb). Because the vehicle loads may be extremely heterogeneous, care is needed to ensure that the samples taken are representative. Typically, the waste from randomly selected loads is dumped in a separate area. A front-end loader scoops representative portions at least four times as large as the samples to be taken. These portions are mixed, placed into a conical pile, and one-quarter is removed as a sample. The total number of samples to be obtained in this manner depends upon the accuracy needed or desired.

A statistical analysis of some initial samples provides information about the variability in the wastes (i.e., standard deviation) and can be used to determine the quantity of waste that must be included in the study in order to achieve results with a desired precision. Alternatively, an initial estimate for the number of samples needed can be calculated using the field data shown in Table 2.15. After the preliminary sampling has provided a preliminary value for the composition of the mixture (expressed as decimal fractions) and the standard deviation, s, of these fractions, the number of samples, N, required can be calculated by the following procedure:

(1) Select a component to govern the precision of the composition measurements (i.e., the "governing component").

**Table 2.15 Estimates of the Mean Fractional
Composition (by Weight) and
Standard Deviations Reported for
Municipal Solid Waste Sampled
During Weekly Sampling Periods at
Several Locations in the United States**

Component	Mean Fractional Composition	Standard Deviation
Mixed paper	0.22	0.05
Newsprint	0.10	0.07
Corrugated	0.14	0.06
Plastics	0.09	0.03
Yard waste	0.04	0.14
Food waste	0.10	0.03
Wood	0.06	0.06
Other organics	0.05	0.06
Ferrous	0.05	0.03
Aluminum	0.01	0.004
Glass	0.08	0.05
Other inorganics	0.06	0.03
	—	
	1.00	

Source: Standard Test Method for Determination of
the Composition of Unprocessed Municipal Solid
Waste, ASTM Method D 5231-92, American Society
for Testing and Materials, Philadelphia, PA, 1992.
Copyright ASTM. Reprinted with permission.

(2) Calculate the number of samples required to achieve the desired precision using the formula

$$N = (t^* s / e x)^2$$

where
  N = the number of samples to be sorted,
  e = the desired precision, expressed in decimal form,
  x = the estimated mean fraction of the governing component in the
     mixture,
  s = the estimated standard deviation determined from an initial sam-
     pling,
  t^* = student t statistic corresponding to the desired level of confidence.

For this initial calculation, use the t^* value corresponding to $N = \infty$ (i.e.,
1.96 for a 95% confidence level or 1.645 for a 90% confidence level) and

values for the mean fraction and standard deviation from Table 2.15 or the preliminary sampling to find N.

   (3) Go to Table 2.16, but rather than using the t^* value corresponding to $N = \infty$, select the t^* corresponding to the value of N calculated in step 2. Enter this new value of t^* into the formula and calculate a new value of N. Return again to Table 2.16 for a value of t^* corresponding to the new N and recalculate. Repeat this process until the values of N differ by less than 10%. (Note: After actual sorting of the waste has been done, the measured mean fraction and measured standard deviation will provide better estimates for the governing component than those found in Table 2.15.)

   The formula used in the calculations above assumes a normal distribution for the mass fractions of the components, which may not be the case for small samples. Klee (1993) addresses the issue of sampling with skewed distributions using an iterative sampling method and a computer program PROTOCOL to perform the necessary calculations.

Example 2.2: Suppose we wish to determine the amount of glass in our municipal solid waste to a precision of 10% with a confidence level of 90%. Samples of about 130 kg are to be used. How much waste must be sorted?

Solution: Use Table 2.16 to find an estimate for the mean fraction of glass and the standard deviation. Using the formula for sampling precision, with $t^* = 1.645$, $e = 0.10$, $x = 0.08$, and $s = 0.05$

$$N = (t^* s / e x)^2 = [(1.645) \ (0.05)/(0.10)(0.08)]^2$$
$$N = 106$$

Referring again to Table 2.16, the value of t^* corresponding to 106 samples is about 1.659. Inserting this into the formula yields

$$N = [(1.659) \ (0.05)/(0.10)(0.08)]^2$$
$$N = 108$$

This is within 10% of the previous value of 106. Therefore, the number of samples required would be 108. At 130 kg per sample, about 14,000 kg of waste must be sorted.

Table 2.16 Values of t Statistics as a Function of the Number of Samples and Confidence Interval

Number of Samples	Confidence Interval	
	90%	95%
2	6.314	12.706
3	2.920	4.303
4	2.353	3.182
5	2.132	2.776
6	2.015	2.571
7	1.943	2.447
8	1.895	2.365
9	1.860	2.306
10	1.833	2.262
11	1.812	2.228
12	1.796	2.201
13	1.782	2.179
14	1.771	2.160
15	1.761	2.145
16	1.753	2.131
17	1.746	2.120
18	1.740	2.110
19	1.734	2.101
20	1.729	2.093
21	1.725	2.086
22	1.721	2.080
23	1.717	2.074
24	1.714	2.069
25	1.711	2.064
26	1.708	2.060
27	1.706	2.056
28	1.703	2.052
29	1.701	2.048
30	1.699	2.045
31	1.697	2.042
36	1.690	2.030
41	1.684	2.021
46	1.679	2.014
51	1.676	2.009
61	1.671	2.000
71	1.667	1.994
81	1.664	1.990
91	1.662	1.987
101	1.660	1.984
121	1.658	1.980
141	1.656	1.977
∞	1.645	1.960

Source: Standard Test Method for Determination of the Composition of Unprocessed Municipal Solid Waste, ASTM Method D 5231-92, American Society for Testing and Materials, Philadelphia, PA, 1992. Copyright ASTM. Reprinted with permission.

2.8 SEASONAL VARIATION OF SOLID WASTE QUANTITIES

Solid waste quantities may vary significantly throughout the year. Typically, in temperate climates, the daily, weekly, and monthly MSW quantities are above average in summer. This is attributable in part to the increase in yard and garden waste and to construction, home improvement, and community cleanup activities. The MSW quantities tend to be below average in winter when the level of these activities is considerably less. There may be local or regional anomalies in the generation pattern such as those arising in resort and vacation areas where the influx of seasonal residents may greatly increase the residential and commercial waste load. In localities with large quantities of industrial waste, the generation pattern will be determined by the type of industries present. For example, food processing plants may only operate at harvest time, but pulp and paper industries tend to operate with constant production throughout the year.

Table 2.17 shows the variation of the monthly waste generation for Brown County, Wisconsin, using data for the years 1984–1989. The entries in the table were calculated in the following manner. For each month and year and for each type of waste, the daily average quantity for each month was calculated by dividing the monthly quantity by the number of days in the month, then the percentage by which this average daily quantity for each month differs from the daily average for the year (i.e., annual waste quantity divided by 365, or 366 in leap year) was calculated. Finally for each month and type of waste, the mean value and standard deviation of the daily average percentages for the five years 1985–1989 were calculated. A small standard deviation value indicates that the generation pattern is consistent from one year to the next. The residential and commercial waste have much smaller standard deviations compared to the other categories, indicating that the monthly generation patterns for residential and commercial wastes can be predicted more accurately than the other types of waste.

The quantity and composition of the waste will exhibit daily or weekly fluctuations, as well as monthly or seasonal variations. These fluctuations may be important in waste management planning. Processing and disposal facilities must be designed with a capacity large enough to accommodate all of the waste most of the time, with allowances made for future growth of the waste quantities. It is uneconomical to design a disposal facility for a much greater capacity than is needed, but also imprudent to undersize the facility so that the current and future growth of waste inputs cannot be accommodated. One sizing method is to design a facility for a weekly delivery that is the average of the highest four-week period projected for the year (Research and Education Association, 1978). A second method is to design the facility with a capacity that will be

Table 2.17 The Mean Percentage by Which the Daily Waste Quantity Calculated for Each Month Differs from the Annual Average Daily Value Using Data from Brown County, WI for the Years 1985-1989. The Standard Deviations of the Values Used to Compute the Mean Are in Parentheses

Month	Residential	Commercial	General Industrial	Categories of Solid Waste Process Residue/ Sludge	Ash	Tires	Construction/ Demolition
1	-21.3 (4.0)	-13.0 (3.4)	4.4 (12.5)	4.7 (16.2)	11.2 (17.6)	-23.4 (14.3)	-48.5 (12.6)
2	-28.2 (3.0)	-9.2 (7.2)	3.4 (4.3)	6.7 (31.1)	16.4 (23.3)	-17.8 (17.2)	-32.9 (32.6)
3	-17.0 (4.8)	-2.4 (4.3)	0.5 (6.7)	19.5 (40.9)	8.8 (16.2)	-19.3 (20.3)	-25.7 (17.3)
4	14.8 (2.9)	3.8 (5.0)	-0.8 (7.1)	19.8 (18.9)	-1.1 (14.3)	-5.9 (17.7)	-0.6 (14.3)
5	21.5 (2.3)	-0.5 (5.0)	-9.9 (7.0)	0.8 (15.4)	-12.6 (18.0)	-7.1 (14.7)	11.9 (21.1)
6	13.2 (12.2)	4.9 (3.3)	-2.1 (10.6)	-6.2 (22.2)	-18.6 (12.6)	5.2 (11.2)	21.1 (9.7)
7	2.6 (6.3)	5.0 (6.3)	-4.0 (8.7)	-13.1 (32.6)	-12.3 (10.7)	11.8 (21.5)	28.3 (10.3)
8	11.7 (3.4)	11.2 (3.3)	3.7 (7.3)	27.8 (47.7)	-3.79 (13.5)	20.8 (17.9)	29.6 (12.7)
9	10.0 (6.1)	3.9 (5.8)	2.1 (5.9)	-31.3 (28.2)	-1.4 (13.3)	17.9 (33.5)	13.5 (27.7)
10	11.2 (7.1)	2.3 (7.5)	7.2 (6.0)	-17.0 (35.0)	-1.9 (15.5)	13.2 (19.7)	44.4 (30.8)
11	-3.5 (7.0)	-1.0 (4.3)	2.3 (10.6)	-22.3 (18.0)	4.1 (22.6)	-3.0 (15.6)	-14.4 (12.3)
12	-16.6 (5.3)	-6.1 (2.6)	-6.3 (9.8)	9.6 (28.5)	11.9 (17.9)	6.4 (30.8)	-28.9 (21.7)

Source: Rhyner, C.R., The monthly variations in solid waste generation, *Waste Management & Research*, 10, 67, 1992. Reprinted with permission.

sufficient 95% of the time. In other words, the capacity is exceeded in only about 2 weeks of the year.

2.9 MUNICIPAL AND INDUSTRIAL WASTEWATER

In developed countries, residential, commercial, and industrial wastewater is collected and treated prior to returning the effluent to a river or other surface water body. The material removed during treatment is a sludge, a solid material which typically contains a high amount of moisture. Dewatered solids may be landfilled, applied to land as a soil amendment, or incinerated. Sludge disposal will be discussed in more detail in later chapters.

There are 15,300 publicly-owned wastewater treatment plants in the United States treating approximately 80 billion liters of water daily (*U.S. Federal Register*, 1989). In addition, there are over 70 million people using septic tanks for wastewater disposal. Seventeen million housing units discharge approximately 8 billion liters of wastewater daily through septic systems (Cantor and Knox, 1985).

Wastewater treatment processes are designed to remove organic and inorganic constituents and to kill pathogens. Municipal wastewater treatment generates 7–9 million dry metric tons of sludge annually. The quantity is expected to increase in the future because of population increases and more stringent treatment requirements. Table 2.18 illustrates the amounts of sludge that might be generated when the best current technology is employed in wastewater treatment.

Table 2.18 Sludge Generation from Best Current
Wastewater Treatment Technology

Process Step	Liters of Sludge from One Million Liters of Municipal Wastewater	Percent Solids
Primary treatment	2,500-3,500	3-7%
Secondary treatment	15,000-20,000	0.5-2%
Tertiary treatment	10,000+	0.2-1.5%

Source: *U.S. Federal Register*, 54(23), 5745–5902, Feb. 6, 1989.

SOURCES AND FLOW RATES OF MUNICIPAL WASTEWATER

The primary sources of municipal wastewater are residential units and retail and commercial business establishments. Other sources may include institutional and recreational facilities. In some communities substantial amounts of wastewater may be contributed by large industries. Wastewater arising from industrial sources is often pretreated to reduce the waste load and to remove specified hazardous materials before final treatment in a municipal wastewater treatment plant.

Estimates of the quantity of wastewater from residential areas can be determined on the basis of population and average per capita contribution. Examples of flow rates from residential sources are shown in Table 2.19. Such data are useful in the planning of sewer and wastewater treatment plant construction. However, since there can be considerable variation in flows among communities, design rates should be based on actual flow data from selected residential areas similar to those under study.

Commercial wastewater flow rates can vary greatly among the various sources, as depicted in Table 2.20. Therefore, in planning for wastewater treatment from commercial sources it is important that flow records be obtained from similar facilities.

Table 2.19 **Wastewater Flow Rates from Residential Sources**

	Unit	Range of Daily Flow Rates (L/unit)
Apartment	Person	130–280
Cottage (Seasonal)	Person	95–190
Mobile home	Person	110–190
Single family home	Person	110–380

Conversion: L × 0.264 = gal

Sources: Metcalf & Eddy, Inc., *Wastewater Engineering: Treatment, Disposal, and Reuse*, McGraw-Hill, Inc., New York, NY, 1991.

Salvato, J.A., *Environmental Engineering and Sanitation*, John Wiley and Sons, New York, NY, 1982.

**Table 2.20 Wastewater Flow Rates from Commercial and
              Institutional Sources**

Source	Unit	Range of Daily Flow Rates (L/unit/day)
Airport	Passenger	11–19
Automobile service station	Vehicle served	27–50
	Employee	34–57
Department store	Toilet room	1500–2300
	Employee	30–45
Hospital	Bed	470–910
	Employee	19–57
Hotel	Guest	190–210
	Employee	27–50
Laundry (self–service)	Machine	1700–2500
	Wash	170–210
Milk processing plant	50 L of milk	45–100
Motel	Rental unit	290–680
Office	Employee	27–61
Restaurant	Meal served	8–16
School	Student	57–114
Shopping center	Employee	27–50
	Parking space	4–8

Conversion: L x 0.264 = gal

Sources: Metcalf & Eddy, Inc., *Wastewater Engineering: Treatment, Disposal, and Reuse*, McGraw-Hill, Inc., New York, NY, 1991.

Salvata, J.A., *Environmental Engineering and Sanitation*, John Wiley and Sons, New York, NY, 1982.

CHARACTERIZATION OF MUNICIPAL WASTEWATER

Municipal wastewater is complex and highly variable. It typically contains a wide range of organic and inorganic compounds both in soluble and particulate (suspended solids) forms which can cause serious environmental degradation if inadequately treated. Municipal wastewater also contains high populations of microorganisms, some of which have the potential for causing disease.

Because of the complex nature of wastewater it is, for all practical purposes, impossible to perform a complete chemical and microbial analysis. It is for this reason that wastewater is described in terms of three broad categories—organic and inorganic materials and microbial content.

Organic Compounds

Organic compounds possess at least one carbon atom and one or more hydrogen atoms. Oxygen is often part of the structure of organic compounds and, in addition, sulfur, nitrogen, and phosphorus are part of certain organic structures. Organic compounds can range from relatively simple structures such as methane (CH_4) to very large organic molecules such as starches, proteins, and others which may have molecular weights of many thousands of atomic mass units. In nature, organic molecules may be oxidized biologically by many microbes using the organic compounds as a source of carbon and energy for growth. For example, in the presence of oxygen, glucose can be oxidized according to the following equation:

$$C_6H_{12}O_6 + 6O_2 \rightarrow 6CO_2 + 6H_2O + energy$$

The ability of many microbes to utilize organic materials and oxygen for growth may lead to undesirable changes in streams and lakes to which wastewater high in organic content is discharged. The oxygen-utilizing microbes rapidly deplete oxygen during the degradation of the organic material in discharged waste. Oxygen depletion may lead to the death of fishes and other organisms. On the other hand, in properly designed and maintained wastewater treatment systems these same microbes can be utilized to efficiently degrade the organic matter in wastewater prior to its discharge.

A major need in wastewater management, then, is for a test to easily determine the organic content of a waste and to do so in a manner that provides a useful measure of the efficiency of a given treatment process. Several methods of measuring organic content are in common use. These include biochemical oxygen demand (BOD), chemical oxygen demand (COD), and total organic carbon (TOC).

Biochemical Oxygen Demand (BOD)

The BOD test takes advantage of the ability of microorganisms to oxidize organic matter using oxygen as an electron acceptor (oxidizing agent). A typical BOD test consists of a 300 mL bottle filled with a highly aerated sample of wastewater to be tested, plus a "seed" containing a large number of microbes. The bottle is then tightly sealed, so no oxygen can enter from the atmosphere. The sample is incubated in the dark at 20°C. for five days. The oxygen content of the bottle is measured at the start of the test with an oxygen electrode and again after five days of incubation. The change in oxygen levels is then used to calculate the BOD of the original sample and is typically expressed as mg of BOD per liter. The BOD measurement based on a five-day incubation time and denoted as BOD_5 is commonly used in the United States, but in European countries, a seven- day incubation (BOD_7) is often used. The BOD_5 of a typical raw sewage might be 350 mg/L, whereas after efficient treatment, the BOD_5 may be less than 20 mg/L. It is important to recognize that the BOD_5 test measures the amount of those organic materials that are readily degraded by microbes within the five-day period of incubation. There are some complex organic compounds that may be resistant to breakdown or may degrade slowly over a much longer period of time.

Chemical Oxygen Demand (COD)

The COD test uses a mixture of potassium dichromate and sulfuric acid at 150°C in the presence of a silver catalyst which acts as a strong oxidizing agent. Under these conditions most of the organic carbon and hydrogen are oxidized to carbon dioxide and water. At the same time the dichromate is reduced to trivalent chromium. Any remaining dichromate is converted to a compound which can be precisely measured using a titrimetric technique or by spectrophotometric means. The organic content of a sample is expressed as milligrams of COD per liter (mg/L). While the COD test provides a strong oxidizing environment, there are certain aromatic compounds such as pyridines, benzene, toluene and others which are not readily oxidized.

Table 2.21 presents the results of a typical analysis of municipal wastewater showing COD and BOD concentrations. The concentrations of total and suspended solids and certain inorganic compounds, such as ammonia, nitrogen, and phosphate, are also listed.

Total Organic Carbon (TOC)

The TOC test is based on the use of an automated apparatus, called a total carbon analyzer, which oxidizes the carbon (organic and inorganic carbon) in a sample in the presence of a catalyst in a high temperature stream of air. The process is rapid and reproducibility is excellent.

Inorganic Compounds

The inorganic components of wastewater cannot be measured in single tests similar to the BOD or other tests used to measure the organic content of wastewater. However, the number of inorganic compounds of major concern in municipal wastewater is limited largely to nitrogen and phosphorus compounds and certain heavy metals such as mercury (Hg), lead (Pb), zinc (Zn), and cadmium (Cd). The concentrations of these compounds are determined on an individual basis.

Table 2.21 Example Analysis of a Municipal
Wastewater

Component	Concentration (mg/L)
Total solids (TS)	850
Volatile TS	250
Nonvolatile TS	600
Total suspended solids (TSS)	220
Volatile TSS	170
Degradable volatile TSS	90
Nonvolatile TSS	50
Chemical oxygen demand (COD)	400
Suspended COD	210
Soluble COD	190
Biochemical oxygen demand (BOD_5)	220
Suspended BOD	120
Soluble BOD	100
Ammonia nitrogen	30
Phosphate	20
Sodium	75
Magnesium	25
Sulfate	70

The principal forms of nitrogen which are of concern in waste treatment are ammonium (NH_4^+), nitrite (NO_2^-), and nitrate (NO_3^-). Excessive amounts of these forms of nitrogen can cause undesirable changes in the environment in a number of ways. Ammonia may be oxidized to nitrate by nitrifying bacteria which use dissolved oxygen in water for this conversion. Hence, this microbial activity may deoxygenate water just as in the case of excess organic matter described earlier. In addition, because of its solubility in water, ammonia is toxic to many aquatic organisms at low concentrations. For example, the USEPA has established 0.02 mg/L non-ionized ammonia (NH_3) as the lower concentration limit of this species of nitrogen in receiving waters.

While nitrate does not present the same level of toxicity as ammonia, it may be a problem due to the fact that it is an excellent nutrient for green plants such as algae and other aquatic plants. Its presence in water may lead to excessive growth of such plants. This condition is further exacerbated when excess phosphate, another plant nutrient, is present. Under such nutrient-rich (eutrophic) conditions, algae will thrive, producing oxygen as well as depleting dissolved carbonates during photosynthesis. High rates of photosynthesis lead to increased oxygen and pH during the daylight hours, while at night the algae respire using oxygen and producing carbon dioxide, thereby lowering the pH. This diurnal fluctuation in dissolved oxygen and pH is very stressful to many fishes and other forms of aquatic life. As a consequence, eutrophic conditions caused by excess nitrate and phosphate tend to reduce species diversity as the more sensitive species of plants and animals gradually die out.

Microbes

Municipal wastewater and sludges by their very nature contain large numbers of microbes. The population of microbes can vary greatly even within a given community but many tens of millions of microbes per milliliter of wastewater are common. While many of these organisms are nonpathogenic there are many that can be the cause of serious diseases in humans. Table 2.22 lists some of the more common bacterial, viral, and intestinal parasite organisms that may be present in municipal wastewater.

Waterborne diseases are transmitted as the result of fecal contamination of waters used for drinking, bathing, or recreation. The potential for disease transmission is enormous, especially in developing countries. The World Health Organization (Koning, 1987) estimates 1.25 billion people suffer from waterborne infections.

Table 2.22 Principal Bacterial, Viral and Intestinal Parasites in Human and Animal Wastes

Bacteria	Viruses	Intestinal Parasites
Salmonella	Enteroviruses	Protozoa
Shigella	Poliovirus	Entamoeba
Vibrio	Echovirus	Giardia
Mycobacterium	Hepatitis A	Cryptosporidium
Campylobacter	Norwalk	Toxoplasma
Staphylococcus	Adenovirus	Nematodes (round worms)
Klebsiella	Reovirus	Ascaris
Clostridium	Rotavirus	Necator
Brucella		Strongyloides
Enterobacter		Toxocara
Escherichia		Trichuris
Leptospira		Cestodes (tapeworms)
Listeria		Echinococcus
Pseudomonas		Taenia
Yersinia		

Because the populations of microbes are so great and so varied, it is not practical to try to identify all the microbes that may be present in a wastewater. As a consequence of this difficulty, much emphasis has been placed on the development of tests which use **indicator organisms** to provide guidance as to the effectiveness of a particular wastewater treatment process. One of the best known of such indicator systems is the fecal coliform test which is based on the detection of coliform bacteria (*Escherichia coli* is the best known member of the group) in water used for drinking and bathing. The fecal coliform test is based on the premise that the organisms which respond positively to the test are those common to the human intestinal tract. While a positive test does not mean pathogens are also present it is an indication that the particular water sample has been exposed to fecal contamination and therefore should not be used for human consumption without some type of treatment to eliminate the contaminating organisms.

The fecal coliform test is not the perfect indicator tool (there is none), and in some situations other tests may also need to be used to assure the safety of a particular water or a sludge for application to agricultural land. The processing of sewage sludges and wastewaters for pathogen destruction is discussed in Chapters 7 and 11, particularly as it relates to the application of wastewater and sludge to land used to grow food crops.

INDUSTRIAL WASTEWATER

Industries consume large quantities of water for processing. The characteristics of the discharged waters from industrial sources are very different from municipal wastewater in that the concentrations of pathogens are typically low or absent. Table 2.23 provides a summary of water quantities discharged by selected industries. The untreated amounts may, in some cases, include uncontaminated cooling water.

The composition of industrial wastewater is varied and tends to be industry specific. For example, wastewater from food processing industries is often high in organic content (high BOD), while metal processing wastewater has almost no organic content but may be very acidic and/or contain toxic compounds such as cyanide.

DISCUSSION TOPICS AND PROBLEMS

1. Using reasonable choices of generation factors for residential, commercial, and industrial solid waste, prepare an estimate for the quantity of waste generated annually by a city of 100,000 people.

Table 2.23 Water Quantities Discharged by Selected Industries in the United States

Industry	Water Discharges (millions of cubic meters)		
	Treated	Untreated	Total
Food	745	1,353	2,098
Textile	205	194	399
Chemical	3,739	7,585	11,323
Paper	4,898	1,801	6,699
Primary metal	884	1,228	2,112
Totals of all industries	15,291	18,582	33,873

Conversion: 1 m^3 = 1000 liters = 264.2 gallons

Source: U.S. Dept. of Commerce, Census of Manufacturers, 24/12:MC82–5–6, 1982.

2. The typical composition of the municipal solid waste generated in the United States by weight is shown in the table below with typical moisture values. Calculate (a) the composition of the waste, percent by weight, with the moisture removed and (b) the moisture content of the waste.

Component	Percent by Weight	Typical Moisture Content (%)
Paper and paperboard	37.5	6
Yard waste	17.9	60
Food	6.7	70
Glass	6.7	2
Ferrous metal	6.3	3
Nonferrous metal	2.0	2
Plastics	8.3	2
Wood	6.3	20
Rubber and leather	2.4	2
Textiles	2.9	10
Miscellaneous	3.0	8

3. Using municipal solid waste composition data from the table in Problem 2, estimate what fraction is combustible.

4. What volume would the 147.3 million tonnes of discarded U.S. municipal solid waste occupy if it were landfilled? Use the composition data from Table 2.5 and the landfilled densities from Table 2.13.

5. Using population data and reasonable assumptions about generation rates, estimate the quantity of each of the following types of wastes in your community or solid waste service region: (a) automobile tires; (b) waste automobile motor oil; (c) junked automobiles.

6. How much municipal solid waste must be sampled to determine the fraction of plastics to a precision of 10% (i.e., e = 0.10) at a confidence level of 90%. Use the data in Table 2.15 for a preliminary estimate of the mean and standard deviation. Assume a nominal sample size of 130 kg.

7. Prepare an estimate of the quantity of solid waste that will be disposed in the year 2000. Explicitly state all assumptions used to develop the estimate.

8. According to United States census data, the average household size in the U.S. decreased from 2.75 persons in 1980 to 2.63 persons in 1990. During this decade, MSW generation increased at an annual rate of 2.8%. The population was 227 million persons in 1980 and 249 million persons in 1990. Prepare an estimate for the portions of this increase attributable to population growth, to changes in household size, and increased per capita generation.

9. Using 1992 U.S. production data and the ratios of mining wastes from Table 2.11, prepare an estimate for the annual quantity of mining wastes generated in the United States.

	Production (tonnes)
Bauxite	600,000 (estimated)
Coal	938,000,000
Copper	1,157,000
Gold	290
Iron	57,000,000
Lead	474,000
Phosphate rock	46,300,000
Silver	2,170
Uranium	4,040
Zinc	515,000

10. Both the BOD and COD tests are measures of the amount of organic material in a wastewater, but the values measured are often different. Which of the tests tends to indicate a larger amount of organic material? Why?

11. A community generates an average daily wastewater flow of 50,000,000 liters. Using data from Table 2.20, estimate the daily quantity of solids (dry mass basis) which will result, using best current wastewater treatment technology.

12. What is meant by the term "indicator organisms" as used in waste-water treatment?

13. A potato processing plant generates wastewater which contains sugars, starch, some proteins, other miscellaneous organic compounds, ammonia, nitrate, and phosphate. A sample of this wastewater is analyzed using the BOD method. Which of the waste components will contribute to the BOD value measured? Explain your answer.

14. A major new housing development is added to a community. Approximately 1000 people will be living in single family residences and another 600 will live in apartments. Calculate the amount of additional wastewater which will be generated by this increase in population.

15. Three new restaurants are part of the housing development described in Problem 14. How would you estimate the additional wastewater which these businesses would contribute to the wastewater flow?

REFERENCES

Anderson, L.A. and Tillman, D.A., *Fuels from Waste*, Academic Press, New York, 1977.

Alter, H., The origins of municipal solid waste: The relations between residues from packaging materials and food, *Waste Management & Research* 7:103–114 (1989).

Alter, H., The future course of solid waste management in the U.S., *Waste Management & Research*, 9, 3–20, 1991.

Alter, H., The origins of municipal solid waste: II. Policy options for plastics waste management, *Waste Management & Research*, 11, 319–332, 1993.

ASTM Method D5231–92, Determination of the Composition of Unprocessed Municipal Solid Waste, American Society for Testing and Materials, 1916 Race Street, Philadelphia, PA, 19103, 1992.

Black, R.J., Muhich, A.J., Klee, A.J., Hickman, H.L. and Vaughn, R.D., The National Solid Wastes Survey: An Interim Report, National Technical Information Service, NTIS Report PB 260 102, 1968.

Cantor, L. and Knox, R., *Septic Tank Effects on Ground Water Quality*, Lewis Publishers, Chelsea, MI, 1985.

Denison, R.A. and Ruston, J. *Recycling and Incineration*, Island Press, Washington, DC, 1990.

Eller, R., Solid waste management in Japan, *Waste Age*, 23(5), 53–64, May 1992.

Golueke, C.G. and McGauhey, P.H., Comprehensive Studies of Solid Waste Management, NTIS Report PB 218 265, National Technical Information Service, U.S. Department of Commerce, Washington, DC, 1970.

Illinois Department of Energy and Natural Resources, Office and Commercial Waste Reduction, prepared by Camp, Dresser & McKee Inc. and Holt, Ross, & Yulish Inc., Report ILENR/RR-91/10, Springfield, IL 62704–1892, 1991.

INVENT '94 1994, Ashact Ltd., Bridge House Station Approach, Great Missenden, Bucks, HP16 9AZ, Phone 0494 891100; FAX 0494 890320.

Klee, A.J., New approaches to estimation of solid waste quantity and composition, *J. Env. Eng.*, 119(2), 248–261, 1993.

Koning, H. W., Setting Environmental Standards—Guidelines for Decision Making, World Health Organization, 1987.

LeRoy, D., Giovannoni, J.M., and Maystre, L.Y., Sampling method to determine a household waste stream variance, *Waste Management & Research*, 10, 3–12, 1992.

Metcalf & Eddy, Inc., *Wastewater Engineering: Treatment, Disposal, and Reuse*, McGraw-Hill, Inc., New York, NY, 1991.

MI-WEIGH Software, Inc., P.O. Box 510168, St. Louis, MO 63151, (314) 772–4144.

Neal, H.A. and Schubel, J.R., *Solid Waste Management and the Environment: The Mounting Garbage and Trash Crisis*, Prentice-Hall, Englewood Cliffs, NJ, 1987.

Niessen, W.R. and Alsobrook, A.F., Municipal and Industrial Refuse: Composition and Rates, *Proceedings: 1970 National Incinerator Conference*, American Society of Mechanical Engineers, NY, 1972, pp. 319–337.

Office of Management and Budget, *Standard Industrial Classification Manual*, U.S. Printing Office, Washington, DC, 1987.

Powelson, D.R. and Powelson, M.A., *The Recycler's Manual for Business, Government, and the Environmental Community*, Van Nostrand Reinhold, New York, 1992.

Rampacek, C., Mining and Mineral Waste, *Accomplishments in Waste Utilization: A Summary of the Seventh Mineral Waste Utilization Symposium*, U.S. Bureau of Mines Circular: 8884, U.S. Printing Office, Washington, DC, 1980.

Rattray, T., Source reduction—an endangered species? *Resource Recycling*, 9(11), 64–65, 1990.

Research and Education Association, *Modern Pollution Control Technology, Vol. II*, Research and Education Association, New York, NY, 1978.

Rhyner, C.R., Wenger, R.B., Raridon, R.J., and Westphal, J.M., Domestic solid waste and household characteristics, *Waste Age*, 7, 29–39, 50, April 1976.

Rhyner, C.R., The monthly variations in solid waste generation, *Waste Management & Research*, 10, 67–71, 1992.

Salvato, J.A., *Environmental Engineering and Sanitation*, John Wiley & Sons, New York, NY, 1982.

Smith, F.A., Comparative Estimates of Post Consumer Wastes, National Technical Information Service, NTIS Report PB 256 491, 1975.

Smith, F.L., Jr., A Solid Waste Estimation Procedure: Materials Flows Approach, U.S. Environmental Protection Agency, EPA/SW-147 U.S. Printing Office, Washington, DC, 1975.

Steiker, G., Solid Waste Generation Coefficients: Manufacturing Sectors, Regional Science Research Institute Discussion Paper Series No. 70, Regional Science Research Institute, P.O. Box 8776, Philadelphia, PA, 19101, 1973.

Tchobanoglous, G., Theisen, H., and Eliassen, R., *Solid Wastes: Engineering Principles and Management Issues*, McGraw-Hill Book Company, New York, NY, 1977.

U.S. Congress, Office of Technology Assessment (USCOTA), Facing America's Trash: What Next for Municipal Solid Waste?, OTA-O-424 U.S. Printing Office, Washington, DC, October 1989.

U.S. Congress, Office of Technology Assessment (USCOTA), Managing Industrial Solid Wastes From Manufacturing, Mining, Oil and Gas Production, and Utility Coal Combustion—Background Paper, OTA-BP-O-82 U.S. Printing Office, Washington, DC, February 1992.

U.S. Dept. of Commerce, Census of Manufacturers 24/12:MC82-5–6, U.S. Printing Office, Washington, DC, 1982.

U.S. Dept. of Commerce, *Statistical Abtracts of the United States 1992*, U.S. Printing Office, Washington, DC, 1992.

U.S. Dept. of Transportation, Federal Highway Administration, Highway Statistics 1990, PHWA-PL-91–003, U.S. Printing Office, Washington, DC, 1990.

U.S. Environmental Protection Agency, Report to Congress: Wastes from the Extraction and Beneficiation of Metallic Ores, Phosphate Rock, Asbestos, Overburden from Uranium Mining and Oil Shale, U.S. Environmental Protection Agency, USEPA/530–SW–85–033, U.S. Printing Office, Washington, DC, 1985.

U.S. Environmental Protection Agency, U.S. Environmental Protection Agency Subtitle D Study—Phase I Report, EPA/530–SW 86–054, U.S. Printing Office, Washington, DC, 1986.

U.S. Environmental Protection Agency, Office of Solid Waste and Emergency Response, Report to Congress: Solid Waste Disposal in the United States, Vols. 1–2, EPA/530–SW–88–011, U.S. Printing Office, Washington, DC, 1988.

U.S. Environmental Protection Agency, Characterization of Municipal Waste in the U.S.: 1990 Update, EPA/530–SW–90–042, U.S. Printing Office, Washington, DC, 1990.

U.S. Environmental Protection Agency, Markets for Scrap Tires, EPA/530–SW–90–074A, U.S. Printing Office, Washington, DC, 1991.

U.S. Environmental Protection Agency, Characterization of Municipal Waste in the U.S.: 1992 Update, EPA/530–R–92–019, U.S. Printing Office, Washington, DC, 1992.

U.S. Environmental Protection Agency, National Biennial RCRA Hazardous Waste Report, EPA/530–R–92–027, U.S. Printing Office, Washington, DC, 1993.

U.S. Federal Register, 54(23), 5745, U.S. Printing Office, Washington, DC, Feb. 6, 1989.

Washington State Department of Ecology, *Best Management Practices for Solid Waste: Recycling and Waste Stream Survey*, 1987.

Weston, R.F., Inc., New York Solid Waste Management Plan: Status Report 1970, National Technical Information Service, NTIS Report PB 213 557, 1971.

Wisconsin Department of Natural Resources, Report on the State of Wisconsin Solid Waste Management Plan, Madison, WI, 53707, 1974.

World Resources Institute, *World Resources 1992–93*, Oxford University Press, New York, NY, 1992.

CHAPTER 3

Collection and Transportation
of Solid Waste

3.1 OVERVIEW

Garbage collection was one of the first municipal sanitation services developed in the United States. In the late nineteenth century, concerns about the impacts on human health from garbage and other types of solid waste piled up in the streets and gutters of the nation's towns and cities was one of the issues which stimulated the development of municipal programs for "public cleansing." Ever since, collecting municipal solid waste and transporting it to disposal or processing sites has been a major activity in departments of public works.

Local governments have not been the only collectors of solid waste, however. Over the centuries "trash men" have made their way through the streets picking up items that retained some economic value, or, for a fee, have hauled a load of waste into the countryside where it was out-of-sight to city dwellers. In recent decades "trashmen" have evolved into entrepreneurs whose business is solid waste collection. Companies have been formed which have as their major business the collection and transport of solid waste. Such companies are responsible for collecting the majority of the commercial, institutional, and industrial waste. While the public sector retains the major responsibility for collecting residential solid waste, private firms provide a significant portion of this service as well. Even though both the public and private sectors are involved in solid waste collection and transport, the major focus of this chapter is on municipal programs.

The topic of municipal solid waste collection merits special consideration if for no other reason than it is the most expensive part of the entire municipal solid waste management system. In recent decades, collection costs have typically accounted for two-thirds to three-fourths of the total cost of municipal solid waste management systems. For example, it has been estimated that collection costs in the United States in 1986 were $10.4 billion out of a total municipal solid waste cost of $13.8 billion. In 1990 the collection and total municipal solid waste costs were estimated to be $13.7 and $20.3 billion, respectively (Solid Waste Management Economics Report, 1991). Thus, in 1986 collection costs constituted about 75% of the total solid waste management cost, while in 1990 the figure was approximately 67%.

The five-year period mentioned above has witnessed a more rapid increase in disposal costs than in collection costs. In that time period, collection costs increased by approximately 32% while disposal costs nearly doubled. Nevertheless, collection remains as the costliest component of a solid waste management system. Therefore, a major preoccupation of those in charge of municipal solid waste collection systems is providing the level of service demanded by the public at the least possible cost.

Until recently, the solid waste collection and transportation component was viewed as an independent subsystem within the larger solid waste collection and disposal system. Except for large bulky articles, there was no need to segregate different components of municipal solid waste; all of it went to the same place: a landfill or a central processing facility. Thus, the goal of an efficient collection system could be sought in isolation from other solid waste management goals, and usually meant choosing the most appropriate trucks, designing balanced collection districts and good routes, and organizing and administering productive crews.

With the current emphasis on recycling and composting, there is a need to segregate materials and, in some cases, transport them to different sites. Therefore, collection systems have become much more complex and designing efficient ones provides a difficult challenge to managers. Collection systems must be integrated with the other components of a community's solid waste management program and support its overall goals. If, for example, different agencies are responsible for the collection and subsequent processing of materials, conflicts could arise. Whereas a community's goal might be to separate materials for recycling, the agency responsible for collection may be interested in maximizing the commingling of materials in order to minimize collection costs.

To meet current challenges, Pferdehirt, O'Leary, and Walsh (January, 1993) call for the development of an "integrated collection strategy" which incorporates the following goals in designing a collection system:

1. The system should provide locally appropriate level(s) of service designed to meet political, health, and regulatory requirements

2. The system should accomplish its defined service requirements at the lowest possible cost
3. The system should develop locally appropriate partnerships between the private and public sectors
4. The system should build in flexibility to meet changing demands
5. The system should support achievement of waste reduction/ diversion policies

In order to attain these goals a number of specific issues must be addressed and a variety of tools used in planning for or modifying solid waste collection systems.

3.2 PRODUCTIVE COLLECTION SYSTEMS

Solid waste collection systems which are cost-effective, responsive to customer's needs, and consistent with and supportive of the overall goals of a community's solid waste management program, can be operated by municipal authorities or managers of private companies. Within a given community, solid waste collection usually involves a combination of public and private sector activity. Private companies provide collection services to a community through a contractual or licensing arrangement.

Planners of collection services, whether the operation is run within the public or private sector, need to have good estimates of the solid waste quantities generated within the service area. Poor estimates can result in the purchase of too much or too little equipment, or in the assignment of an inappropriate number of persons to specific tasks. Estimates of generation rates can be obtained by using data of the types discussed in Chapter 2. However, if they are available, it is better to use previous records of measured solid waste quantities for a given community. It is important that reliable estimates are obtained for different segments of the solid waste stream, those arising from residential, commercial, and industrial sources.

Once reliable estimates of waste quantities are in hand and goals have been determined, a number of specific factors must be addressed. They include frequency of collection, set-out locations, type of containers, type of collection vehicles, and crew size.

FREQUENCY OF COLLECTION

Most communities provide collection service once or twice a week. Once-a-week service is the most common; however, in warmer climates twice-a-week service is sometimes provided because of health and odor concerns. When the collection of recyclables is involved, matters become considerably more complicated. Decisions must be made on which components of the solid waste stream to group together, and on the manner and frequency of collection for each group. Specific alternatives for collection of recyclables will be described in detail in the next section.

As collection frequency increases, productivity, measured on a tonne per hour basis, goes down because smaller amounts of waste are collected in a given time period. Conversely, less frequent collection will result in greater tonnages collected in a given time period; however, fewer collection locations will be serviced in that time. Increasing collection from once-a-week to twice-a-week requires approximately 50% more crew members and equipment (Hickman, 1986).

SET-OUT LOCATIONS

In urban areas customers are usually required to place the waste materials by the curbside for pickup on collection day. Alternatives to curbside pickup are alley and backyard pickup. Alley pickup is usually possible only in older residential or commercial areas.

Backyard pickup is the most convenient for the customer but also the most costly. In one study (Hickman, 1986), it was estimated that crew productivity for backyard pickup was about half that of curbside or alley pickup. Because of this low productivity and high cost, backyard pickup, which was once quite common, is now rarely available. Alley pickup in residential areas allows citizens to set out the waste containers for pickup when it is convenient, not necessarily on the day that pickup occurs.

TYPES OF CONTAINERS

Three types of containers are commonly used in solid waste collection: metal or plastic cans, bags, or special containers for mechanized collection. Each type has advantages and disadvantages.

When cans are the option, customers are usually allowed to choose a size ranging from 75 to 120 liters (20 to 32 gallons). They are readily available for purchase and are convenient for use and storage. Their use may also conserve resources. When placed at curbside or when stored in backyards the contents of the containers are secure from foraging animals.

Bags, on the other hand, are susceptible to breakage and their contents relatively easy prey for animals with sharp claws or teeth. Bags offer the important advantage of convenience to collection crews; they can be grabbed quickly and tossed into the collection vehicle without need to take the time to return an empty container to the curb. Studies have shown that this convenience leads to productivity improvement. In some cases customers are allowed to use a variety of plastic or paper bags, provided they conform to minimum thickness and maximum weight specifications. In other communities a specific bag must be purchased from the collection agency.

Containers which are used for mechanized collection are large plastic bins, usually mounted on wheels. They can be wheeled to the curb or alley where their contents are dumped into the collection vehicle with the aid of a mechanized lift. They have the advantage of reducing fatigue and injuries to collection workers. These containers have come into more frequent use in recent years to assist with the collection of recyclables. They have the disadvantages of high capital cost, the need for collection vehicles with special loading mechanisms, and operational difficulties which may arise due to physical barriers or steep grades.

COLLECTION VEHICLES AND CREW SIZE

Municipal solid waste is usually collected in compactor trucks equipped with rear-, side-, or top-loaders. The trucks may be loaded manually or with lifts or arms which allow for automated loading using the specially designed containers described above. Compactor trucks have hydraulic equipment for compacting the waste, thereby increasing the payload. Solid waste at curbside has a typical density of 118 kg/m^3 (200 lb/yd^3), while a top-of-the line compactor truck will compress the waste to a density of 593 kg/m^3 (1000 lb/yd^3). The capacity of these compactor trucks ranges from 11 to 23 m^3 (15 to 30 yd^3) or more.

Trucks with special compartments for collecting recyclables or a combination of recyclables and mixed solid waste are now in use. Trailers have also been designed which can be pulled behind pickup or compactor trucks to assist with the collection of recyclables. A more detailed description of this type of equipment will be provided in the next section.

The capacity of a solid waste collection vehicle has an important bearing on overall productivity. If a vehicle has a small capacity, more loads are collected in a workday; therefore, a greater amount of time is spent hauling the waste from the collection district to the disposal site than is the case for a larger vehicle. Since labor costs make up a large portion of the expense budget, time spent off-route lowers productivity.

The number of crew members is determined by the type of collection vehicle employed and the nature of the service provided. The high cost of labor acts as an incentive to keep crew sizes as small as possible. Whereas three-member crews were once common, most crews now consist of one or two members. Collection vehicles with a right-hand-side drive and a side loader are specifically designed to be operated by one person.

How can the various factors described above interact to create a productive collection system? It is not only a matter of crew members working harder and faster, but equipment and methodologies must be employed which will allow workers to devote more time to actually collecting solid waste. A number of mathematical models have been developed which managers can use to measure productivity of solid waste collection systems. Examples of productivity measures are tonnes of waste collected per day, cost per tonne of waste collected, cost to collect a household unit, and collection time required per stop. These models can be used as tools for identifying factors which offer the greatest opportunity for productivity improvement.

An example of a productivity model is one developed by Stearns (1982) to measure the productive collection time per stop of a two-person crew working with a rear-loading compactor:

$$P = 0.0033D + 0.16N + 0.09T + 0.03S + 0.02$$

where

P = productive collection time required per stop, in minutes, including the driving time from the previous stop
D = distance between stops in meters
N = number of refuse containers at a service stop
T = total number of throw-away items (including paper or plastic garbage bags) serviced at a stop
S = number of services collected at each stop

A similar model for one operator with a rear-loading compactor is:

$$P = 0.0165D + 0.15N + 0.08T + 0.08$$

These models include route-related factors only.

Example 3.1: Suppose on a given collection route serviced by a two-person crew with a rear-loading compactor the average distance between stops is 50 meters; the average number of containers and the average number of throw-away items at each service stop is 1 and 2, respectively; and the average number of services collected at each stop is 1.5. What is the crew's productive collection time per stop? How does its productivity compare with that of another two-person crew collecting in an area where the average distance between stops is 75 meters?

Solution: The equation given above for a two-person crew must be applied:

$$P = 0.0033(50) + 0.16(1) + 0.09(2) + 0.03(1.5) + 0.02 = 0.57 \text{ min/stop}$$

The answer to the second question is given by:

$$P = 0.0033(75) + 0.16(1) + 0.09(2) + 0.03(1.5) + 0.02$$
$$= 0.65 \text{ min/stop}$$

Thus a 50% increase in the distance between stops results in a 14.5% increase in the time required to service a stop.

A more sophisticated model developed by Stone and Stearns (1969) includes route-related and other factors associated with the set of activities occurring during an entire workday of a collection crew. It is:

$$X_1 = Vt\rho/Q + B + K + D$$

where
- X_1 = the total time in minutes to complete one trip (collect and dispose of one full load)
- V = vehicle volumetric capacity (m^3)
- t = average time per collection stop plus travel time to the next stop (min)
- ρ = average density of refuse in the vehicle (kg/m^3)
- Q = average quantity of refuse per collection stop (kg)
- B = one-way average driving time between route and disposal site (min)
- K = total nonproductive time (min); includes dispatch, lunch and relief, yard to route time, and disposal site to yard time
- D = average disposal time (min/load)

After the first load has been discharged at the disposal site, the crew must decide whether there is sufficient time to return to the collection route for a second load. Factors which must be considered when making this decision are the minimum partial load which the crew is allowed to collect and the policy on overtime. If, for example, it is assumed that the minimum partial load that may be collected is one-eighth of the volumetric capacity of the truck and a maximum of a half hour of overtime may be utilized (for a maximum working day of 510 minutes), the decision is made as described in the following paragraph.

If $X_1 + 2B + D > 510$, the crew's workday ends after collecting and transporting one load to the disposal site. If, however, $X_1 + 2B + D < 510$, the crew will return for a second or third load as time permits. In general, the truck makes a total of n trips, where

$$X_n = (n + a - 1)Vt\rho/Q + (2n - 1)B + K + nD$$

provided $X_n < 510 < X_{n+1}$ and $a > 1/8$. If $a < 1/8$, only $n - 1$ trips are made.

When the above equations have been used to estimate the number of full and partial loads to be collected, measures of efficiency, such as the number of tonnes collected per truck, the total services collected per truck, and the labor cost per tonne of waste collected can be calculated. Equations which can be used for this purpose are:

$$N = n + a - 1$$

$$T = NV\rho/1000$$

$$SC = NV\rho/Q$$

$$L_T = 480 \ CS \ (\text{if } X_n < 480)$$

$$L_T = CS \ \{1.5 \ (X_n - 480) + 480\} \ (\text{if } X_n > 480)$$

$$M_H = L_T/T$$

$$L_c = M_H(b)$$

$$V_c = V_T/T$$

$$C = L_c + V_c$$

In these equations N = daily number of loads per truck, T = total number of tonnes of refuse collected per truck, SC = total services collected per truck, L_T = total person-minutes of labor time, CS = crew size (including driver), M_H = person-minutes per tonne, b = labor cost in dollars per minute, L_c = labor cost per tonne, V_T = vehicle cost, V_c = vehicle cost per tonne, and C = total cost per tonne.

Example 3.2: Suppose a two-person collection crew is collecting in an area where the average quantity of waste per collection stop is 18 kg and the average time required to service a collection stop is 0.55 minutes. The collection vehicle has a capacity of 12.2 m³ and the average density of the waste in the vehicle is 384 kg/m³. The one-way driving time between the collection route and the disposal site is 25 minutes, the average disposal time per load is 15 minutes, and the daily nonproductive time is 60 minutes. Use the Stone/Stearns model to determine the number of loads the crew is able to collect per day.

Solution: We must first determine the time required to fill the vehicle to capacity. This is given by

$$V t \rho / Q = (12.2)(0.55)(384)/18 = 143 \text{ min.}$$

Next, successively determine the values of X_1 and X_2.

$$X_1 = Vtp/Q + B + K + D = 143 + 25 + 60 + 15 = 243$$
$$X_2 = (2 + a - 1) \, Vtp/Q + [2(2) - 1]B + K + 2D$$
$$= (2 + a - 1) \, 143 + 3(25) + 60 + 2(15)$$
$$= (1 + a) \, 143 + 75 + 60 + 30 = (1 + a) \, 143 + 165$$

Setting the latter expression equal to 510 and solving for "a" yields a value of approximately 1.4. Thus a second full load is possible. When a is set equal to 1 in the above equation, it gives the value $X_2 = 451$. Since this number is less than 510, it is necessary to calculate X_3.

$$X_3 = (3 + a - 1) \, 143 + [3(2) - 1] \, 25 + 60 + 3(15)$$
$$= (2 + a) \, 143 + 230$$

Again setting the latter expression equal to 510 and solving for a, the result this time is approximately –0.04. Thus two full loads can be collected, but the time needed to collect and dispose of an additional 1/8 load would exceed the allowable limit of 510 minutes. Formally,

$$N = n + a - 1 = 2 + 1 - 1 = 2$$

Example 3.3: Continuing with the previous example, suppose the labor costs (including fringe benefits) for the two-person crew are $21.00 per hour for the driver and $15.00 per hour for the helper, a total of $36.00 per hour for the two. What is the labor cost per tonne of waste collected?

Solution: The number of tonnes collected can be calculated from the equation:

$$T = NVp/1000$$

This yields

$$T = 2(12.2)(384)/1000 = 9.37 \text{ tonnes}$$

Even though only 451 minutes are required for the crew to do its work, both members must be paid for the full 480 minutes or 8 hours. Thus, the labor cost per tonne is

$$(\$36.00/\text{hr})(8 \text{ hr/day})/(9.37 \text{ t/day}) = \$30.74$$

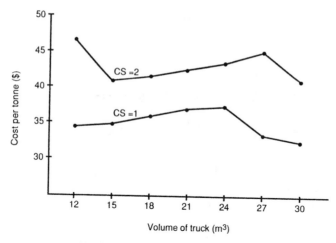

Figure 3.1 Collection cost as a function of truck volume and crew size.

With the assistance of models such as these, simulations can be conducted to help in developing strategies for efficient collection. For example, if one wishes to study the relationship between vehicle volumetric capacity and the cost of collection for different crew sizes, the result is likely to be a set of curves similar to those shown in Figure 3.1. These curves were constructed by applying the Stone/Stearns model to a typical residential collection district. The collection cost data used in this simulation were taken from an article by Miller (1993). From results like those displayed in Figure 3.1, it is clear why there has been a push in recent years for one-member crews.

Studies have been conducted to determine the relationship between the average number of containers at a collection stop and the time in minutes required to service that stop (including travel to the next stop). For collection in typical residential districts, these relationships are those shown in Figure 3.2.

A number of general observations can be made from simulation studies based on the Stone/Stearns model. Among them are:

1. Generally, the use of larger vehicles enables the collection of more services by each crew because with fewer trips to the disposal site a greater portion of the day is spent on the collection route.

2. The ability of the two- and three- person crews to collect a greater number of services per day reduces the total equipment

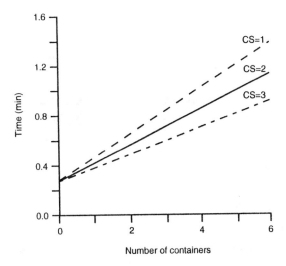

Figure 3.2 Average time per collection stop as a function of the number of containers (Stone and Stearns, 1969).

requirements for the refuse collection operation, thereby reducing total equipment costs. But increased labor costs of the multi-person crews more than offset the reduced equipment costs.
3. The one-person operation is less costly than either the two- or three-person crew, regardless of the truck size. As the total nonproductive time and driving time to the disposal site decrease, the cost differential between crews and the effect of truck size on cost both become less.

3.3 COLLECTION OF RECYCLABLES

Most of the principles discussed in the previous section apply to the collection of recyclables. However, there are a number of specific problems and issues pertaining to the collection of recyclables which go beyond the broad guidelines that apply to collection systems for municipal solid waste. As noted earlier, the effective planning and implementation of collection systems for recyclables requires their integration within the total solid waste management system; a decoupling of the collection component from the processing component is likely to lead to problems and result in ineffective programs.

A number of key issues that must be considered when designing a collection system for recyclables have been identified by Pferdehirt, O'Leary, and Walsh (March, 1993). They are:

1. *Distribution of Sorting Responsibility*:
   How are responsibilities for sorting materials distributed among households, collection crews, and central sorting facilities?

2. *Containers*:
   What types of containers should residents use to set out materials? This decision must be coordinated with the types of materials selected for collection, the distribution of sorting responsibilities, collection frequency, and collection truck type.

3. *Collection Trucks*:
   What types of vehicles will be used to collect materials? This decision must be coordinated with set-out requirements, driver sorting responsibilities, and delivery requirements at the materials recovery facility or market. Capital funding limitations may require the development of an interim strategy focusing on use of existing equipment.

4. *Collection Schedule*:
   How often should materials be collected? This decision must be coordinated with the size of the containers and the schedule for solid waste collection.

5. *Education and Promotion*:
   Methods must be developed to build and maintain participation and compliance with set-out quality standards. Feedback to participants and residents must be built into the program in order to generate enthusiasm.

6. *Funding Mechanism*:
   How will the program be funded? How can funding of recycling and solid waste be coordinated to provide incentives for reduction and recycling? Can savings in waste collection and disposal be used to reduce the cost of recyclables collection?

7. *Links with Markets*:
   What materials are marketable in the region under consideration? What facilities should be developed or used to sort or consolidate collected materials?

8. *Integration of Program with Other Solid Waste Services*:
   The following must be coordinated: population served, materials collected, integration with processing facilities, funding of services, location of facilities, and collection schedules.

COLLECTION ALTERNATIVES

The key issues listed above must be addressed when choosing a recycling collection system. Several basic choices are available. Among them are: drop-off centers, curbside collection of recyclables, mixed waste collection and processing, and buyback centers.

Drop-off Centers

In a drop-off system, residents are required to deliver recyclables to a location where containers are provided for depositing the materials. In some cases the container equipment is permanently located at a given site; in others it is mobile and moved from one location to another. Drop-off centers may have staff members available to provide quality control and assist with unloading, or they may be unattended and rely on the users themselves to place materials in the appropriate containers. Unattended drop-off centers sometimes experience problems with contamination resulting from inappropriate commingling of materials. Use may be limited to certain days or hours, or it may be unrestricted, with open access available around the clock. In most cases separate containers are provided for each type of material. Drop-off centers are generally located at sites to which citizens regularly travel, such as shopping mall parking lots, or at solid waste facilities, such as transfer stations or landfills.

Drop-off centers are often used in rural areas where residential collection services are unavailable. However, they are also used in suburban and urban areas to supplement curbside collection programs. They provide an outlet for citizens who may have missed a recyclables collection day, or serve as a means for collecting certain specialized items, such as used tires or batteries. Drop-off centers for collecting recyclables work best when citizens are accustomed to using drop-offs for solid waste. Often economic analyses of drop-off centers ignore the costs to citizens in delivering the material to the drop-off location.

Curbside Collection

There are numerous options for curbside collection based on the degree of sorting generators must provide, the type of collection vehicle used, whether or not the remaining solid wastes are collected simultaneously, and whether or not a centralized sorting facility is part of the system.

1. *Source Separation/Multibin Truck*:
   This system relies on the free labor of generators to separate materials from one another and place them in separate containers prior to pickup. The vehicle used for collection is specifically

designed to accommodate recyclables and has as many compartments as material types, usually three or four.

2. *Commingled Recyclables/Multibin Truck*:
   In this system recyclables are set out commingled in a container and collection crews sort the materials as they collect them. Crew members then toss them into separate compartments in the collection truck. Obviously this approach requires more labor on the collection route than the source separation/multibin system because of the extra time required for the sorting of materials by crew members. Whether this leads to an overall increase in cost for the solid waste management system depends on the nature of the changes induced throughout the system. For example, since collection workers can screen materials for contaminants, this approach offers the important benefit of obtaining high quality recyclables. The sale of these materials may yield a greater income, thereby offsetting, at least in part, the higher labor costs on the collection route. A variation in this approach requires generators to divide materials into two or three major components—containers and paper, for instance—to make the sorting job by workers easier.

3. *Commingled Recyclables/Central Processing*:
   In this approach residents set out commingled recyclables in containers in the same manner as in the previous option. However, crew members toss them onto the truck in commingled form and transport them to a centralized materials recovery facility for sorting. This approach enables collection crews to spend their time collecting rather than sorting materials. Instead of placing all commingled materials in a single compartment, a variation is sometimes used where materials are segregated into two major components and placed in separate compartments. In this way, for example, it is possible to prevent newspapers from becoming contaminated with broken glass.

4. *Co-Collection System:*
   In a co-collection system recyclables and solid wastes are collected simultaneously in the same truck. In some cases the recyclables and refuse are placed in the same compartment; in others separate compartments are available for the two components. When the two components are mixed in the same compartment, bags with different colors may be used to distinguish the contents. An obvious disadvantage of placing recyclables and solid wastes in the same compartment is the danger of contamination. Also, bags containing recyclables may break open, resulting in the loss of their contents. Yet another disadvantage is a loss in collection

efficiency because compaction cannot be as high. An advantage is that existing equipment can be used and flexibility in the assignment of trucks to collection routes can be maintained. If recyclables and solid wastes are to be placed in separate compartments, new multicompartment vehicles must be purchased or existing trucks must be modified.

5. *Post-Collection Separation:*

In this approach, solid waste is collected without any separation of materials by residents or collection crew members. Collection occurs in the same manner as when no sorting is planned. All separation takes place at a central processing facility where sorting is usually accomplished through handpicking of materials from conveyor belts supplemented with various types of mechanical sorting methods.

As noted in the above paragraphs, a variety of vehicles are available for curbside collection of recyclables. Trucks with adjustable compartment sizes can be manually loaded from the side. Some of them have low profiles for easier loading and to assist in minimizing worker back strain. Other trucks designed specifically for collection of recyclables have side- or front-mounted bins. When the bins are filled they are mechanically raised overhead and the materials dumped into compartments located in the interior of the truck. Manually loaded vehicles generally have less capacity than those which are loaded at the top by mechanical means. Top-loaded vehicles have more broken glass than side-loaded ones.

Trailers are also available for collecting recyclables. They can be towed with a pickup truck or a regular compactor collection truck. As noted above, regular compactor trucks—rear-loading or side-loading—can be used to collect commingled recyclables. In addition, there are specially designed trucks with separate compartments that allow recyclables and solid waste to be collected simultaneously.

The collection of recyclables from multifamily housing units may require special consideration. Planning must take into account the existing procedures for solid waste collection. Commonly used procedures are an indoor trash room, an indoor chute system, and an outdoor collection bin. If an indoor depot is used, recyclables can be collected in the same room or in a nearby area. The containers used for recyclables must be clearly marked or color coded to prevent confusion with the general refuse containers. A high-rise condominium in Miami Beach, Florida, is experimenting with the use of a chute system for collection of recyclables. Residents would dial the type of waste to be dropped, then collection bins would move into place to receive the material (Walsh, Pferdehirt, and O'Leary, 1993). Assuming enough space is available, an outdoor collection bin could accommodate recyclables in a manner similar to an indoor collection area.

Home storage of recyclable materials may be a problem for residents in multifamily housing units. Kitchen areas and storage spaces are typically more limited than in single family units. For this reason, collection systems which allow commingling of wastes may be easier to implement in multifamily units.

Buyback Centers

Buyback centers provide yet another avenue for diverting recyclables from landfills and using them as raw materials for new products. Buyback centers are establishments to which participants can deliver materials in return for a cash payment. Common types of buyback establishments include multimaterial centers operated by private or not-for-profit recyclers, and aluminum can banks. All have as their goal the generation of revenue by processing recyclable materials for sale to industries to be used for manufacturing new products. Aluminum cans and ferrous/nonferrous scrap metals are the backbone of most buyback centers. Other types of materials such as glass and paper are sometimes accepted for little or no cash return. Buyback centers do not offer a means for communities to implement comprehensive recycling programs. Often they are not conveniently located for large numbers of residents, and the range of items they accept is limited.

SOCIAL AND POLITICAL FACTORS

The overall goal in implementing a recycling program is to maximize the recovery of materials in a cost-effective manner. Economic factors and financial issues play a crucial role in the successful implementation of a recycling program. However, a number of social and political issues need to be addressed as well. Participation rates can be significantly affected by giving attention to social and behavioral issues.

A basic decision which must be made is whether a recycling program should be voluntary or mandatory. A voluntary program relies on a "carrot" approach and seeks to encourage participation through education or appeals to support the enhancement of environmental quality. In a mandatory program, citizens are required to participate. In a study of 357 residential curbside recycling programs in the United States, Everett and Peirce (1993) found that mandatory programs collect more material than those in which participation is voluntary. The same study uncovered no clear evidence that mandatory programs accompanied with enforced punishment of recalcitrants results in the collection of more material than when punishments are absent. Generally, mandatory recycling is utilized only with curbside collection programs. Participation is usually not mandated for drop-off programs unless a drop-off system is used for the collection of solid waste as well.

Participation rates are also affected by the frequency and nature of the collection schedule. Collection of recyclables on a weekly basis seems to work best. In cases where schedules have been changed from twice-monthly to weekly collection, the amount of material collected has increased, sometimes substantially. If recyclables are collected on the same day as the remaining solid waste, residents have an easy reminder of when to set them out. However, there is no clear evidence that same-day collection results in larger quantities of materials collected (Everett and Peirce, 1993). Another scheduling option which is sometimes used is to have a weekly pickup of recyclables, but to collect a different type of material each week. Such a program has a built-in disadvantage: residents are required to maintain a calendar as a reminder to set out the appropriate material each week. This can result in confusion and lower participation rates.

Residents, public officials, and collection workers often have strong feelings about the type of container to be used in a curbside recycling program. Color, shape, size, purchase, and replacement policy are factors that are debated when residents consider the appearance and convenience of different types of containers. Collection workers want containers which facilitate the collection task and minimize fatigue and injuries. Rectangular bins, buckets, and plastic bags are the types of containers most commonly used in curbside recycling programs. Paper bags and large wheeled containers are also sometimes used.

If recycling programs are to be successful, an effective education program is a necessity. Educational campaigns targeted at residents, school children, commercial and retail business operators, and other sectors of the population will help any collection program become more effective. Special care must be given to tailor materials to the educational level of the targeted audience.

ECONOMIC ANALYSIS OF COLLECTION ALTERNATIVES

Given the large number of issues that arise when planning a recycling collection program, the selection of a cost-effective alternative is a major challenge. The specific objective is to hold collection costs to a minimum while meeting the overall goals of the recycling program.

In a study conducted by the National Solid Wastes Management Association, a "full-cost" accounting methodology was used to identify and allocate all costs associated with the collection of recyclables (Miller, 1993). All the operating and capital costs, including those for equipment, labor, buildings, land, overhead, and administration, were included in the study. The operating and capital costs were determined on an annual basis for a hypothetical collection route. In designing the hypothetical collection route an attempt was made to mimic a "typical" suburban route. The assumptions on which the simulation was based included the following: newspaper is set out in a bundle; commingled glass bottles, aluminum and steel cans, polyethylene terephthalate (PET) soft drink bottles, and high-density polyethylene (HDPE) milk cartons are placed in a bin; the start-up to on-route time is 30 minutes, with 60 minutes for getting to and from the processing site; residential units are predominantly single family with a distance between households of 45 meters; and the standard work-week is five days, 8.5 hours per day with a base hourly wage of eight hours plus a half hour of overtime.

Costs were tallied for one- and two-person crews and for vehicles with capacities of 17.5 m^3 (23 yd^3) and 23.5 m^3 (31 yd^3). The results are displayed in Table 3.1. The distribution of costs among the cost elements

Table 3.1 Costs for the Curbside Collection of Recyclables

Cost Element	One Person Crew ($/y)	Two Person Crew ($/y)
Labor	33,317	54,736
Fringes	11,597	21,273
Vehicle operation and maintenance	31,983	31,983
Building and utilities and other expenses	8,083	10,442
Vehicle depreciation		
17.5 m^3 capacity	7,857	7,857
23.5 m^3 capacity	13,986	13,986
Subtotal		
17.5 m^3 capacity	92,837	126,291
23.5 m^3 capacity	98,966	132,420
Administrative (12%)		
17.5 m^3 capacity	11,140	15,155
23.5 m^3 capacity	11,876	15,891
Total		
17.5 m^3 capacity	103,977	141,446
23.5 m^3 capacity	110,842	148,311

Source: Miller, C., The cost of recycling at the curb, *Waste Age*, 24(10), 46–54, October 1993. Copyright © 1993 by the National Solid Waste Management Association. Reprinted with permission.

Table 3.2 Cost Distribution in the Curbside Collection of Recyclables

Cost Element	One Person Crew	Two Person Crew
Labor and fringes	38–46%	47–56%
Vehicle operation and maintenance	30%	22%
Building and utilities	4%	3%
Other expenses	4%	5%
Vehicle depreciation	7–13%	5–10%
Administration	12%	12%

Source: Miller, C., The cost of recycling at the curb, *Waste Age*, 24(10), 46–54, October 1993. Copyright © 1993 by the National Solid Waste Management Association. Reprinted with permission.

for one- and two- person crews are displayed in Table 3.2. Depending on the recycling participation rates, the cost per tonne ranged from $90 to $135. Obviously, the specific cost information obtained from this study cannot be applied directly to a given situation, but the results from the study give an indication of the ranges in the costs associated with the collection of recyclables.

A recent innovation in the planning process has been the use of computer models to assist in making decisions about recycling alternatives and collection options. These models cannot replace thoughtful planning, but they are useful tools for exploring "what if" questions, identifying those assumptions which have significant impacts on program effectiveness, and narrowing the range of options to be considered. Kohrell et al. (1992) have studied a number of commercially available computer software packages (kNOwWASTE, RECYCLE, RecycleWare, PRIDE, and WastePlan) and used them to conduct several simulations based on various assumptions concerning collection programs for recyclables, city sizes, and participation rates. The computer simulations produced a number of results which led to the following observations:

1. All recycling alternatives cost more per household than the no-recycling alternative when there are negligible revenues from the sale of recovered materials and the tipping fees are low (i.e., in the range of $10–20 per tonne). For a given revenue from the sale of recyclables, the cost difference tends to decrease linearly as the tipping fee increases.

2. Because the cost difference decreases as the tipping fee increases, at some value of the tipping fee each of the recycling collection alternatives becomes more economical than the no-recycle alternative. For larger cities this point of economic feasibility is attained at a lower tipping fee value than that for medium-sized or smaller cities.

3. Drop-off systems tend to be the lowest cost recycling alternative (on a $/tonne basis) when there are negligible revenues from the sale of recyclables and the tipping fees are low. As the value of recyclables or the tipping fee increase, the economics improves with greater participation in the program and, therefore, curbside collection becomes more viable.

4. Of the curbside collection alternatives, the use of one truck for the simultaneous collection of recyclables and solid waste tends to be the least costly. However, if a co-collection system is used, the recyclables may be difficult to market, which may make this system more expensive than initial conclusions would suggest. The weekly collection of recyclables with a multibin collection truck tends to be the most costly. However, materials collected under this system are likely to be the most marketable, thereby yielding greater revenues which offset, at least in part, higher collection costs.

3.4 COLLECTION ROUTING AND SCHEDULING

Systems for collecting municipal solid waste or recyclables should include routes which minimize travel distances and travel times during pickup. A variety of tools are available to assist with the development of well-balanced and efficient collection routes. These tools range from heuristic or "rule-of-thumb" techniques to sophisticated computer programs designed specifically for the purpose of selecting efficient collection routes and schedules.

Well-balanced collection routes are those for which all collection crews spend approximately the same amount of time in productive work each day without accumulating significant amounts of overtime. Here, productive work means the time spent on all tasks, including collection, travel, unloading, and dispatch times. In short, the buyer's goal is to receive "a day's work for a day's pay" (Bodin et al., 1993a). The process of determining well-balanced collection routes is sometimes called macro-routing.

Figures 3.3, 3.4, and 3.5 illustrate the concept of a well-balanced collection schedule (Bodin et al., 1993a). In Figure 3.3 the workload is well-balanced, no overtime has accrued, but "the gap" indicates that there has

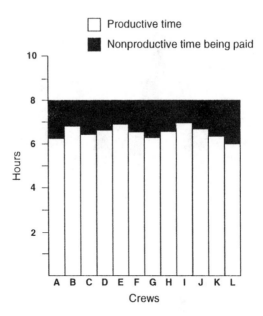

Figure 3.3 A well-balanced collection schedule with a considerable amount of non-productive time (Bodin, L., Fagan, G., and Levy, L., The RouteSmart ve-hicle routing and scheduling system, paper presented at Conference on Cost-Effective Collection of Recyclables and Solid Waste, Madison, WI, June 2–4, 1993a). Reprinted with permission.

been significant nonproductive time. Thus, "a day's work for a day's pay" has not been achieved. Figure 3.4 displays a situation where workloads are out of balance, some overtime has been accumulated, and the question of whether the buyer is receiving "a day's work for a day's pay" cannot be clearly answered. The goal is to achieve the type of schedule displayed in Figure 3.5. Workloads are balanced, overtime is modest, and the em-ployer receives "a day's work for a day's pay."

A second goal which is sought in tandem with balanced routes and schedules is to minimize the number of routes and the total deadhead time accumulated when traveling over a street twice. This goal provides the pathway to significant savings in solid waste collection. The process of designing routes for achieving this goal is often referred to as micro-routing.

An important heuristic technique for macro-routing is to divide the total area into districts, each of which is to be serviced by all crews in a given day. Railroad embankments, rivers, heavily traveled streets, and other types of barriers need to be taken into account when determining such districts. Once these districts are identified they can be divided into routes for individual crews.

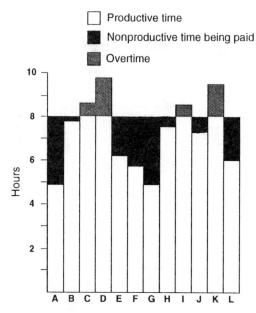

Figure 3.4 A poorly balanced collection schedule (Bodin, L., Fagan, G., and Levy, L., The RouteSmart vehicle routing and scheduling system, paper presented at Conference on Cost-Effective Collection of Recyclables and Solid Waste, Madison, WI, June 2–4, 1993a). Reprinted with permission.

Schur and Shuster (1974) have identified a number of heuristic principles which can be applied to both macro-routing and micro-routing. They are:

1. Routes should not be fragmented or overlapping. Each route should be compact, consisting of street segments clustered in the same geographical area.
2. Total collection plus haul times should be reasonably constant for each route in the community (equalized workloads).
3. The collection route should be started as close to the garage or motor pool as possible, taking into account heavily traveled and one-way streets (see Rules 4 and 5).
4. Heavily traveled streets should not be collected during rush hours.
5. In the case of one-way streets, it is best to start the route near the end of the street with the highest elevation, working down it through the looping process displayed in Figure 3.6.
6. Services on dead-end streets can be considered as services on the street segment that they intersect, since they can only be

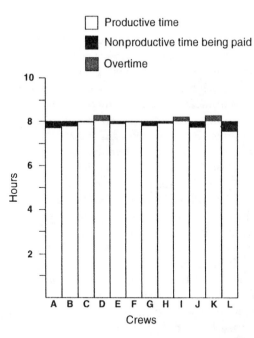

Figure 3.5 A well-balanced and productive collection schedule (Bodin, L., Fagan, G., and Levy, L., The RouteSmart vehicle routing and scheduling system, paper presented at Conference on Cost-Effective Collection of Recyclables and Solid Waste, Madison, WI, June 2–4, 1993a). Reprinted with permission.

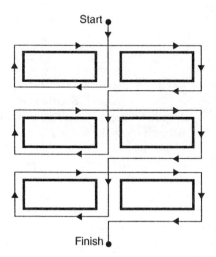

Figure 3.6 Routing pattern for one-way street collection (Schur, D.A. and Shuster, K.A., Heuristic Routing for Solid Waste Collection Vehicles, USEPA Office of Solid Waste Report SW-113, 1974).

collected by passing down that street segment. To keep left turns at a minimum, collect the dead-end streets when they are to the right of the truck. They must be collected by walking down, backing down, or by making a U-turn.

7. Waste on a steep hill should be collected, when practical, on both sides of the street while vehicle is moving down hill. This facilitates safety, ease, and speed of collection. It also lessens wear on vehicle and conserves gas and oil.

8. Higher elevations should be at the start of the route.

9. For collection from one side of the street at a time, it is generally best to route with many clockwise turns around blocks. (Heuristic Rules 8 and 9 emphasize the development of a series of clockwise loops in order to minimize left turns, which generally are more difficult and time-consuming than right turns. Especially for right-hand-drive vehicles, right turns are safer.)

10. For collection from both sides of the street at the same time, it is generally best to route with long, straight paths across the grid before looping clockwise.

11. For certain block configurations within the route, specific routing patterns should be applied.

Figures 3.6, 3.7, and 3.8 provide examples of heuristic tools which can be applied, depending upon the type of block patterns which are encountered. Working with street maps, it is possible to combine various approaches to systematically plot timesaving routes.

In addition to heuristic techniques, more sophisticated approaches based on mathematical and computer models are available to assist with the development of good routes and schedules for solid waste collection vehicles. These tools address two classes of problems: node routing and arc routing problems. An example of a node routing problem in solid waste management is the servicing of drop-off facilities situated at specific, discrete locations within a region. Routing of vehicles to deliver goods or to provide services at a specific set of locations within a geographical region is a common example of a node routing problem not in the field of solid waste management. Arc routing and scheduling problems are those for which each street requires a specific service. Household solid waste collection and door-to-door delivery of telephone books are specific examples.

A classical mathematical problem which underlies the node routing problem is the **traveling salesman problem**. In the traveling salesman problem the objective is to find a minimum distance tour a salesman would follow in visiting N cities, returning at the end to his point of origin. It is assumed that the distance between each pair of cities is known. If the distance from city A to city B is the same as the distance from city

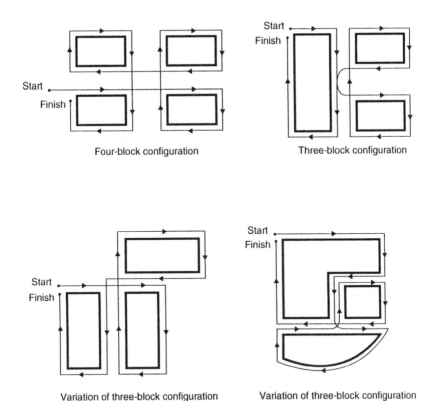

Four-block configuration   Three-block configuration

Variation of three-block configuration   Variation of three-block configuration

Figure 3.7 Routing patterns for three- and four-block configurations (Schur, D.A. and Shuster, K.A., Heuristic Routing for Solid Waste Collection Vehicles, USEPA Office of Solid Waste Report SW-113, 1974).

B to city A, the traveling salesman problem is said to be symmetric. If these two distances are different, which might be the case, for example, when one-way streets are involved, the problem is said to be an asymmetric traveling salesman problem.

Similarly, a classical mathematical problem underlies the arc routing and scheduling problem. Called the **Chinese postman problem**, its objective is to find a minimum distance tour a postman would traverse through a street network in order to deliver the mail to the residents living on those streets.

The traveling salesman and Chinese postman problems can both be envisioned with the aid of a graph, a mathematical construct consisting of a set of nodes, and a set of arcs. The arcs join some or all of the nodes. Two examples of graphs are displayed in Figure 3.9. In the traveling salesman problem, the nodes represent the cities and the arcs are the

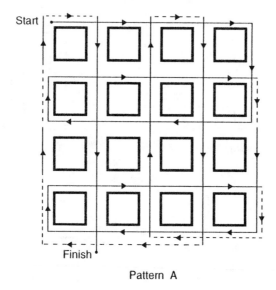

Pattern A

Pattern B

Figure 3.8 Routing patterns for multiblock configuration (Schur, D.A. and Shuster, K.A., Heuristic Routing for Solid Waste Collection Vehicles, USEPA Office of Solid Waste Report SW-113, 1974).

highways joining them. Associated with each arc is a number representing the distance from one city to another. The objective is to construct a tour which includes all the cities, beginning and ending at the salesman's home base, so that the total distance is a minimum. In the Chinese post-

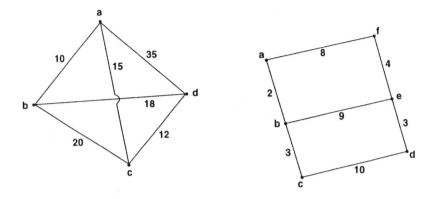

Figure 3.9 Example of graphs. (Note that graphs are not drawn to scale.)

man problem the arcs are the streets and the nodes are the intersections in the street network. Numbers associated with the arcs indicate the lengths of the streets. The objective is to construct a tour which includes all the arcs; again, the total distance is to be a minimum.

Exact and approximate solution procedures exist for solving both of these problems, even when the number of cities or streets is quite large. A few of the simplest types of solid waste collection and scheduling problems can be modeled as either a traveling salesman or Chinese postman problem and, in many of these cases, solved using one of these solution procedures. Gottinger (1991) provides references where these models have been used to analyze solid waste management problems. However, in most cases the solid waste management problems are considerably more complex, involving important aspects which cannot be modeled as a simple traveling salesman or Chinese postman problem.

Problem characteristics which typically arise in solid waste collection and which require more sophisticated problem-solving approaches include the following:

1. Collection fleets consist of multiple vehicles of several different types.
2. Service locations or areas may require certain vehicle types.
3. The vehicle fleet may be housed in several depots. Each depot may have a limit on the number of vehicles which can be housed there.
4. Some service locations may have time windows; that is, they must be serviced within a given time period.

    5. Specific street or road designs must be taken into account when computing travel distances. For example, in some cases U-turns may be allowed, while in others they may be prohibited.

    6. Multiple objectives must be taken into account.

Computer software packages have been developed which take into account some or all of these problem elements. The most elementary ones require manual entry of customer locations and use euclidean distances (crow's flight distances) to estimate travel distances and times.

More sophisticated ones incorporate a geographic information system (GIS) and advanced routing algorithm components. A GIS is a computer software system which has capabilities for storing, managing, analyzing, and displaying spatially referenced data. In the last decade or so GISs have become available which incorporate geographical data at the block level. For example, data can be purchased from the United States Census Bureau which can be converted into a digital street map of a given region. Such information is available for most parts of the United States. These databases allow the user to work with actual distances traversed when a vehicle travels from one location to another in a street network. When combined with good routing algorithms, similar to those which have been developed to solve traveling salesman and Chinese postman problems, a computer package of this type can be used to generate detailed travel paths to meet the objectives of a solid waste collection system.

RouteSmart (Bodin, Fagan, and Levy, 1993b) is an example of a computer package which incorporates many of the features described above. The developers of this package state that it offers the potential for communities to save substantial amounts of money. In one application, Oyster Bay, New York, was able to save three out of 40 vehicles and $750,000 a year. In the near future, computerized routing and scheduling systems similar to RouteSmart are likely to be a part of an organization's management information system.

3.5 TRANSPORT OF SOLID WASTE

Solid waste is usually transported from the collection district to the disposal site in a collection vehicle. However, sometimes the distance to the disposal site is so great that hauling the waste there in collection vehicles would require crews to spend great amounts of time away from the fundamental task of collection, thereby lowering productivity to unacceptable levels. In such cases it may be feasible to construct a transfer station to which collection vehicles can deliver waste for transfer to large transport vehicles for the long haul to the disposal site. The transport vehicles are able to haul the waste to the disposal site in bulk quantities in a cost-effective manner because the only labor requirement enroute is that of the driver.

Clearly the question of whether a transfer station is economically feasible depends very strongly on the distance between the waste generation area and the location of the disposal facility. While the question of feasibility must be answered on a case-by-case basis, most experts suggest that transfer stations are difficult to justify unless this distance, at a minimum, is in the range of 16-24 kilometers (Walsh, Pferdehirt, and O'Leary, 1993). In a given case, a careful analysis must be conducted in which the cost of direct haul to the disposal site is compared with the cost of bulk transport by way of a transfer station. Studies have shown that the relationship between the two, as a function of the one-way haul distance from the transfer station to the disposal site, is described by curves of the type displayed in Figure 3.10 (Wilson, 1981). The intersection point of the direct haul and transfer station curves identifies the distance at which a transfer station becomes feasible.

The overall cost of the bulk delivery alternative is dominated by the fixed cost of the transfer station. This is seen in Figure 3.10 by noting that the transfer station curve intersects the vertical axis at a point quite far from the origin. While the cost value represented by this point includes the fixed portion of the haul cost and the haul cost in the collection

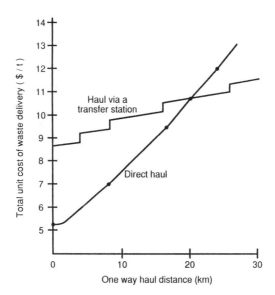

Figure 3.10 Comparative costs of waste delivery to a disposal site by direct haul and via a transfer station (Wilson, D.C., *Waste Management: Planning, Evaluation, Technologies*, Oxford University Press, New York, NY, 1981).

vehicle to the transfer station, the major component is the fixed cost of the transfer station. The comparative advantage in the operating cost of transport in bulk quantities over that for direct haul is depicted by the fact that the cost curve for the former rises less steeply than that for the latter. The step function behavior of the transfer station curve is caused by changes in the number of trips each vehicle can make per day. It should also be noted that the point where the direct haul curve intersects the vertical axis represents the cost of transporting the waste in the collection vehicle to the transfer station location.

Transfer stations provide several potential benefits which may not show up clearly in equations derived from cost analyses. These include an increased flexibility in choosing disposal and processing facilities, reduced maintenance costs for collection vehicles, improved operating efficiency at disposal facilities, and additional opportunities for the recovery of recyclables. If direct haul is the mode of operation, the choice of disposal facilities is usually very limited. A transfer station may open up new alternatives. A wider range of choices may enable a community or company to take advantage of lower rates from competing facilities.

Since roadways at transfer stations are nearly always paved, the wear and tear on collection vehicles is usually considerably less than when the waste is hauled directly to a landfill. Roadways at landfill sites are much more difficult to maintain because they extend into the landfill working areas. Greater numbers of flat tires occur when collection vehicles are required to traverse areas where recently deposited waste is merely compacted or, at best, is covered with a thin layer of fill material. Also, landfill access roads are hard on the transmissions of collection vehicles.

Operating efficiencies at disposal sites can be significantly improved when large transport vehicles are used, simply because fewer of them are required to enter the working area. The result is less traffic congestion and less interference with the daily working operations at the facility than is the case when all collection vehicles are required to enter the site.

Transfer station facilities also provide an opportunity for the recovery of materials prior to disposal. Wastes can be sorted to recover certain types of recyclables or to remove hazardous materials which may cause problems at a disposal facility. Sorting of recyclables or compostable materials can be done manually or with the assistance of mechanical equipment. Transfer stations can also provide locations for recycling drop-off containers or as sites for dropping off household hazardous wastes, used oil and tires, and used batteries.

TYPES AND SIZES OF TRANSFER STATIONS

At their very simplest, transfer stations consist of little more than a concrete slab and a front-end loader. Collection vehicles dump the wastes on the slab and the front-end loader places them in a semitrailer or truck for transport to the disposal site. Another simple possibility is for the waste to be dumped into a pit. Stationary or mobile equipment can then be used to load the waste into a semitrailer with or without compaction.

A second major design type is a dump station in which collection vehicles dump waste directly into open-top trailers. Hoppers direct the waste into the trailers and clam-shell equipment can be used to distribute it and, perhaps, provide a minimal amount of compaction. A disadvantage of this approach is that it does not allow the waste to be stored at the transfer station for a time, should bottlenecks occur at some point in the system.

Many transfer stations include stationary compaction equipment. Collection vehicles dump the waste into a pit from which a hydraulically operated stationary ram pushes the waste onto a trailer and compacts it. The greater density creates maximum payloads for transporting the waste to the disposal site.

Transfer stations are typically located in areas zoned for industrial use. It is important that they be located near intersections which allow quick access to major highway networks. They are usually housed in simple buildings which provide cover for unloading, storing, and loading of wastes and restrict litter to their inner confines.

To determine the appropriate size of a transfer station, accurate information is needed on present and projected waste quantities within the service region. A number of other factors need to be considered as well. According to Walsh, Pferdehirt, and O'Leary (1993) these include: the capacity of collection vehicles using the facility, the desired number of days of storage space on the tipping floor or pit area, the time required to unload the collection vehicles, the number of vehicles that will use the station and their expected days and hours of arrival, waste sorting and processing to be accomplished at the facility, transfer trailer capacity, hours of station operation, availability of transfer trailers waiting for loading, time required to attach and disconnect trailers from tractors, or to attach and disconnect trailers from compactors, and time required to load trailers. They go on to say that owners of transfer stations often complain of overly cramped quarters and wish that tipping and storage areas had been made larger.

TYPES OF TRANSPORT SYSTEMS

Thus far in the discussion of solid waste transport systems it has been assumed that semitrailers provide the mode of transportation for distance hauling. While trucks and semitrailers are most commonly used, railroad cars and barges are also utilized.

Compaction semitrailers usually have a capacity in the 50 to 96 m^3 range and are equipped with an unloading blade which is powered by the tractor's hydraulic system or by a separate engine provided for this purpose. Compaction trailers are enclosed and are specially reinforced to withstand the added stress from the compaction operation. For this reason they are heavier than most noncompaction trailers even though their volume is smaller.

Noncompaction trailers are constructed with lighter materials than those used for compaction trailers and are often open at the top. Some noncompaction trailers are equipped with push-out blades for unloading; others are unloaded with the aid of a hydraulic lift which tips the trailer to an angle sufficiently large to enable the waste to slide off.

As distances to landfill sites have increased in recent years, railroad transfer has become more common. With distances of 80 km (50 mi) or more to disposal sites, rail transport becomes more and more economically attractive. A variety of equipment options are available, including dedicated boxcars, containerized freight systems, and boxcars with removable roofs. In some cases, waste is baled to facilitate handling and increase payloads. The payload for a railroad boxcar is in the range of 80 to 90 tonnes, while that for a semitrailer is 18 to 23 tonnes. Railroad transfer stations are more expensive to construct than truck transfer stations designed to accommodate similar quantities of waste because of the added cost of rail lines and the special equipment needed to load the rail cars.

A barge transport system exists in New York City, but on the whole they are quite rare in the United States. Such systems are more common in Europe. When navigable waterways provide a passage from a source of waste generation to a disposal site, barge transport can be economically viable. Payloads are large; as much as 500 to 600 tonnes of waste can be accommodated on a single barge. Environmental concerns arising from the construction of marine transfer stations and from the siting of landfills near waterways are likely to lead to a decline of barge transport in the future.

3.6 FACILITY LOCATION MODELS

As noted in the previous section, the location of solid waste facilities can have an important effect on collection and hauling costs. Solid waste systems which include facilities located at a distance from the sources where the waste is generated incur greater equipment and labor costs than those where the facilities are close at hand. Many factors must be taken into account when deciding where to locate a facility. For example, when a site for a sanitary landfill is to be chosen, hydrogeological conditions are of paramount importance. For almost any type of solid waste facility, access from major highways and number of nearby dwellings are also important factors to consider. Nevertheless, in most solid waste planning problems, precise information on the cost of transporting solid waste to alternative sites can be very helpful in the decision-making process.

Mathematical models exist which can identify an optimal location for a solid waste facility based on estimates of costs incurred when waste is transported from the sources where it is generated to the facility location. With a knowledge of this optimal solution and its sensitivity to changes, a decision-maker is able to effectively incorporate this information into a larger decision framework where the full range of factors having a bearing on the choice of a facility location is considered. In this section we discuss these models and illustrate the way in which the information derived from them can be used to assist in identifying locations for solid waste facilities.

The models are based on the assumption that transportation costs are proportional to distance (or travel times). Construction of the models requires that a coordinate system be superimposed on a map of the geographical region in which a solid waste management system is being developed. If there are m sources of solid waste in the region, the coordinate system can be used to identify centroid coordinates (x_1,y_1), (x_2,y_2), . . . , (x_m,y_m). The facility location problem is to find the location (x,y) such that

$$z = f(x,y) = \sum_{i=1}^{m} w_i \, d_i(x,y)$$

is a minimum. In this formulation w_i is a weight which measures, at least in a relative sense, the quantity of waste arising at source i and $d_i(x,y)$ is the distance from source i to the unknown location, (x,y), of the solid waste facility.

In formulating this model one has to determine an appropriate metric for measuring distances. The **euclidean** or **straight-line metric**,

$$d_i(x,y) = \sqrt{(x - x_i)^2 + (y - y_i)^2}$$

by itself is hardly appropriate because garbage trucks do not travel from one point to another in a road network as the "crow flies." The **metropolitan metric**,

$$d_i(x,y) = |x - x_i| + |y - y_i|$$

may be appropriate if travel occurs over a road or street network which is laid out in a grid-like fashion, similar to that in lower Manhattan in New York City. In this regard it is interesting to note that the metropolitan metric is sometimes called the Manhattan metric.

In many applications a multiple of the euclidean metric can be used to adjust crows-flight distances so that they approximate actual distances traversed over a road network. Statistical studies have been conducted by Love and Morris (1979) to determine appropriate constants for different types of road networks. If a constant multiple, k_i say, of the euclidean metric is used to approximate travel distances, the structure of the model given above remains intact because the constant $k_i\, w_i$ can be replaced with a new constant a_i. This change will alter the value of z in the optimal solution but not the optimal location (x,y).

Using an iterative algorithm developed by Weiszfeld (1937) it is possible to solve facility location problems with a large number of sources. After selecting an initial estimate (x^0, y^0), successive values are calculated using the formula

$$T(x,y) = \frac{\displaystyle\sum_{i=1}^{m} \frac{w_i}{d_i(x,y)}(x_i,y_i)}{\displaystyle\sum_{i=1}^{m} \frac{w_i}{d_i(x,y)}}$$

In nearly every case the algorithm converges to the optimal solution. The only exception occurs when one of the iterates coincides with a centroid. When two successive iterative values are the same, or within a specified tolerance, an optimal point can be identified.

When the metropolitan metric is used, the facility location problem is easy to solve. In that case the function $z = f(x,y)$ can be separated into two parts, one a function of x and the other a function of y. Each is a piece-wise linear function whose segments have increasing slopes from left to right. To obtain a solution one looks for the point where the slopes of successive segments change from negative to positive or for a segment with zero slope. In the first case a unique optimal solution can be identified from the point of intersection of the segments, and in the second case the points on the horizontal segment yield multiple optimal solutions.

By itself, knowledge of an optimal solution is often insufficient to assist the decision-making process in determining a location for a solid waste facility. When other criteria are taken into account, the optimal location determined by the model may be precluded from consideration. If the optimal solution is augmented with results from sensitivity analyses; that is, information which describes the manner in which the value of the objective function changes as deviations from the optimal location occur, the location model is much more likely to be of value. Sensitivity analyses can be accomplished nicely through the use of a computer-generated contour mapping program which constructs a map displaying equal cost contours. These contours display at a glance the increase in transportation

Table 3.3 Municipalities, Locations, and Waste Quantities for the Brown County Application

Municipality	Centroid Coordinates	Annual Waste Quantities (tonnes)
1	(11.3, 17.1)	6286
2	(8.3, 15.8)	8545
3	(13.0, 15.5)	1979
4	(9.6, 14.3)	3224
5	(7.9, 14.0)	2860
6	(13.5, 18.4)	17444
7	(10.0, 19.0)	21236
8	(7.5, 22.3)	4657

Source: Wenger, R.B., and Rhyner, C.R., Solving facility location problems with computer contour mapping, *Journal of Environmental Management*, 27(4), 429–436, 1988.

costs that occur if a location is chosen which differs from an optimal one. They enable a decision-maker to examine trade-offs when alternative locations are under consideration. A more complete discussion of the techniques used to construct cost contours can be found in Wenger and Rhyner (1988).

As an illustration, an application to Brown County, Wisconsin, is presented. A site is to be found for a sanitary landfill that is to serve eight municipalities which generate varying quantities of solid waste. The data are displayed in Table 3.3.

Using Weiszfeld's algorithm, the optimal location for the sanitary landfill is found to be near the point (11, 18). The cost contours are displayed in Figure 3.11. The results derived from models of this type seldom provide "the answer" to the question of where solid waste facilities should be located. However, they often provide significant insights to those who have the responsibility for making decisions.

3.7 OTHER MATHEMATICAL MODELS IN WASTE MANAGEMENT

In the last two or three decades, a number of mathematical models have been developed to assist decision-making processes and planning in the waste management field. By themselves, these models seldom provide specific answers to solid waste management problems but, as noted in the previous section, they often provide insights to assist with the broader decision-making process.

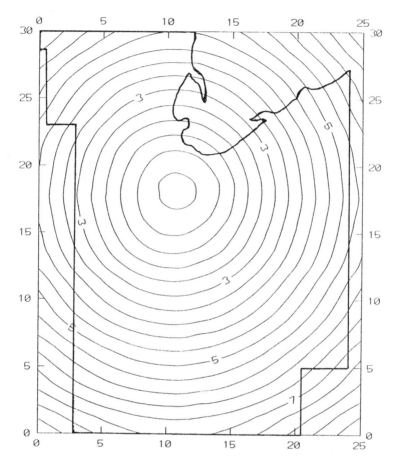

Figure 3.11 Equal cost contours for facility location example (Wenger, R.B. and Rhyner, C.R., Solving facility location problems with computer contour mapping, *Journal of Environmental Management*, 27(4), 429, 1988). Reprinted with permission.

Most of the models which have been developed pertain to collection, routing, and scheduling problems. Some examples have been discussed in previous sections of this chapter. These models focus primarily on day-to-day management problems and issues.

Another area of model development has been concerned with long-range planning issues on a regional basis. The objective of regional planning models can be described as follows: Given the potential locations of intermediate processing facilities and landfills; the location and capacities of existing facilities; collection, transportation, processing, and fixed costs of facilities; and the quantities of waste generated at the sources; determine which facilities should be built or retained and how the waste should be routed, processed, and disposed of so that the overall cost of the system is a minimum. In some models the collection system is decoupled from the processing and disposal system, but in others the entire system is taken into account.

In the following paragraphs a specific example of such a model is presented. It is based on the diagram displayed in Figure 3.12. The notation used in the model is described first.

Indices:

$i = 1, 2, \ldots, I$ identify the sources of solid waste

$j = 1, 2, \ldots, J$ identify the processing facilities (materials recovery facilities or incinerators)

$k = 1, 2, \ldots, K$ identify the transfer stations

Variables:

x_{ij} = amount of waste sent from source i to processing facility j (tonnes)

x_{ik} = amount of waste sent from source i to transfer station k (tonnes)

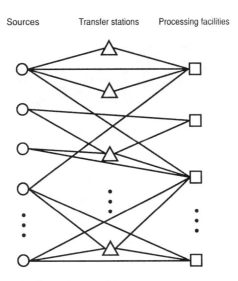

Sources   Transfer stations   Processing facilities

Figure 3.12 Schematic of a regional solid waste management system.

x_{kj} = amount of waste sent from transfer station k to processing facility j (tonnes)

Z_j = total amount of waste entering processing facility j (tonnes)

Y_k = total amount of waste shipped through transfer station k (tonnes)

W_j = a binary variable which is 1 if processing facility j is to be constructed and 0 if not

V_k = a binary variable which is 1 if transfer station k is to be constructed and 0 if not

Parameters:

E_j, E_k = operating costs of processing facility j and transfer station k, respectively ($/tonne)

F_j, F_k = fixed costs of processing facility j and transfer station k, respectively ($/tonne)

C_{ij}, C_{ik} = cost of transporting waste from source i to processing facility j and from source i to transfer station k, respectively ($/tonne)

C_{kj} = cost of transporting waste from transfer station k to processing facility j ($/tonne)

Q_i = quantity of waste generated at source i (tonnes)

C_j, C_k = maximum capacity of processing facility j and transfer station k, respectively (tonnes)

The model then takes the form

$$\text{Minimize } Z = \sum_j (E_j Z_j + F_j W_j - I(Z_j)) + \sum_k (E_k Y_k + F_k V_k)$$
$$+ \sum_i \sum_j C_{ij} X_{ij} + \sum_i \sum_k C_{ik} X_{ik} + \sum_k \sum_j C_{kj} X_{kj}$$

subject to the conditions in the following constraint set:

$$\sum_j X_{ij} + \sum_k X_{ik} = Q_i, \quad \text{for all } i$$

$$\sum_i X_{ik} - Y_k = 0, \quad \text{for all } k$$

$$\sum_j X_{kj} - Y_k = 0, \quad \text{for all } k$$

$$\sum_i X_{ij} + \sum_k X_{kj} - Z_j = 0, \text{ for all } j$$

$$Y_k - C_k \ V_k \le 0, \text{ for all } k$$

$$Z_j - C_j \ W_j \le 0, \text{ for all } j$$

$$x_{ik} \ge 0, \text{ for all } i, \ k$$

$$x_{ij} \ge 0, \text{ for all } i, \ j$$

$$x_{kj} \ge 0, \text{ for all } j, \ k$$

$$W_j = 0 \text{ or } 1, \text{ for all } j$$

$$V_k = 0 \text{ or } 1, \text{ for all } k$$

The expression which is to be minimized is called the objective function. The first sum in this expression represents the cost of constructing and operating processing facilities minus the income received from the sale of products generated by the facilities (in the case of an incinerator the product could be heat in the form of hot water). The term $I(Z_j)$ represents this income; the units are dollars/tonne. The second sum in the objective function represents the cost of constructing and operating the transfer stations. The third sum contains the cost of transporting wastes from sources to facilities and from transfer stations to processing facilities.

The inequalities and equations which occur after the "subject to" statement are referred to as constraints. They represent conditions which must be satisfied by the solid waste system. The first four equations in the constraint set are conservation of flow requirements. The first equation ensures that the waste from each source goes somewhere. The second equation requires that the sum of the waste arriving from different sources equals a transfer station throughput, and the third ensures that the throughput goes to a set of processing facilities. The fourth equation requires that the sum of the waste arriving from sources and transfer stations equals the throughput for a processing facility.

The fifth and sixth constraints are inequalities which ensure that the maximum capacity of each facility is not exceeded. The inequalities in lines seven through nine prevent negative flows from occurring in the system. The last two constraints ensure that the binary variables take on only the values of 0 and 1.

To be complete, landfills should have been included in the system because it is never possible to recover all materials for useful purposes. Incinerators generate ash and residues remain after recyclables have been recovered at a materials recovery facility; both are normally landfilled.

These items can be accounted for in the model, in part at least, by considering any residues as additional products. The cost of disposal of these residues can be included in the $I(Z_j)$ term as a negative income.

A version of the model described above was used by Wenger and Hansen (1983) to study the income derived from the sale of heat in Danish incinerators. Gottinger (1991) discusses a number of case studies based on models similar in structure to the one given above. He also provides an excellent overview of mathematical models that have been developed over the last two or three decades in the solid waste management field.

The development of sophisticated mathematical models in the solid waste management field has occurred in parallel with the appearance of improved computer hardware and software resources. The fields of management science and operations research have contributed much to this overall developmental effort. An excellent textbook in the operations research field which contains some examples from solid waste management is *Introduction to Operations Research* by Hillier and Lieberman (1990). Mathematicians have also contributed by discovering new algorithms which can be used to solve large problems in an efficient manner. It is not likely the mathematical models will ever replace decision-makers in the solid waste management field, but they can be important tools for aiding decision-making processes.

DISCUSSION TOPICS AND PROBLEMS

1. How is the solid waste collected in your community? Is residential waste collected by a municipality or by private haulers (or perhaps a combination of the two)? Who collects the commercial wastes?

2. If your community has a municipal collection system, what is the amount in the municipal budget to cover its costs? How have these costs changed over the years? What are the costs on a per tonne basis? If the latter figure is not available, estimate it on the basis of per capita generation rates from Chapter 2.

3. As products become lighter (for example, glass containers are replaced with plastic ones), what problems arise in making comparisons on a cost/tonne basis?

4. After observing the collection services available in your community, what recommendations would you make to improve efficiency or lower costs?

5. Does your community have a program for collecting recyclables? If so, is participation required or is it voluntary? What recyclables are collected and how are they collected? Where do the recyclables go once they are collected?

6. If your community has a recycling program, how many tonnes of the various materials are collected in a given time period? What is the participation rate and how might it be improved?

7. Are there any buyback centers in your community? What types of materials do they accept and how much do they pay for them? Check the prices on several different days over a period of time. What kind of variation do you observe in the prices?

8. Apply the Stearns productivity model to the case of a one-person crew by assuming that there are two containers and two throw-away items at each service stop. Construct a curve which displays the productive collection time per collection stop, P, as a function of the distance, D, between stops. Take values for the distance between collection stops in increments of 10 ranging from 30 to 100.

9. Take a map of a region in a city and use the heuristic routing guidelines to design a good collection route. Find out what method the waste collection agency in your community uses to route collection trucks.

10. Suppose a single crew member is collecting solid waste in a neighborhood where the average quantity of waste per collection stop is 22 kg and the average time required to service a collection stop is 0.6 minutes. The vehicle capacity is 15 m^3 and the average density of the waste in the vehicle is 390 kg/m^3. The one-way driving time between the collection route and the disposal site is 30 min, the average disposal time per load is 10 min, and the daily nonproductive time is 60 min. Use the Stone/Stearns model to determine the number of loads the worker is able to collect per day. If the labor costs are $25.00 per hour and the vehicle costs are $150.00 per day, including depreciation, what is the cost per tonne of waste collected?

11. Suppose the first network in Figure 3.9 represents a traveling salesman problem. Identify all possible tours starting at a and such that b, c, and d are all visited exactly once. Is there a tour of shortest length?

12. Find a shortest tour which will cover all the "streets" in the second network displayed in Figure 3.9.

13. Use a programmable calculator or a computer and write a program for Weiszfeld's algorithm. Use it to verify the stated result in the example in Section 3.6.

14. Apply the program developed in the previous problem to find an optimal location for a transfer station when sources of solid waste are located at the points (10,10), (20,25), (50,15), (60,40), (70,30), and (75,20). Assume the relative quantities of solid waste arising from these sources are 10, 20, 15, 40, 10, and 50, respectively. Plot the source points and the location of the optimal solution on graph paper. Does the result seem reasonable?

15. Solve the previous problem by using the metropolitan metric instead of the euclidean metric.
16. Identify two or three sites in your community which could serve as locations for a transfer station. Explain your choices.
17. How might a mathematical model like that discussed in Section 3.7 be used to assist in analyzing a regional solid waste management problem in your area?

REFERENCES

Bodin, L., Fagan, G., and Levy, L., The RouteSmart vehicle routing and scheduling system, paper presented at Conference on Cost-Effective Collection of Recyclables and Solid Waste, Madison, WI, June 2–4 1993a.

Bodin, L., Fagan, G., and Levy, L., Vehicle routing and scheduling problems over street networks, paper presented at Conference on Cost-Effective Collection of Recyclables and Solid Waste, Madison, WI, June 2–4 1993b.

Everett, J.W. and Peirce, J.J., Curbside recycling in the U.S.A.: Convenience and mandatory participation, *Waste Management and Research*, 11(1), 49–61, 1993.

Gottinger, H.-W., *Economic Models and Applications of Solid Waste Management*, Gordon and Breach Science Publishers, New York, NY, 1991.

Hickman, H.L., Collection of residential solid waste in *The Solid Waste Handbook: A Practical Guide*, Robinson, W.D., Ed., John Wiley & Sons, New York, NY, 1986.

Hillier, F.S. and Lieberman, G.J., *Introduction to Operations Research*, Fifth Edition, McGraw-Hill Publishing Company, New York, NY, 1990.

Kohrell, M.G., Rhyner, C.R., Wenger, R.B., and Katers, J.F., Choosing the Best Method for Collecting Recyclables, pamphlet written for and circulated under The Cooperative Extension System, National Initiatives Program, University of Wisconsin Extension, Green Bay, WI, 1992.

Love, R.F. and Morris, J.G., Mathematical models of road travel distances, *Management Science*, 25(2), 130–139, 1979.

Miller, C., The cost of recycling at the curb, *Waste Age*, 24(10), 46–54, October, 1993.

Pferdehirt, W., O'Leary, P., and Walsh, P., Developing an integrated collection strategy, *Waste Age*, 24(1), 25–38, January, 1993.

Pferdehirt, W., O'Leary, P., and Walsh, P., Alternative methods for collection of residential recyclables, *Waste Age*, 24(3), 63–74, March, 1993.

Schur, D.A. and Shuster, K.A., Heuristic Routing for Solid Waste Collection Vehicles, USEPA Office of Solid Waste Report SW-113, 1974.

Solid Waste Management Economic Report, SWMER Communications, April, 1991.

Stearns, R.A., Measuring productivity in residential solid waste collection systems, in *Residential Solid Waste Collection*, GRCSA:3–1/3–19, 1982.

Stone, R. and Stearns, R.A., For more efficient refuse collection try analyzing your system with a mathematical model, *American City*, 84(5), 98–100, May 1969.

Walsh, P., Pferdehirt, W., and O'Leary, P., Transfer stations and long haul transport systems, *Waste Age*, 24(12), 57–65, December, 1993.

Weiszfeld, E., Sur le point pour lequel la somme des distances de n points donnes est minimum, *Tohoku Mathematics Journal*, 43, 355–386, 1937.

Wenger, R.B. and Hansen, J.A., The development of a heat income function for regional solid waste management, *Waste Management and Research*, 1(1), 69–82, 1983.

Wenger, R.B. and Rhyner, C.R., Solving facility location problems with computer contour mapping, *Journal of Environmental Management*, 27(4), 429–436, 1988.

Wilson, D.C., *Waste Management: Planning, Evaluation, Technologies*, Oxford University Press, New York, NY, 1981.

CHAPTER 4

Solid Waste Source Reduction and Recycling

4.1 OVERVIEW

Integrated waste management, a concept introduced in Chapter 1, includes the following elements: (1) reduction of the quantity and toxicity of the waste generated; (2) reuse of materials; (3) recycling; (4) composting of organic materials; (5) incineration with energy recovery; (6) landfilling; and (7) incineration with no energy recovery. The first three elements address waste prevention and the diversion of materials from the waste stream. They are the focus of this chapter. The last four involve either the transformation, destruction, or permanent storage of the materials once they enter the waste stream. These will be discussed in later chapters.

Reductions in the quantity and toxicity of waste (also called source reduction) and reuse of materials before they enter the waste stream are actually practices implemented by manufacturers and consumers, as distinguished from waste management techniques that waste management professionals use to reduce the quantity of waste slated for disposal. Educational programs directed toward both consumers and manufacturers to increase awareness of environmental issues in general, and waste management issues in particular, may encourage both groups to examine their waste reduction and reuse practices. Consumers need to be conscious of the environmental consequences of their buying habits and particular lifestyle choices. Manufacturers need to be aware not only of the possibilities for waste reduction in their manufacturing processes, but also how their product design and packaging will impact waste disposal.

The third element in the integrated waste management approach, recycling, captured the imagination of environmental groups and the public in the early 1970s. (It is important to note that recycling has actually been a common business practice in the United States for several hundred years.) Unfortunately, recycling in the 1970s was viewed by many people as a singular activity—the segregation and collection of paper, metals, glass, and plastics—rather than a holistic dynamic system that closes the loop as symbolized by the familiar recycling symbol consisting of the three chasing arrows shown in Figure 4.1. In many cases, well-meaning volunteers and organizations devoted their time and energy to collecting materials, only to see the materials landfilled because they had no market value. They failed to understand the complexity of the secondary materials market and the reluctance of manufacturers to commit themselves to increasing their use of reclaimed materials as raw materials for the manufacture of new products. Perhaps the most important lesson learned from the experiences in the early 1970s is that markets for the collected materials must be developed to make recycling successful.

By the late 1980s, interest in recycling by the public, industries, and local governments was revived as new incentives appeared. These incentives included rising disposal costs, social and political difficulties in identifying sites for constructing new disposal facilities, and both actual and threatened federal and state legislation targeting the problems of waste disposal. By the 1990s, most states had enacted some type of mandatory source reduction or recycling legislation, as had most other developed countries in the world. By 1992, 42 states had established numerical goals requiring some combination of source reduction, recycling, or composting. Several states do not mandate numerical goals, but instead ban certain materials from disposal. The state mandated MSW management goals, waste reduction targets, disposal bans, and recycling laws are summarized in Appendix E. Along with the "stick approach" of requiring that certain types of materials be diverted from the waste stream, some states also employ the "carrot approach" by initiating programs designed to stimulate markets for these materials.

Collection/processing

Consumption   Remanufacture

Figure 4.1. The recycling symbol.

To fully understand the problems and opportunities recycling presents, it is necessary to have a basic knowledge of the individual commodities, the manufacturing processes, the dynamics of the industries, and the role of the secondary materials industry in matching the supplies of reclaimed materials to the demands of manufacturers. Questions the recycling professional must keep in mind are: (1) What are the specific requirements for the raw materials used by an industry?; (2) How can a particular commodity compete, technically and economically, with virgin raw material?; (3) What incentives or disincentives encourage or discourage the use of reclaimed materials by manufacturers?; (4) What is the demand for the reclaimed materials?; and (5) Can new uses be identified or new products developed that will increase the demand? The answers to these questions are different for each commodity. In this chapter, the nature of the secondary materials markets for each commodity is discussed, along with a brief overview of each of the industries.

4.2 SOLID WASTE SOURCE REDUCTION

The USEPA (1989) defines source reduction as "the design, manufacture, and use of products so as to reduce the quantity and toxicity of waste produced when the products reach the end of their useful lives." This section focuses primarily on reducing the quantity of material entering the waste stream, but this should not be construed as slighting the importance of reducing the toxicity of the waste as well. Indeed, many manufacturers are aware of and sensitive to the public's concern about product toxicity. Many have responded by changing the design or formulation of products prior to manufacturing so they contain less or none of the substances that pose risks when the products become part of the waste stream.

One prominent example is the change in the construction of household batteries, a major source of mercury and cadmium in municipal solid wastes (MSW). Until recently, batteries accounted for 88% of the mercury and 54% of the cadmium in the waste stream (Hurd et al., 1992). Battery manufacturers eliminated intentionally-added mercury from zinc-carbon cell batteries in the early 1990s. They also reduced the amount of mercury in alkaline cells by over 90% between 1982 and 1990, and then eliminated all intentionally-added mercury by the early 1990s (Johnson and Hirth, 1990). Cadmium-free rechargeable batteries have been developed for certain uses, with lithium and other materials used to replace cadmium. Another example is the substitution of water-based inks and paints for solvent-based counterparts and a reduction in the use of pigments formulated with heavy metal compounds.

Source reduction is a concept that applies to both manufacturers and consumers. In the case of manufacturers, source reduction is a planned approach to minimizing both the consumption of raw materials in producing and delivering products and waste generation that results at the end of the product life. The results of waste reduction efforts may be quantified from production data and waste disposal data by determining how much raw material is needed to produce a unit of product, how much waste results, and how much and what type of packaging is used for marketing the product.

Source reduction by consumers or households is a more difficult topic to address. People engage in diverse activities, have different habits, incomes, lifestyles, and hold different attitudes about possessions. Educational literature and programs can convey ideas about the choices and actions that reduce waste, but it is difficult to measure or quantify the results of such programs.

SOURCE REDUCTION BY MANUFACTURERS

Manufacturers have several opportunities to address waste reduction. These include reducing production waste, designing products with longer useful lives or that are repairable, designing products using less material, and packaging their products using less material or using a recyclable material. These possibilities are explored below.

Production Waste

The current thrust in manufacturing is to increase production efficiency by automating fabrication processes and reducing waste. The first of these reduces labor costs and the second, material costs. One way to reduce waste is to lower the amount of trimming scrap, perhaps by improving the layout of a pattern used to cut parts from a sheet of material. Another way to reduce waste is to improve the quality of production and reduce the number of defective parts discarded.

Improved Product Designs

One way to reduce the amount of consumer discards is to design products with a longer useful life. A conscious decision by manufacturers and designers is usually required to make goods more durable and repairable. Repairability is important because appliances, automobiles, and other consumer items are discarded when the cost of repair is a significant fraction of the cost of a replacement. One design triumph has been the steel-belted radial passenger car tire which has a tire life of 64,000 km (40,000 mi) or more—often twice that of previous types of tires. On the other hand, a number of products can be identified which are designed counter to this "increased lifetime" philosophy; namely, disposable or "single-use" products. Examples are disposable cameras, flashlights, razors, lighters, diapers, and beverage containers.

A second way to reduce consumer discards through improved design is to design products to perform a given function, but with less material than was used previously. An example of this is the reduction in the weight of beverage containers and other similar products, a practice called lightweighting. Between 1972 and 1989, the weight of an aluminum beverage can was reduced by 26%. In the same period, the weight of a polyethylene terephthalate (PET) 2-liter soft drink container was reduced by 21%. A continuing trend toward lightweighting in beverage containers is evident from the actual and projected weights displayed in Table 4.1.

Product Packaging

Packaging is used for multiple purposes (USCOTA, 1989), including: (1) protection of products during shipping and shelflife; (2) prevention of food spoilage; (3) display of consumer information; (4) compliance with government regulations; (5) prevention of tampering; (6) prevention of theft; (7) consumer convenience; and (8) attractive presentation of products to consumers.

Generally, the need for packaging is not the issue per se. One or more of the purposes cited above can be used to justify most packaging, but because packaging accounts for 29.2% of MSW by weight and 32.7% by volume, it is an obvious target for source reduction efforts. Attention should be focused on selecting appropriate packaging materials (i.e., recyclable or low volume when disposed), minimizing excessive packaging, and designing new packages using less material or with easily recyclable components.

The choices of packaging materials are not always straightforward. In the past, consumers have been advised to avoid plastics because they do not biodegrade and to select recyclable packaging instead, but this advice may not always be accurate. Rattray (1990) compares the "brick-pack" and steel can packages for ground coffee. The weight of the steel can is 113 g, or 11.3 t for 100,000 cans which occupy a volume of 35 m³ when compacted in a landfill. On the other hand, the weight of the multilaminate pouch (polyester, nylon, aluminum foil, low-density polyethylene) is 17 g. The same amount of coffee is delivered in 1.7 t of packages which occupy 4.3 m³ in a landfill, only one-eighth of the space occupied by the steel cans. The steel cans are recyclable, whereas the pouches are not; however, a recycling rate of over 85% must be achieved before the disposed volumes are comparable. In addition, the manufacture of "brick-packs" requires only about one-third as much energy as that for steel cans. "Brick-packs" can also be transported and stored more efficiently.

Table 4.1 Weights of Soft Drink Containers as a Result of Lightweighting

Type of Container	Weights of Containers (g)		
	1987	1990[a]	1995[a]
0.47 L (16 oz) refillable glass bottle	398.8	297.7	255.1
0.47 L (16 oz) nonrefillable glass bottle	205.5	184.3	170.1
1 L nonrefillable glass bottle	510.3	425.2	396.9
0.29 L (10 oz) nonrefillable glass bottle	155.9	141.7	124.7
0.47 L (16 oz) PET bottle			
1-piece	28.6	27.0	25.0
2-piece (with HDPE base)	29.4	29.4	29.4
1 L PET bottle			
1-piece	42.0	40.0	38.0
2-piece (with HDPE base)	50.5	49.0	47.5
2 L PET bottle			
1-piece	56.1	54.5	54.0
2-piece (with HDPE base)	68.0	67.5	67.5
3 L PET bottle			
1-piece	84.7	78.5	78.5
2-piece (with HDPE base)	95.5	93.7	93.7
0.35 L (12 oz) aluminum can	16.6	15.7	14.4

[a]Projected.

Source: Franklin Associates, Ltd., Comparative Energy and Environmental Impacts for Soft Drink Delivery Systems, prepared for The National Association for Plastic Container Recovery (NAPCOR), Charlotte, NC, March, 1989.

There is no unanimity of opinion about what constitutes excessive packaging, but if one were to ask a group of people for examples of excessive packaging and waste, many would identify the food wrappings and containers provided with the meals served at fast-food restaurants. Another expected response to the question is the "blister-packs" used to sell small household items (e.g., batteries, screws, razor blades) from self-serve display racks in stores. These packages have two major drawbacks: (1) they are constructed of two or more dissimilar types of materials which make them difficult to recycle, and (2) they often enclose two or more items when only one may be needed. Both of these examples represent opportunities to reduce packaging or to make it more amenable to recycling.

SOURCE REDUCTION BY HOUSEHOLDS

Environmental and public interest groups commonly provide brochures and pamphlets with advice suggesting ways to reduce the amount of discarded household waste. Some typical suggestions (US-COTA, 1989) are:

- Buy items that are reusable instead of disposable.
- Reuse product containers and purchase beverages in refillable bottles.
- Select products that are durable or repairable.
- Buy in bulk or larger sizes.
- Avoid containers made of mixed materials.
- Compost yard waste and food waste in residential backyards.
- Buy fresh rather than prepackaged fruits and vegetables.
- Donate usable but unwanted items to friends or charities.
- Buy products that contain fewer toxic substances.

The first four items are readily implemented by any frugal consumer. It makes good economic sense as well as good environmental sense to buy goods that are designed and constructed with long useful lives and to buy larger packages that provide more product for the money and for the amount of packaging material.

The suggestion that containers made of mixed materials should be avoided needs careful consideration. The objective in designing containers of homogeneous construction is to reduce the amount of nonrecyclable materials in MSW; but, as was illustrated in the comparison of the steel coffee can with the brick-pack pouch, the environmental impact of the pouch is probably less, despite its multimaterial construction.

Some of the suggestions require changes in shopping habits. Factors such as price, brand recognition, and convenience may dominate buying decisions, but as the public becomes more environmentally-aware, incremental reductions in waste quantities may result.

Some communities, such as Seattle, Washington, have instituted volume-based or weight-based fees for waste collection and disposal to provide an incentive for households to reduce waste. A more detailed discussion of this topic is found in Chapter 13.

4.3 PRE-CONSUMER AND POST-CONSUMER RECYCLABLE WASTE

Recycling, the third element in the integrated waste management list, is the topic of this section and the remainder of this chapter. In discussing this topic it is important to distinguish between "post-consumer" and "pre-consumer" recyclable materials. The classification is based on the source of the materials. Post-consumer recyclables, as defined in the federal Resource Recovery and Conservation Act, are those products generated by a business or consumer that have served their intended end uses. Pre-consumer materials are materials and by-products generated from, and commonly reused within, an original manufacturing process. Sometimes pre-consumer recyclable material is called "home" or "prompt" scrap.

The importance of the distinction between pre- and post-consumer recyclables becomes apparent when purchase requirements specify that certain products must contain post-consumer recyclables. In 1993, President Clinton issued an Executive Order mandating that certain government paper purchases consist of 20% post-consumer paper by the end of 1994, with higher goals set for future years. Furthermore, as of 1992, over 40 states and 200 local governments had mandatory or voluntary procurement laws specifying the purchase of paper products, plastics, and other goods which contain recycled materials by governments (Raymond, 1992). Many of these state and local mandates specify post-consumer recycled content; therefore, if companies wish to sell recycled products to governments, post-consumer recycled content is essential. There are few laws mandating businesses or consumers to purchase recycled products, so their purchases are usually on a voluntary basis. The National Recycling Coalition (Washington, D.C.) has established the Buy Recycled Business Alliance, with many state chapters, to assist businesses in buying recycled products.

4.4 MARKETS FOR RECYCLABLES

A market is an institution through which price-making forces operate, and which serves as a link between buyers and sellers of a particular material. In recycling, the market infrastructure includes two tiers: intermediate markets and end-use markets.

Intermediate markets—the secondary materials industries—serve as an important link between the suppliers of collected materials and the demands of the manufacturers who comprise the end-use markets. Persons employed in the intermediate market sector are commonly categorized as collectors, processors, brokers, and converters. The exact terminology varies by recyclable material and according to specific function. Functions include acquiring recyclable materials by buying or salvaging, separating and classifying materials to conform to standard grades for raw materials, processing the materials into a specified physical form (e.g., baling, shredding, crushing), and selling the resulting commodities. Brokers often do not handle or process material, but provide an important service to producers and consumers of recyclables through their detailed knowledge of supply and demand and their willingness to use this knowledge in making purchases and sales.

Many large communities choose to do some or all of this processing at recycling centers or materials recovery facilities (MRFs) which are operated by the public or private sector. Estimates indicate that there were approximately 900 such processing facilities in the U.S. in 1992 (Steuteville and Goldstein, 1993).

As with most commodities, the supply of a recyclable material and its corresponding demand tend to be cyclical over time. At times, demand exceeds supply and at others, supply exceeds demand. These cycles influence pricing. For example, the supply of used aluminum cans is typically higher during the summer months when beverage consumption (and the supply of aluminum cans) is up, which results in decreased prices for aluminum cans. On the other hand, demand by can-makers for used aluminum cans is very strong during the spring months when additional quantities are needed to meet summer orders; prices then rise accordingly.

An illustrative list of prices for recyclables is contained in Table 4.2. The reader needs to keep in mind the volatility of prices for recyclables. At a different time or in different locations, they are likely to be significantly different.

When governmental actions are taken to alter the supply or demand of recyclables, serious imbalances may result in the marketplace. This has occurred when state governments pass recycling laws that mandate collection of recyclables without allowing for a corresponding remanufacturing demand or technology to keep pace. The markets for newspaper, office paper, green glass, and plastics are among the materials affected most by legislation. For example, in the late 1980s, as a result of recycling laws mandating collection, recyclable newspaper was in such oversupply in the northeastern part of the United States that most recyclers were paying remanufacturers (paper mills) to accept their paper. As a result of experiences like these, most people involved with recycling now recognize the importance of implementing policies to stimulate demand as supplies are being increased, or of mandating supply increases at a pace reasonably consistent with demand.

Manufacturers have both environmental and economic incentives for using recyclable materials instead of virgin resources. As shown in Table 4.3, environmental benefits such as reduced energy and water use requirements are attainable when aluminum, steel, paper, and glass are

Table 4.2 Range of Prices Paid by Midwest Dealers and Processors for Collected Materials and by Manufacturers for Processed Secondary Materials During the Period April 1993-March 1994

	Dealer Prices ($/t)	Manufacturer Prices ($/t)
Paper	Unbaled[a]	Baled[b]
Newspaper (#6)	0–22	11–33
Corrugated cardboard (#11)	0–33	17–33
Magazines	0–11	0–22
Mixed office paper	0–38	6–17
White ledger	33–88	99–176
Metals	Unbaled[c]	Baled[b]
Aluminum cans	440–660	750–880
Steel cans	58–98	85–95
Lead batteries	88–143	—
Plastics	Loose[d]	Flaked[b]
PET – code 1 (clear)	44–220	836–990
PET – code 1 (green)	—	660–750
PET – code 1 (color, mixed)	44–176	—
HDPE – code 2 (clear)	44–264	265–500
HDPE – code 2 (color)	55–132	220–330
Glass[b]		
Clear	0–11	55
Brown	0–11	17–44
Green	0–11	0–17
Mixed color	0–11	—

[a] *Paper Stock Report.*
[b] *Recycling Times.*
[c] *American Metal Market.*
[d] *Plastics News.*

Table 4.3 Environmental Benefits Derived from Substituting Secondary Materials for Virgin Resources

Environmental Benefit	Percentage Reduction			
	Aluminum	Steel	Paper	Glass
Reduction of:				
Energy use	90–97	47–74	23–74	4–32
Air pollution	95	85	74	20
Water pollution	97	76	35	—
Mining wastes	—	97	—	80
Water use	—	40	58	50

Source: Pollock, C., Mining Urban Wastes: The Potential for Recycling, Worldwatch Paper 76, Washington, DC, Worldwatch Institute, 1987.

manufactured using recycled materials. Manufacturers can realize corresponding economic benefits through reduced energy and pollution control costs, along with decreased raw material costs. A disincentive for using recycled materials arises from the fact that many manufacturers have major investments in facilities and equipment which utilize virgin materials and, often, either own sources of these raw materials or have long-term contracts to assure an adequate supply.

Remanufacturing technologies and trends vary considerably depending on the recyclable material. The remaining sections of this chapter focus on the processes for remanufacturing several common recyclables.

4.5 PAPER RECYCLING

Paper manufacturing is a competitive, international industry. The United States, with its abundant forest resources and low-cost production facilities, plays a major role in this industry. There are well over 600 paper and paperboard mills scattered throughout the country, with facilities located in nearly every state. The southern part of the United States dominates the industry, followed by the northeast, northcentral, and extreme northwest regions (USCOTA, 1989). These mills produce such items as newsprint, printing and writing papers, tissue, kraft and packaging paper, and unbleached kraft paperboard.

Wastepaper and paperboard is comprised of nearly a hundred different specialized grades that fall into five major categories: old newspaper (ONP), old corrugated cardboard (OCC), high-grade deinking papers, pulp substitutes and mixed papers. These items account for a larger fraction of MSW, and concomitantly have a higher recovery rate than any other category of material. Using a strict definition, recovered paper and paperboard includes pre-consumer waste fiber generated internally during the manufacturing process in the form of trimmings, roll ends and re-pulped rolls, as well as materials generated in households, offices, and other post-consumer sources. In practice, internally-generated paper and paperboard is usually not included in most recovered paper statistical reports.

As shown in Table 4.4, in 1992 waste paper and paperboard recovery was 30.5 million tonnes (33.6 million tons), 38.4% of the total paper and paperboard supply for that year (American Forest and Paper Association, 1993). The recovery rate has grown annually since 1970. The goal of the paper and paperboard industry is to achieve a recovery rate of 50% by the year 2000. As Figure 4.2 shows, most of the paper and paperboard recovered in 1992, 78%, was utilized for the manufacture of products similar to those from which the recycled fiber was derived. This manufacturing took place in nearly 300 U.S. mills (Andover International Associates, 1992).

Table 4.4   Recovered Paper and Paperboard by Grade in 1992

Grade	Quantity (tonnes)
Old newspapers	6,483,000
Old corrugated containers	13,958,000
Mixed papers	3,609,000
Pulp substitutes	3,436,000
High grade deinking	3,012,000
Total	30,498,000

Conversion: tonnes × 1.102 = tons

Source: American Forest and Paper Association Recovered Paper Statistical Highlights 1992, Washington, DC, 1993.

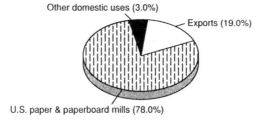

Figure 4.2. Outlets for recovered paper in 1992 (American Forest and Paper Association, 1993).

Exports and other domestic uses, such as cellulose insulation or animal bedding, accounted for the remainder.

PAPER MANUFACTURING AND RECYCLING

The first authentic papermaking was practiced in China as early as 100 AD, utilizing a suspension of bamboo or mulberry fibers. The U.S. paper industry started when the first papermill was built near Philadelphia in 1690. From those early days until 1860, paper was made exclusively from recycled fiber derived from cotton and linen rags and waste paper. As demand for paper and paperboard grew, techniques were developed for utilizing wood fiber in papermaking (American Paper Institute, 1990).

Coniferous (softwood) and deciduous (hardwood) woods are the most commonly used plant fibers in U.S. papermaking. These fibers are composed of cellulose and hemicellulose; lignin, a complex material that provides interfiber bonding; and extractives, such as fatty acids, alcohols, and resin acids. The key to papermaking, whether using virgin or wastepaper pulp, is that fibers must be conformable; that is, capable of being matted into a uniform sheet. They must also be capable of forming strong bonds at the points of contact. Proper conformability and bonding begin with the pulping process, when the bonds in the wood structure are ruptured.

There are three methods of pulping virgin fiber: (1) mechanical pulping, where the fibers are liberated by the application of pure mechanical energy; (2) chemical pulping, which utilizes added chemicals to degrade and dissolve lignin and leave cellulose; and (3) semichemical pulping, which combines the first two forms.

Recovered paper and paperboard require the use of a different pulping process. The key to the recovered fiber pulping process is a continuous pulper, much like a blender, where the material is beat into a pulp and rejects and other extraneous materials are removed. Most recovered pulps are then "deinked," through a process in which inks are removed by mechanical disintegration or chemical treatment. Depending on the type of recovered fiber and its intended end-use, washing and/or flotation steps may follow to further clean the pulp. In the washing step, pulp is washed free of ink and other contaminants on a papermaking screen. In the flotation process, chemicals are added in the pulper to promote air bubbles that will float ink particles off the pulp. If necessary, the pulp may also be bleached.

The resultant pulp is subjected to a number of processing steps which vary, depending on the pulping method utilized and the intended end-use. Virtually all pulp grades require screening, thickening, and storage operations. Cleaning is usually required only when appearance is important (Smook, 1989). Once processed, pulp from virgin or secondary sources enters the paper manufacturing stage. Figure 4.3 presents a schematic

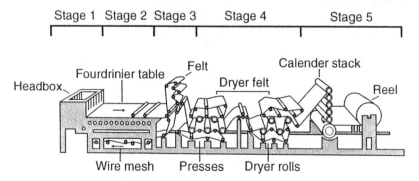

Figure 4.3. A papermaking machine.

diagram of a simple paper machine showing these five basic stages in paper manufacture (Smook, 1989):

1. Sheet formation in the headbox—the flowspreader distributes pulp slurry (1% pulp, 99% water) uniformly into the headbox and onto the moving forming wire.
2. Forming—the Fourdrinier wire forms the fibers into a sheet; pulp is dewatered by about 20%.
3. Pressing—fibers are pressed together, with another 20% of water removed.
4. Drying—the sheet is dried to about 90–95% solids, and fiber bonding develops.
5. Converting and finishing—the sheet is pressed between calender rolls to reduce thickness, and is wound onto reels.

There are various incentives for mills to use wastepaper. In some cases significant energy savings occur when products are manufactured from wastepaper, or a mixture of wastepaper and virgin fiber, instead of 100% virgin fiber. This is particularly true in the case of paper products such as newsprint, printing paper, packaging paper, and tissue paper. However, several studies indicate that many paperboard products, especially linerboard and boxboard, require more energy if produced using recycled fiber (USCOTA, 1989). In other cases, mills use wastepaper because abundant supplies are available at cheaper rates than wood pulp. Some mill technologies require the use of wastepaper as a raw material. A final reason for using wastepaper is the incentive provided by government mandates for paper and paperboard with recycled content.

WASTEPAPER AND PAPERBOARD USES

The total amount of wastepaper used and the proportion of the total comprised by each grade vary considerably by the type of final product. For some paper products, wastepaper is used in conjunction, and sometimes competes directly, with wood pulp. In other instances, the paper product is manufactured from 100% wastepaper. Table 4.5 shows the amounts of wastepaper and paperboard used in the major product types, including exports, in 1992.

Newspaper

The grade of wastepaper categorized as old newspaper (ONP) includes newspapers originating from households (with advertising slicks), over-issues, news blanks, and groundwood computer printout paper. In 1992, 55% of all newsprint paper produced was recovered. As Table 4.5

Table 4.5 Consumption of Wastepaper and Paperboard in the Manufacture of Various Paper Products in 1992

Product Type	Quantity (tonnes)				
	ONP[a]	OCC[b]	High Grades	Mixed Papers	Total
Containerboard	202	7,758	262	172	8,394
Recycled paperboard	1,345	3,163	639	1,532	6,679
Newsprint	2,122	—	—	214	2,336
Tissue	484	84	2,165	440	3,173
Printing & writing papers	46	—	1,863	66	1,975
Others	1,117	447	231	429	2,224
Exports	1,186	2,509	1,330	802	5,827
Total	6,502	13,961	6,490	3,655	30,608

Conversion: tonnes x 1.103 = tons

[a]ONP = old newspaper.

[b]OCC = old corrugated cardboard.

Source: American Forest and Paper Association, Recovered Paper Statistical Highlights 1992, Washington, DC, 1993.

indicates, approximately one-third of ONP is recycled into newsprint, with recycled paperboard, exports, and other uses also important. Other uses include building products, molded pulp products, animal bedding, and hydromulch.

Perhaps one of the most important changes in the paper industry has been the development and implementation of flotation deinking technology. Garden State Paper Company, Inc., a New Jersey newsprint company, is generally credited with developing the flotation deinking process, first used around 1960 to deink newsprint and other low grade paper. However, following its development, the use of the flotation technology became much more prevalent in European countries, and only the recent push for increased recycling has stimulated further interest in this technology in the United States. Today, most new newspaper deinking operations include some type of flotation, as this process has proved to be one of the most economical methods for removing inks and fillers to produce recycled newsprint. Flotation requires that a mixture of old newspaper and old magazines be used, which has also stimulated markets for the latter.

A number of different building product manufacturers use ONP as a raw material for manufacturing applications. Cellulose insulation is the most important of these so-called "dry-process" recycled products made from newspaper. Cellulose insulation became popular in the early 1970s, but has been on the decline since late in that decade. Other building materials that use ONP include fiberboard, insulation board, roofing, siding, flooring and other construction materials. On average, about 45% of the feedstock for these products is wastepaper.

The molded pulp industry is a small-scale user of old newspaper. Products from this industry include egg cartons, nursery pots, meat trays, packaging materials, and decorative pieces. This industry experienced growth beginning in the late 1980s, after a ten-year period of declining sales caused by the penetration of the plastics industry. Environmental concerns have caused people to look for substitutes for products typically made of plastic, hence the resurgence of interest in molded pulp.

In rural areas, turning old newspaper into animal bedding is a viable small-scale use. The low price of old newspaper bedding has made it a very attractive substitute for traditional straw bedding, particularly during seasons with weather extremes. Research has shown that ONP does not adversely affect animals using the bedding, nor the land where the bedding is applied as part of manure.

Shredded newspaper can be mixed with grass seed to produce a hydromulch product. The hydromulch product can then be blown on an area to be seeded for grass. Straw has been the traditional material used for this purpose, but newspaper can be used less expensively.

Corrugated Cardboard

Corrugated and solid fiber containers, plant cuttings, and kraft paper and bags are all categorized as old corrugated cardboard (OCC). In 1992, approximately 60% of all OCC generated was recycled, a rate that has been increasing annually. Most OCC originates from nonresidential sources. The majority of OCC is used to manufacture containerboard and recycled paperboard, such as the facing and inner fluting of cardboard boxes (see Table 4.5). Most mills using OCC are located in the eastern half or far west coast of the United States. Utilization of OCC also depends heavily on export markets.

High-Grade Deinking Papers/Pulp Substitutes

High-grade deinking papers are generally defined as office type papers of high quality. Print free grades are classified as pulp substitutes. The bulk of these high quality papers are recycled into tissue products and fine printing and writing papers, as Table 4.5 indicates. The major suppliers of these grades of paper are commercial operations and institutions, such as government offices. Mills have either added or upgraded deinking facilities to accommodate the growing supplies of these paper grades, and to allow them to meet growing government mandates for tissue and writing papers with recycled content.

Mixed Papers

Mixed papers represent a catchall category for lower grade papers recycled from residential or commercial sources. These lower quality grades of paper are recycled into numerous products, with recycled paperboard leading the field (see Table 4.5). Recycling of these types of paper, particularly from residential sources, represents an area of growth potential that will likely accelerate when the needs of mills for wastepaper cannot be satisfied by available supplies of higher quality wastepapers.

4.6 METALS RECYCLING

FERROUS METALS

Ferrous metals are those metals which contain iron. Ferrous metals are used primarily to manufacture durable iron and steel products such as vehicles, structural steel and appliances, and nondurable steel packaging material for food, beverage, and other types of containers.

Significant quantities of ferrous metals are recovered to be used as raw materials for the manufacture of new iron and steel products. As Table 4.6 indicates, over 50 million tonnes (55 million tons) of scrap iron and steel were recovered in the United States in 1992. Since these products are not always tracked as part of MSW, obtaining accurate data on recovery rates can be a challenge. Nevertheless, it is clear that the ferrous recycling industry is a major industry in the United States, reaching a

Table 4.6 U.S. Ferrous and Nonferrous Metals Recovery, 1992

Category	U.S. Consumption ($\times 10^3$ t)	Exports ($\times 10^3$ t)	Total Recovery ($\times 10^3$ t)
Ferrous			
Iron and steel	41,695	8,708	50,403
Nonferrous			
Aluminum	2,852	295	3,147
Copper	1,274	246	1,520
Lead	889	60	949
Nickel and stainless	589	223	812
Zinc	250	57	307

Conversion: tonnes $\times$ 1.102 = tons

Source: Garino, R.J., 1992 commodity wrap–up, *Scrap Processing and Recycling*, 50(3), 53–60, May/June, 1993.

level of $19 billion in 1992 (Institute of Scrap Recycling Industries, 1993c).

Junked vehicles are a major source of ferrous scrap. Approximately seven million automobiles are recycled annually, in addition to trucks, buses, and motorcycles, making automotive recycling the sixteenth largest industry in the U.S. and the single largest segment of the recycling industry (Powelson and Powelson, 1992). The U.S. Department of Transportation (U.S. DOT, 1990) estimates there are 143 million automobiles and 44 million trucks on the road in the United States each year. About 75% of an average automobile can be recycled, a proportion which continues to grow. Historically, most of that amount was ferrous metal, but increasingly, metals are losing share to plastics.

White goods are large, bulky household appliances such as refrigerators, freezers, stoves, washers, dryers, and air conditioners. The annual discard rate is estimated to be about 2.7 million tonnes (3 million tons) (USEPA, 1992). Appliances typically contain large amounts of ferrous metals along with some nonferrous metals such as copper and aluminum. Their recycling rate is estimated to be 30% to 50% (Powelson and Powelson, 1992). Federal mandates imposed in the late 1980s and early 1990s drastically altered the white goods recycling system by requiring salvagers to deal with polychlorinated biphenyl compounds (PCBs) and chlorofluorocarbon gases (CFCs). PCBs are present in electrical capacitors manufactured in the U.S. prior to 1979 or later in other countries, and must be removed from appliances and disposed of as toxic materials. CFCs must be recovered from appliances prior to processing to prevent their release into the atmosphere. Prior to 1990, CFCs were often released into the atmosphere.

The so-called "tin cans" commonly used as food and beverage containers are another source of ferrous scrap. These cans are constructed primarily of steel. A very thin coating of tin is applied to prevent corrosion from damaging the containers' contents. In order that tin-coated steel cans can be recycled, steel manufacturers' specifications may require that the tin be removed. This can be done through a chemical process called detinning. The tin can then be recovered as metal through electrolysis. There are about 3.75 kilograms of tin in one tonne of steel cans (Powelson and Powelson, 1992). Using this process, both high-quality tin and steel are recovered for reuse as raw materials. In some cases, the detinning process can be skipped because impurities from tin do not interfere with the manufacture of new steel products from steel can scrap.

Despite the fact that steel is recycled more than any other material and its magnetic properties enable steel-based materials to be easily separated from solid waste, a substantial portion of steel food and beverage containers are not recovered. However, efforts are underway, spearheaded by the Steel Recycling Institute, to convince scrap dealers to process steel cans, including aerosol and paint cans, and to encourage industrial end-users to utilize steel can scrap. The steel industry's goal is to recover and recycle 66% of all steel cans by 1995 (Lund, 1993).

The majority of ferrous scrap consumed in the United States (nearly three-fourths) is utilized by the steel industry. Approximately one-fourth is used by the ferrous castings industry and a small portion is consumed in such uses as copper precipitation and the production of ferro-alloys. Exports of ferrous scrap to such countries as the Republic of Korea, Japan, Turkey, Spain, and Taiwan also account for significant scrap consumption (USCOTA, 1989).

FERROUS METALS MANUFACTURING AND RECYCLING

The sources of iron used in steel manufacturing are virgin ores or scrap iron and steel. Scrap materials have been utilized in ironmaking from the time this craft was first mastered approximately 5,000 years ago. In the United States, wartime scrap drives in which citizens were encouraged to donate iron and steel materials to be used in the manufacture of weaponry, occurred from the American Revolution through the Korean War. The growth in demand for metal during the Industrial Revolution resulted in higher prices for scrap metals and stimulated the development of a scrap metal processing industry which led to the major industry it has become today.

The backbone of the iron and steel industry in the United States is comprised of steel mills and iron foundries. The three basic types of furnaces used in steelmaking are the open hearth furnace, the electric arc furnace, and the basic oxygen furnace.

Twenty to thirty percent of the melt capacity in a basic oxygen furnace must be composed of steel scrap. The remaining 70% to 80% consists of molten pig iron which is produced in a blast furnace from iron ore, limestone, and coke. A small quantity of scrap is often charged with this mix as well. The molten pig iron is combined with steel and aluminum scrap in the basic oxygen furnace to produce a variety of flat rolled products such as steel sheet and coil for the automotive, appliance, and canning industries.

The electric arc furnace is designed to operate almost entirely on steel scrap. It is in essence a scrap melting furnace. Advanced technologies used in electric arc furnaces since the mid-1970s have reduced the time needed to produce a given batch of steel from 180 to 70 minutes, and have dramatically reduced energy consumption as well (USCOTA, 1989). Electric arc furnaces typically produce steel structural beams, reinforcement bars, and wire. Approximately 40% of new steel produced in the United States is made from electric arc furnaces.

Although once the dominant type of furnace in steelmaking, the open hearth furnace has declined substantially in recent years because of the

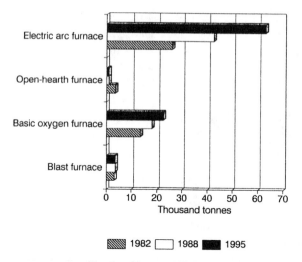

Figure 4.4 The consumption of iron and steel scrap by different types of furnaces (Leading Edge Reports, U.S. Bureau of Mines).

length of time required to make a batch of steel. As can be seen from Figure 4.4, open hearth furnaces do not have a prominent role in the use of scrap steel for manufacturing new steel products. Basic oxygen furnaces and electric arc furnaces consume most of the iron and steel scrap.

The United States steel industry underwent a major shift during the 1980s. It invested over $20 billion in modernization programs to become more competitive internationally and replace outdated technology. These technology improvements have also reduced the amount of in-house scrap steel. This has been a factor in creating an increased demand for post-consumer scrap steel, especially steel food and beverage containers.

The iron foundry industry utilizes a variety of furnaces to produce castings for the automotive and agricultural industries. Most foundries consume varying quantities of steel scrap in their manufacturing processes. Based on research conducted at the University of Wisconsin during the early 1990s, it is believed that iron foundries which utilize a pearlitic base metal (i.e., mixture of iron and iron carbide) to produce gray and ductile castings should also be able to utilize steel cans as a scrap source because the tin content of the steel cans appears to improve product performance (Steel Recycling Institute, 1992).

NONFERROUS METALS

Nonferrous metals are those metals which contain little or no iron. Of the many types of nonferrous metals in use, durable and nondurable aluminum is far and away the most common, followed by copper, lead, nickel, stainless steel, and zinc (Institute of Scrap Recycling Industries, 1993a). Table 4.6 indicates the amount of these major nonferrous metals collected for recycling in the United States in 1992.

ALUMINUM

Aluminum scrap consists of industrial (new) scrap that is a by-product of an aluminum manufacturing process and obsolete (old) scrap consisting of such post-consumer items as used aluminum beverage cans, building siding, and foil. In 1990, industrial and obsolete aluminum scrap made up nearly 31% of the total raw material used in the manufacture of aluminum products. In that same year, for the first time, the amount of old scrap exceeded quantities of industrial scrap, at 57% of total aluminum scrap (Apotheker, 1991).

The United States is the largest aluminum producer in the world, with containers and packaging accounting for the largest share of shipments. Nearly 80% of the aluminum in MSW consists of beverage containers. The national aluminum beverage can recycling rate was nearly 70%, or 889,000 tonnes (979,000 tons) in 1992 (Powell, 1993).

ALUMINUM MANUFACTURING AND RECYCLING

Alumina, a refined product of bauxite, is the raw material for primary aluminum production. Alumina is produced by dissolving powdered bauxite in sodium hydroxide. Few U.S. companies refine bauxite into alumina in this country; most import alumina from Australia, Jamaica, and Suriname. Primary aluminum production begins when alumina is dissolved in a molten bath of cryolite (sodium aluminum fluoride). The bath also contains a small amount of aluminum fluoride and calcium fluoride. Electrolysis is used to recover 99% pure aluminum. This pure metal is then alloyed with various elements to produce the qualities desired for specific end-uses, such as can sheet, transportation uses, and building and construction products.

In secondary aluminum production, scrap aluminum is melted in a furnace, to which alloying elements and primary aluminum are added as needed. It is then cast into ingots or other aluminum products (USCOTA, 1989). A major incentive for using aluminum scrap is the 90% to 95% energy savings compared to that used in the production of aluminum from bauxite.

4.7 PLASTICS RECYCLING

In comparison with paper and steel products which have been part of civilization for hundreds of years, plastics have arrived on the scene only recently. The chemical engineering-based technology used in manufacturing plastics did not become fully established until after World War II. There are more than 12,000 companies in the United States for whom plastics are a major component of their business (Council for Solid Waste Solutions, 1991). In 1990, plastics accounted for 7% of MSW by weight; however, they represented 18% by volume. Table 4.7 shows a breakdown of plastics uses by product type. As shown, packaging and building applications are predominant. Given the newness of plastics, as of the mid-1990s there were no data sources that tracked their recovery rates, which is not the case with other commodities.

There are two main groups of plastics: thermoplastics and thermosets. Thermoplastics comprise about 80% of the plastics market in the U.S. (Institute of Scrap Recycling Industries, 1993b). Thermoplastics can be recycled by being remelted and molded into a new product. Thermoplastics are further divided into commodity and engineering categories. Commodity thermoplastics are low cost, high volume plastics. They include packaging resins such as high-density polyethylene (HDPE) and polyethylene terephthalate (PET). Engineering thermoplastics are produced for specialty markets such as the electronics and transportation industries at a high cost and low volume.

Thermosets, the second major plastics group, comprise 20% of the plastics market in the United States (Institute of Scrap Recycling Industries, 1993b). Thermosets have a long life expectancy and cannot be re-

Table 4.7 Selected Plastics Uses by Product Type, 1989

Product Type	Type of Resin[a] ($\times 10^3$ t)							
	LDPE	HDPE	PS	PP	PET	PVC	Other	Total[b]
Packaging	2038	1847	774	688	480	326	238	6392
Building	143	228	166	15	3	2189	2423	5167
Electronics	180	66	220	21	21	212	279	999
Transportation	0	63	0	140	22	89	684	998
Housewares	188	109	103	120	0	30	66	618
Appliances	0	9	93	77	3	51	310	543
Furniture	0	9	41	44	4	42	400	540
Toys	72	67	94	19	0	17	61	331
Total	2621	2398	1491	1124	533	2956	4461	15,587

[a]Definitions are given in Figure 4.5.

[b]Totals may not sum because of rounding errors.

Sources: Illinois Department of Energy and Natural Resources. Guide for Recyclers of Plastics Packaging in Illinois, prepared by Bottom Line Consulting, Inc.), Report ILENR/RR–90/10, Springfield, IL, 1990.

Modern Plastics, January 1990.

molded into other products. When thermosets are heated more than once, they decompose. However, opportunities for recycling thermosets are available and the recycled material can be used as carpet underlay, fiber-fill, and as a blasting medium for paint removal.

The most commonly recycled plastic packages are PET and HDPE containers, mainly soda bottles and milk jugs. In 1991, 24% of PET soda bottles and 10% of HDPE bottles and containers were recycled (Miller, 1992b and c). It should be noted that soda bottles are often a combination of PET and HDPE materials. The solidly colored base cup (often black) is made with HDPE material, while the remainder is made with PET mater-ial. In the early 1990s, manufacturers also began producing PET soda bot-tles without HDPE bases. Recycling of plastic containers began in the early 1980s, primarily as a response to beverage container deposit legisla-tion which provided a stable supply of bottles. The recycling of other durable and nondurable post-consumer plastics in the United States is negligible. An estimated 95% of pre-consumer industrial scrap is recy-cled, comprising as much as 10% of production (USCOTA, 1989).

Within the thermoplastics category, there are six main types of plastic resins in use. Table 4.8 describes the major properties of each of these resins. In 1988, a voluntary coding system was developed by the Society of the Plastics Industry, Inc. to assist in separating these resins for recy-cling. Figure 4.5 shows this system. Although the coding system is volun-tary, by 1993 at least 36 states passed laws requiring coding labels for certain plastic containers. Concerns over confusion caused by the coding system compelled SPI to examine the symbols used and to make recom-mendations for a possible new system to be implemented in the mid to late 1990s.

PLASTICS MANUFACTURING AND RECYCLING

The primary raw materials for the majority of plastics are natural gas, liquefied petroleum gases, and petroleum. Basic hydrocarbons such as ethane or methane are the building blocks for plastics. From these build-ing blocks, simpler hydrocarbon compounds called monomers are formed, which are then linked together through polymerization to form long chains of repeating molecules with a high molecular weight, called **polymers**. There are hundreds of different polymers used to manufacture plastics products. Each polymer has specific properties that enable it to meet packaging or other product requirements. These properties are deter-mined by the way the monomers are linked together, the molecular weight (which is partially dependent on the length of molecules in the polymer chain), and the type of molecules (Council for Solid Waste Solu-tions, 1991).

Table 4.8 Properties of Selected Packaging

Type of Polymer	Specific Gravity	Water Vapor Barrier	Gas Barrier	Resistance to Grease and Oils	Use Temperature Range (°C)	Applications
Polyester (PET)	1.38-1.41	Good	Good	Good	27 to 205	Bottles, films, thermoforms
Polyethylene						
Low-density	0.910-0.925	Good	Fair	Good	15 to 82	Bottles, tubs, cups, tubing, lids, closures, drums, pails
High-density	0.941-0.965	Good	Fair	Good	5 to 100	Pallets, thermoforms, films, coatings, nettings, foams
Polypropylene	0.900-0.915	Very good	Fair	Good	-18 to 120	Bottles, boxes, cups, film, closures, tubes, trays, vials, thermoforms, foams
Polystyrene						
General purpose	1.04-1.08	Fair	Fair	Fair to good	Up to 90	Cartons, closures, cups, vials, bottles, thermoforms, film, foam
Impact	1.03-1.10	Fair	Fair	Fair to good	-18 to 88	Bottles, tubs, cups, lids, trays, closures, thermoforms
Polyvinyl Chloride						
Plasticized	1.16-1.35	Varies	Good to poor	Good	< -18 to 180	Thermoforms, closures, film
Unplasticized	1.35-1.45	Varies	Good	Good	< -18 to 180	Bottles, thermoforms, closures

Source: American Plastics Council, Plastic Packaging Opportunities and Challenges, prepared by R.F. Testin and P.J. Vergano, Washington DC, 1992.

PETE   HDPE   V   LDPE   PP   PS   Other

PETE – Polyethylene terephthalate (PET)

HDPE – High–density polyethylene

V – Vinyl / polyvinyl chloride (PVC)

LDPE – Low–density polyethylene

PP – Polypropylene

PS – Polystyrene

Figure 4.5 Plastics coding system (Society of the Plastics Industry, Inc.).

There are three main types of manufacturing processes used to produce the most common types of nondurable and durable plastics. They are extrusion, blow molding, and injection molding. Most of these processes begin with the plastic resins as pellets, granules, flakes, or powder, which are then generally subjected to heat and pressure and melted before processing.

Extrusion

In the extrusion process, the feedstock is continuously melted and mixed by a large screw and forced through a die. Profiles are extruded products with a continuous length and constant cross section. Examples of extruded products are window frames, garden hose, plastic lumber, and pipe. Coextrusion allows for different types of plastic to be forced through the same die together. Coextrusion processes are used to produce sheet or film plastic.

Blow Molding

There are three different types of blow molding: extrusion, injection, and stretch blow molding. In extrusion blow molding, the melted plastic is extruded into a tube, called a parison, which is then surrounded by two halves of a bottle mold. Air pressure applied through a blow pin forces the plastic against the mold and then cools the molded product. The bottle is ejected from the mold and trimmed.

Products made by injection blow molding do not require trimming and produce no waste (except for rejects). The melted plastic is injected into a cavity around a rod. This is transferred on the rod to the bottle blow mold, where the bottle is blown and then cooled. This type of processing is appropriate for specialized shapes and closures but not for containers with handles.

In stretch blow molding, preformed plastic is placed into a mold and air is blown through a center rod extension, which expands within the mold. Resin crystalline properties can be used to advantage in this process. Stretch blow molding requires temperature conditioning of the melt form and rapid stretching and cooling. PET bottles are made by stretch blow molding.

Injection Molding

Injection molding is the final common type of plastics molding. Melted plastic is injected into a mold where it is kept under pressure until it has solidified, then the mold opens and the part is ejected. High production rates result in a large volume of solid parts. Small parts that are hard to make with other processes can be manufactured with injection molding. One advantage of injection molding over other molding methods is that any scrap generated can be reground and remolded right at the machine. Food "tubs" used for yogurt, cottage cheese, and similar products are typical examples of injection molded products.

Both pre- and post-consumer plastics can replace or supplement virgin plastic resins in any of the three processes described above. However, much effort is required to minimize contamination. Usually, although not always, the plastics must be separated by resin, and preferably by production method, and many buyers of plastic scrap also require the plastics to be color-sorted with little contamination. The various resins not only have different physical characteristics, they also have different melting points. Mixing of different resins can result in inadequate melting of some materials while other resins might burn. For example, one PVC bottle in a bale of PET can destroy the entire load and damage machinery. Color separation is often important to maintaining the color of the final product. Mixing all of the colors together results in a black resin with fewer commercial applications.

PRODUCTS FROM RECYCLED PLASTIC PACKAGING

Numerous products can be made from secondary plastics. Several of these products are listed in Table 4.9. A few existing products and the potential for newly developed products are discussed below.

Table 4.9 Selected Products Made from Recycled Plastics

Resin	Products
PET	Carpets, fiberfill, strapping, containers, geotextiles, surfboards, sailboat hulls, industrial paints
HDPE/LDPE	Detergent bottles, trash cans, lumber, soda bottle base cups, drainage pipes, animal pens, drums/pails, matting, milk bottle carriers, pallets
PVC	Drainage pipe, fencing, handrails, house siding, sewer/drains
PP	Auto parts, auto batteries, bird feeders, furniture, pails, water meter boxes, bag dispensers, golf equipment, carpets, refuse/recycling containers, geotextiles, industrial fibers
PS	Insulation board, office equipment, office accessories, household products, license plate frames, reusable cafeteria trays
Other	Landscape timber, pens for farm animals, roadside posts, pallets, marine pilings, benches, picnic tables

Source: Council for Solid Waste Solutions, How to Implement a Plastics Recycling Program, Washington, DC, 1991.

Plastic Containers

Plastic containers range from egg cartons to plastic soda bottles, detergent bottles, and tubs for butter. The use of recycled resins in plastic food containers is controlled by the Food and Drug Administration (FDA). The FDA is concerned about possible contaminants from recycled resins affecting the quality of the food. For this reason, little recycled resin was used in food containers prior to the mid-1990s, although technological advances indicate more will be used in the future. Plastic egg cartons and soda bottles are two of the more common exceptions granted by FDA. The FDA has decided that egg shells protect eggs from any possible contaminants. Other exceptions, granted to soft drink companies, allows them to use recycled soda bottles. In one case the bottles are made from recycled PET that has been broken down into its basic chemical monomers and then reprocessed into a new bottle using a process called methanolysis (see chemical processing below). Another process used is the production of a layered soda bottle with a center layer of recycled resin and a virgin plastic outer layer. Finally, the FDA has approved a process whereby recycled PET is cleaned to meet all standards, and can be used directly in the manufacture of new soda bottles.

The use of recycled resins in nonfood container applications is growing. Many of the major detergent companies package their product in bottles containing recycled resins. One of their major concerns has been color integrity. Because products are often recognized by a particular color, companies do not want this color to fade or be inconsistent.

Plastic Lumber

Plastic lumber is slowly growing in popularity. The lumber is made by extrusion and may contain either single resins or mixed resins. Because it is usually not color specific, it provides an important outlet for mixed resins.

Plastic lumber is similar to regular wood lumber in that it can be sawed, nailed, and worked in other ways similar to regular lumber. However, plastic lumber has a number of characteristics that make it superior to wood. One of its positive characteristics is its resistance to weather, water, and insect damage. It also requires little maintenance. The primary negative feature of plastic lumber is its cost, which is usually two to two and a half times that of wood. Another deterrence to the use of plastic lumber has been a lack of performance standards. A process to develop standards was initiated in the early 1990s. The most popular uses of plastic lumber are park benches, sign posts, speed bumps, docks, fencing, and landscaping products.

Other Uses

Companies are developing methods to include recycled resins in the products they currently manufacture and in new products. They are choosing to do this because of the threat of legislation requiring products with recycled content and because of public demand for products made partially or completely from recycled plastics. The products currently manufactured are very diverse. Carpeting and clothing made from PET bottles are the two most rapidly growing uses of recycled plastics. Other examples are trash bags, waste containers, letter trays, food trays, toys, fiberfill for jackets and sleeping bags, and automobile parts.

Chemical Processing

Another form of recycling is the recovery of basic chemicals and fuels from plastics materials. This type of recycling is considered to be a tertiary level of plastics recycling. In primary recycling the plastic material is returned to a product similar to its original. Secondary recycling involves the manufacture of a product different from the original, often an item that can be used as a substitute for wood, concrete, or metal.

Processes for producing chemicals and fuels include pyrolysis, hydrolysis, methanolysis, and glycolysis. These processes break the plastics down into their various chemicals through the use of chemical catalysts and heat. Most tertiary recycling is being conducted by large plastics firms that also make virgin resins.

4.8 GLASS CONTAINER RECYCLING

The glass container industry is the largest component of the glass manufacturing industry; the other three components are flat glass, fiberglass, and specialty glass. Table 4.10 shows the quantities and percentages produced in each of these categories for 1990. An additional 410,000 to 820,000 tonnes (450,000 to 900,000 tons) of glass containers were imported into the United States in 1990 (Miller, 1992a). (Due to the nature of these imports, a closer tonnage estimate is not possible.)

As of 1992, glass containers were manufactured at 73 plants in 26 states. Due to plant consolidation and declining market share, the number of plants has declined somewhat in recent years. Most of these plants produce glass in one or more colors: flint (clear), brown (amber), green, and increasingly, blue. The colors are based on consumer preference and the need to protect certain products (for instance, brown glass is used to shield some beverages from ultraviolet light). Two-thirds of the glass bottles produced in the United States are flint, one quarter are amber, and the remainder are various shades of green and blue. Half of the imported bottles are green. As shown in Figure 4.6, the majority (62%) of glass bottles are used for beverages. Of these containers 31% were used for beer, 22% for other beverages, and 9% for wine and liquor (Miller, 1992a).

The amount of post-consumer glass, or cullet, collected for recycling in the United States has risen every year since 1987, the year the Glass Packaging Institute (GPI) began its data tracking program. GPI statistics indicate the national glass container recovery rate for 1992 was 32.8%, or

Table 4.10 Glass Production by Category, 1990

Glass Category	Production ($\times 10^6$ t)	Percent
Container	9.925	61
Flat glass	3.705	22
Fiber wool	1.459	11
Pressed and blown	0.890	6

Conversion: tonnes x 1.102 = tons

Source: Wisconsin Department of Natural Resources, Waste Glass Container Markets in the Wisconsin Region, prepared by Resource Management Associates, Madison, WI, 1992.

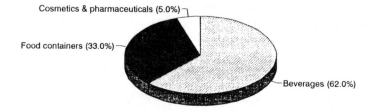

Figure 4.6 Glass bottle markets, 1990 (Miller, 1992a).

3,301,000 tonnes (3,637,000 tons) recovered from 10,048,000 tonnes (11,073,000 tons) of bottles shipped (Apotheker, 1993). Most of the glass recovered in the United States is used by container manufacturers to make new glass bottles and jars.

GLASS MANUFACTURING AND RECYCLING

Silica sand, a very pure form of silicon dioxide, is the most common ingredient in glass manufacturing. Sodium oxide, commonly known as soda ash, is used as a fluxing agent to lower the high melting point of sand. Blending limestone, the third major glass ingredient, with the soda ash reduces water solubility. Silica sand, soda ash, and limestone are all available in abundant supplies. By adding alumina, boric oxides, coloring agents, or other minor ingredients, easily formed durable containers can be produced. In addition to these ingredients, the glass container industry has always used in-house cullet in its batches because cullet melts at lower temperatures than the virgin raw materials used to make glass. If significant quantities of cullet are used, its lower melting results in energy savings in the manufacturing process. As noted in Table 4.3, energy savings are likely to be in the range of 4% to 32%.

Manufacturers started using post-consumer cullet in the late 1960s when improvements in the glass production process caused a decline in the available amount of in-house cullet. Another factor contributing to an increase in the use of post-consumer cullet was an improved and more stable supply of quality cullet resulting from beverage container deposit legislation.

In the glassmaking process, raw materials are first mixed together in a "batch," and then conveyed into the melting furnace where temperatures of 1480°C to 1570°C (2700°F to 2850°F) are reached. After mixing and melting, the resulting molten glass is molded and blown into container shapes. The container is then toughened by cooling it in an annealing oven.

Cullet use levels of 80% to 90% or greater are commonplace in many countries, particularly in Germany and Switzerland. In the United States, cullet utilization rates are lower, commonly 10% to 40%. The lack of consistent supplies of quality cullet inhibits the glass manufacturing industry in the United States from utilizing greater quantities. Contaminants have become a major concern because workers in large recycling programs are unable to carefully sort out unacceptable materials such as ceramics, heat-resistant cookware, light bulbs, metal rings and lids, porcelain, mixed color cullet, and noncontainer glass. Ceramics and heat-resistant cookware, for example, melt at temperatures higher than those in a glass furnace. These materials are partially melted into stones which become foreign objects imbedded in a container, causing weak spots and rendering the container unusable. Metals melt at the same or lower temperatures than glass, and can form molten metal pools at the bottom of the furnace, eventually corroding the furnace and potentially causing millions of dollars in damage. Noncontainer glass differs chemically and melts at higher temperatures than container glass. The result can be stresses, variations in container color, and a finished container which is weak (Central States Glass Recycling Program, 1991).

A polymer coating technology for glass containers was developed in the U.S. in the early 1990s, with test-marketing conducted beginning in 1993. Applied to the outside of glass containers, this thin coating increases their strength by 30% to 100%. Anticipated benefits of this process are the potential for lightweighting containers and enhanced efficiencies in filling lines. Recycling could be favorably impacted by this development as well. Since color can potentially be mixed into the coating, the need for differently colored glass would be eliminated; the glass portion of all bottles could be clear. This would negate problems associated with varying market strength for various glass colors, as well as decreased glass container breakage during collection and processing and the associated problems of mixed color residues.

ALTERNATE GLASS CULLET USES

As the collection rate of post-consumer glass increases, so does the interest in developing noncontainer uses for glass. This is driven by five primary factors: (1) glass cullet is a relatively low-value material, and the cost of transporting it to distant plants further reduces revenues; (2) container manufacturers are sometimes too distant for easy access; (3) there is weak market access for certain color(s), especially green, which is produced at only seven facilities in the United States; (4) breakage of containers yields a mixed color residue not usable in large quantities in container production; and (5) high contamination levels often lead to rejections by glass plants.

A large number of potential applications exist for using recycled glass in the noncontainer sector. The majority of these applications are unproven from an economic or technical standpoint, but some have passed muster.

An example is the use of glass as an aggregate in asphalt road paving material. This material, called **glasphalt**, has been used successfully in a number of cities and counties in the United States. Glasphalt appeared initially in the early 1970s, and after a period of dormancy, a resurgence in interest occurred in the early 1990s. Strict standards, which are generally based on climate and road grade conditions, have been put in place in some states that allow its use; other states simply provide recommendations. These standards tend to vary from one location to another, but a common baseline is a maximum amount of glass in the range of 5% to 10%. Another typical requirement is that the glass particles must be of a size not exceeding 1.0 cm (3/8 in).

In some cases, cullet is used as an ingredient in road base material when traditional gravel is in short supply. For example, the local geology in several Minnesota counties is such that they lack local resources for making Class 5 road base aggregate, a material which has a broader range of particle sizes than the Class 3 material indigenous to that area. Rock to produce Class 5 aggregate was imported from other areas at considerable expense. Through research, the counties found that by incorporating glass into a Class 3 aggregate so that it constituted 10% to 12% of the total, a Class 5 aggregate could be produced that met Minnesota Department of Transportation specifications. The Minnesota example, in which cullet filled a significant resource need in making a road base material, is atypical. In most other instances, glass is incorporated into road base aggregate primarily as a local outlet for post-consumer glass, or as a means for conserving existing gravel supplies.

Another example where cullet derived from post-consumer containers has been utilized in the noncontainer glass sector is in the production of fiberglass. When cullet is used for fiberglass production, it is shipped from recyclers to intermediate glass processors, who clean the glass and reduce it to the approximately 12-mesh size needed by the fiberglass industry. Utilization of cullet for this purpose has been limited, however, because fiberglass manufacturers have very strict cullet requirements in order to assure a quality product. Fiberglass production utilizing cullet has been in existence in the United States since the 1970s, but as recently as 1992 there was only one manufacturer using this post-consumer raw material. Legislation passed in 1991 mandating post-consumer cullet use in California fiberglass manufacturing could affect consumption in this industry. It is unlikely, however, that fiberglass insulation will ever become a dominant market for post-consumer containers on anything but a regional basis.

4.9 TIRE RECYCLING

Approximately 240 million tires were discarded in the United States in 1990 because of wear, damage, or defects (USEPA, 1991). This number does not include the 33.5 million tires that are retreaded each year. The origin of scrap tires and recent generation trends are shown in Table 4.11. The USEPA estimates that less than 7% of the 240 million tires are recycled as products, 10.7% are burned for energy, and 5% are exported.

Almost 79% of the discarded tires are automobile tires which have been replaced with new ones, or have been obtained from junked vehicles. The average automobile tire has a weight of 9 kg (20 lb), which is equivalent to 110 tires/tonne (100 tires/ton). Most of the remaining 21% of the tires come from trucks and buses. These tires are about five times heavier than automobile tires and constitute 22 tires/tonne (20 tires/ton).

Tires are constructed with one of two distinct designs, a nonbelted design which was used for many years or a steel-belted design, which now dominates the tire market. The reason for the popularity of steel-belted radial tires is improved gasoline mileage and a lifespan that is two to three times longer than nonbelted tires. The predominance of steel-belted radial tires results in fewer tires being discarded in a given time period. However, the scrap tires that are produced are more difficult to recycle. Nonbelted tires, on the other hand, are fairly easy to recycle, but today they represent only 10% of all used tires in the waste stream. Figure 4.7 shows the construction of a typical radial tire.

Table 4.11 Scrap Tire Generation in the U.S. (in Thousands)

	1984	1987	1990
Replacement tires			
Passenger	144,580	151,892	152,252
Trucks	31,707	34,514	36,588
Farm equipment	2,592	2,658	2,549
Imported used tires	1,793	2,925	1,108
From scrap vehicles			
Automobiles	26,700	32,412	39,000
Trucks	6,408	9,456	11,000
Total scrap tires	235,961	233,857	242,496
Population (millions)	235.6	242.8	250.0
Scrap tires/person/yr	0.91	0.96	0.97

Source: U.S. Environmental Protection Agency, Markets for Scrap Tires, EPA/530–SW–90–074A, 1991.

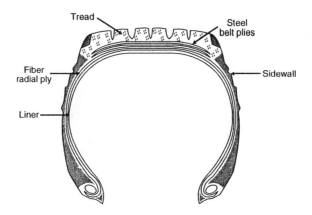

Figure 4.7 Construction of a typical steel-belted tire (adapted from Kohrell, 1993).

DISPOSAL OF TIRES

Figure 4.8 shows the fates of scrap tires. The majority of the scrap tires, 77.6%, are disposed in landfills, stockpiled, or illegally dumped.

Although disposing of tires in a landfill is a common practice, many people view this as a squandering of valuable landfill space because: (1) tires are relatively inert and do not require the safeguards of a sanitary landfill to protect groundwater, (2) tires have a potentially high economic value as a fuel, and (3) tires have potential value as recovered material. By 1991, 36 states had regulated scrap tire disposal, up from only one state in 1985 (USEPA, 1991). Twenty-two states had funded their tire management programs by placing a tax or surcharge on tires or other vehicle-related purchases, such as registration fees. By 1993, 25 states banned tires from landfills. (See Appendix E for a list of those states.)

Landfilling whole tires creates operational problems. Tires have a tendency to rise or "float" to the surface as a result of their density and the air trapped in them when they are covered with soil. The upward movement of the tires compromises the integrity of the protective landfill cover layers. Some landfill operators slice tires in half and nest them to ameliorate this problem. At landfills where there is a shortage of suitable soils for cover material, shredded tires may provide an acceptable substitute.

Current scrap tire stockpiles are estimated to contain 2.5 to 3 billion tires. Not only are stockpiles unsightly, but they represent a fire hazard and are a breeding place for mosquitoes. Once a fire starts in a pile of tires, it is extremely difficult to extinguish.

Burning scrap tires as a fuel has increased significantly in recent years (USEPA, 1991). As Figure 4.8 shows, combustion consumed about

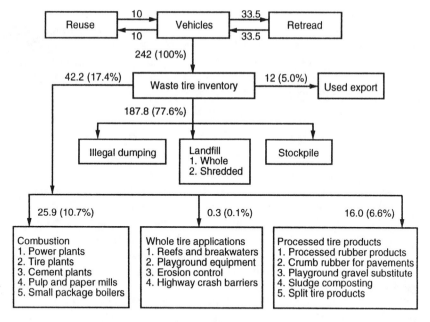

Figure 4.8 Estimated destination for scrap tires in 1990 (U.S. Environmental Protection Agency, Markets for Scrap Tires, EPA/530-SW-90-074A, Washington, DC, U.S. Printing Office, 1991).

10.7% of the scrap tires in 1991. Tires shredded into particles of size 5 cm x 5 cm (2″ × 2″) or smaller, have proved to be an excellent supplementary industrial fuel when used in small percentages, typically 10% or less. Tire-derived fuel (TDF) has more energy per mass or weight than an equivalent amount of most types of coal—28,000 to 37,000 kJ/kg (12,000 to 16,000 BTU/lb) compared to 20,900 to 33,700 kJ/kg (9,000 to 14,500 BTU/lb) for bituminous coal. This energy content is generally equivalent to 9.5 liters of oil (2.5 gallons) per tire (Ansheles, 1991). The use of TDF is increasing in certain industries, including the pulp and paper industry and power utilities. Cement kilns can use shredded or even whole tires for fuel. Furthermore, the cement production can utilize the steel contained in the tires' belts and beads. Despite these advantages, few cement plants in the United States use TDF. This stands in contrast to other countries, including Japan and several in Europe, and where burning tires in kilns is more common (USEPA, 1991). Some plant owners may be reluctant to burn tires because to do so requires repermitting and demonstrating that the plant meets air emission standards.

Closely related to combustion is pyrolysis, a thermal process in which organic materials are heated to high temperatures in the absence of oxygen. The materials decompose into gases, oil, and char, a form of carbon. (Pyrolysis is discussed in Section 8.7.) Typically, one tonne (1.1 tons) of tires yields about 520 L of oil (138 gallons) and 320 kg (700 pounds) of carbon black (Kohrell, 1993). Several pyrolysis projects have operated on an experimental basis (about 10 in 1992), but currently there are no commercial operations.

TIRE RETREADING

Historically, retreading has been the principal method for recycling tires. Retreading is a process of extending the life of a used tire by adding new rubber tread to the used tire casing. Whereas the average nonbelted tire can be retreaded four to six times before the casing is no longer usable, steel-belted tires can be retreaded only two to three times. The retreading of steel-belted radial tires is also a much more expensive process. Airplane and truck tires are the most frequently retreaded tires. There are over 1,900 retreading plants in the United States and Canada, of which 95% are owned/operated by small entrepreneurs whose collective investment is approximately one billion dollars. The remaining 5% are owned/operated by new tire manufacturers (Kohrell, 1993).

USES FOR WHOLE TIRES

With the exception of retreading or burning in a cement kiln, uses for whole tires are limited and account for only 0.1% of the discarded tires. One use that has received considerable attention is the construction of artificial reefs by stringing whole tires together, attaching weights to them, and placing them in the desired locations below the surface of the water. Over 2,000 artificial reefs have been constructed worldwide by government agencies, tire companies, sport fishing groups, civic organizations, and various state agencies. Some of these reefs contain over 3 million tires (Kohrell, 1993). Artificial reefs provide an excellent habitat for a variety of aquatic life in freshwater and saltwater environments. A related use is for the construction of breakwaters. The Army Corps of Engineers has used discarded tires to build more than 20 breakwaters.

Other uses for whole tires include playground equipment, erosion control, construction applications, crash barriers, bumpers on boat docks, and sound barriers.

USES FOR SHREDDED TIRES

Some potential uses require that scrap tires be shredded (e.g., for tire-derived fuel as discussed previously). Shredding, as a unit process, will be discussed in more detail in Chapter 5. Some uses require only a pri-

mary or coarse shredding that produces large chunks of tire (e.g., 5 cm ×
5 cm), while others, such as tire-derived fuel, require smaller pieces produced by secondary shredding. Very finely shredded or ground rubber is
called crumb rubber. It is important to recognize that the cost of shredding generally increases as the required size of product decreases.

Two major uses of primary-shredded tires are daily cover at landfills
(discussed above) and some fabricated products, such as mats used in
blasting or industrial applications. The size of the strips must be uniform
for manufacturing these types of products.

Secondary-shredded tires can be used as tire-derived fuel, or the
shredded pieces can be used at composting facilities as a bulking agent to
increase air flow and optimize composting processes (discussed further in
Chapter 7). In this application, the tire pieces are removed by screening
before the compost is sold.

Crumb rubber is a form of shredded tires in which the particles are
small enough to be reused in molded or mixed products. It can be produced by using two very distinct methods, a cryogenic process or an ambient (room temperature) grinding process. In the cryogenic process, the
pieces resulting from secondary shredding (i.e., about 0.6 cm or less) are
frozen with liquid nitrogen, then ground to a smaller size, while ambient
grinding is done at room temperature. Each year about 2.3 million tires
are converted to crumb rubber for use in rubber and plastic products
(Pillsbury, 1991). Uses for crumb rubber include molded rubber products,
athletic surfaces, and rubberized asphalt.

Crumb rubber utilization is growing in the construction of athletic
surfaces, such as running tracks and tennis courts. Benefits from athletic
surfaces made with rubber include increased flexibility and resistance to
stress; improved adhesive characteristics of asphalt; resiliency retention
in all weather conditions and over the life of the surface; and better trac-
tion for athletes (Kohrell, 1993).

The use with the greatest potential for consuming large quantities of
waste tires is rubberized asphalt. It has been estimated that rubberized asphalt lasts up to 4 times longer than standard asphalt, at about twice the
cost (Arcata Community Recycling Center, 1990). Many states, including
New York, Florida, and Oregon, are now using rubberized asphalt due to
such cost and performance criteria. The use of rubberized asphalt grew by
about 30% per year during the early 1990s. Some additional utilization of
crumb rubber in roads could result from the 1991 Intermodel Surface
Transportation Efficiency Act (ISTEA). Section 1038 of ISTEA requires
states to meet certain minimum recycled rubber utilization requirements
for asphalt in federally-funded road projects. However, Section 1038 has
been widely contested by many groups since its passage, and thus it may
not increase crumb rubber utilization.

4.10 ENVIRONMENTAL IMPACTS OF RECYCLING

Throughout earlier portions of this chapter several observations were made concerning the environmental impacts of recycling. In this section a summary of these impacts is provided as well as a synopsis of the general environmental impacts resulting from recycling.

The most obvious environmental impact of recycling is the conservation of materials. For example, when secondary fiber is used to manufacture newsprint it takes the place of virgin fiber obtained from trees, or when scrap ferrous metal is used in the manufacture of steel, iron ore is saved. It is important to recognize, however, that there are limits to the quantities of materials which can be conserved through recycling.

Recoverability of materials is limited because of physical and economic constraints. Physical limitations arise from the fact that certain materials cannot be recycled after use. Toilet paper is an obvious example. In other cases the lifetime of a product is sufficiently long that the material contained in it does not enter the waste stream or become a candidate for recycling until many years have passed. In addition, materials decline in quantity and deteriorate in quality over time. Paper fibers become shorter and weaker as they are reused, thus there are limits on the number of times they can be recycled. When metals are processed, material is lost through oxidation or is withdrawn as dross during the smeltering process.

Economic limitations may occur at any one of several links in the recycling process. If a reliable supply of secondary material which meets required specifications is not available at a price which is cost-competitive with primary materials, manufacturers are not likely to purchase and utilize this source of raw material. If the potential revenue available to a secondary materials processor does not exceed the acquisition and processing costs, the economic viability of the recycling process breaks down. Communities not mandated by law may be reluctant to institute recycling programs if the cost of collecting and processing recyclables, along with their avoided collection costs, are not matched by the total of the revenue received from the sale of the materials and the cost of the tipping fee avoided at the landfill.

Despite the physical and economic constraints, substantial quantities of materials destined for the waste stream are diverted as a result of recycling activities. An estimated 30 million tonnes were recycled out of a total of 177 million tonnes of solid waste generated in the United States in 1990 (USEPA, 1992). This represents a recycling rate of approximately 16.9%, but also does not include recovery of such materials as sewage sludge, automobiles, and construction and demolition debris. Powelson and Powelson (1992) speculate that it should be possible to recover as much as 50% of the waste generated if obstacles from imagined constraints are removed. Others have concluded that a 50% recycling rate is unrealistic. Alter (1991), for example, carefully analyzes the impediments standing in the way of increasing the recycling rate and concludes that achieving and sustaining a 25% recovery rate is an ambitious goal and one which may not be attainable. However, many communities already report reaching rates of 50% or better. The key factor in all of these debates is exactly what materials are counted when recovery rates are calculated.

Another potential environmental benefit from recycling is energy conservation. Table 4.3 shows that major energy savings occur when aluminum and steel are manufactured from secondary materials instead of virgin materials. Smaller energy savings occur in the case of glass. For most paper products, energy savings occur when wastepaper is used as a raw material instead of 100% virgin fiber; for a few paperboard products, however, more energy may be required. Also on the negative side, it must be recognized that additional energy expenditures may be required in order to collect recyclables and haul them to a processing site than would be the case if these materials were all mingled together with the other refuse components and transported to a landfill or an incinerator.

In some cases pollution reduction may be attributable to recycling. Air pollution levels are lower when a given amount of aluminum, steel, paper, or glass is manufactured from secondary materials instead of primary materials. As can also be seen in Table 4.3, lower levels of water pollution occur in most cases as well. Also, the diversion of recyclables from landfills or incinerators results in smaller quantities of pollutants generated from these disposal processes.

A less obvious environmental benefit resulting from recycling is a reduction in mining wastes when secondary materials are used for manufacturing glass and steel. Each tonne of iron ore or limestone that is mined or quarried produces large quantities of waste. The avoidance of these wastes through the use of secondary materials is a tangible environmental benefit. Finally, it should be noted that the manufacture of glass, paper, and steel from secondary materials requires smaller quantities of water per tonne of material produced than when primary materials are used.

While all of the previously outlined environmental impacts of recycling have been positive, it is important to note there are also negative impacts. For example, recycling at paper mills produces large quantities of sludge, which often contains hazardous materials resulting from deinking and other cleaning processes. On a positive note, many mills are reducing or eliminating the use of hazardous cleaning agents; nonetheless, their presence will be part of paper recycling for some time into the future. In addition, plastics recycling facilities generate large quantities of wastewater through the plastic cleaning process that must be properly disposed of.

As noted in Chapter 1, a comparison of the environmental impacts resulting from alternative manufacturing processes or lifestyle activities is difficult. While there are obvious environmental impacts which can be delineated at certain stages of a product lifecycle, the overall net impact can be determined only on the basis of a full lifecycle analysis.

DISCUSSION TOPICS AND PROBLEMS

1. What quantity and toxicity source reduction techniques are being practiced by governments or industries in or near your community?

2. The ratio of product content to packaging (i.e., the weight of product/weight of packaging) is an indicator that is sometimes useful when thinking about packaging waste. (a) Why is a high value of this ratio desirable? When might it not be desirable? According to the geometric principle of scaling, if the dimensions of an object are increased by some scaling factor, the volume changes by the cube of the factor, whereas the area increases by the square of the factor. (b) Explain how buying a larger quantity in a single package of the same shape affects the product content to packaging ratio? (c) The weights of 1-L, 2-L, and 3-L PET bottles (1987 values from Table 4.1) are 42 g, 56.1 g, and 84.7 g, respectively. Using 1 L = 1000 g of soda, calculate the product content to packaging ratio for these three sizes of bottles. Are the results consistent with the discussion above? If not, can you explain why?

3. Compare the advantages and disadvantages of using plastic pouches, plastic bottles, and paper cartons as packages for milk: (a) if source reduction is the desired outcome; (b) if recyclability is the desired outcome.

4. Describe the recyclables collection and processing system in your community. Which sectors (e.g., residential, commercial, industrial) are served by the program? How many public and private processing centers are there?

5. What kind of recycled product procurement policies are in effect at your school, in your community, or in your state government? Do the policies have an impact in the supply of post-consumer recyclables?

6. Are there any local manufacturing industries (paper mills or iron foundries, for example) that utilize recyclables in their manufacturing process? If so, are pre-consumer or post-consumer materials used? What has their experience been?

7. Many industries have adopted Total Quality Management programs to provide better products and services to their customers. Can you suggest ways in which such programs can contribute to source reduction?

8. Are there local manufacturing industries that could utilize pre-consumer or post-consumer recyclables in their process but choose not to? If so, why?

9. Could any of the alternative uses for newspaper, glass, or tires described in this chapter (animal bedding or glasphalt, for example) be utilized in your community? What would be the impact of these uses on the quantities recycled, and on the processing and transportation requirements?

10. How far can aluminum cans be transported by a citizen before the energy savings from using recycled aluminum rather than virgin materials for the manufacture of new aluminum cans is equal to the energy of the gasoline consumed by the automobile? Assume that the gasoline consumption of the automobile is 12.8 km/L (30 mpg), the energy content of the gasoline is 3.5×10^7 J/L (i.e., joules/liter) and that the energy saved by producing aluminum from recycled cans rather than raw materials is about 2.7×10^8 J/kg of aluminum.

11. An interesting problem encountered in introductory calculus classes that relates directly to waste reduction resulting from product design is this:

> A cylindrical can is to contain a volume V. What are the dimensions of such a can having the least surface area, therefore requiring the least amount of material to make? (The surface area is $A = 2\pi r^2 + 2\pi rh$ and the volume is $V = \pi r^2 h$.)

(a) Show that the most efficient can has height equal to its diameter. (Note: This can be done using calculus or a spreadsheet.)

(b) By what percentage does the amount of aluminum used to construct an actual 0.355 L (12 oz) soda can differ from that needed to construct the ideal can? The dimensions of an actual can are: height = 12.0 cm and diameter = 6.5 cm. (Neglect the material in the seams and deviations from a cylindrical shape near the ends.)

(c) The top and bottom of the can have to be cut from a rectangular sheet of metal and the scrap recycled. How should r and h be chosen so that the least amount of material is used? (Note: Consider inscribing a circle with a diameter d in a square with sides of length d. The area lying outside of the circle corresponds to the scrap.)

(d) Suppose the material for the can costs 3 cents per unit area, and the scrap can be sold for 1 cent per unit area. How should r and h be chosen to minimize the cost of materials?

REFERENCES

Alter, H., The future course of solid waste management in the U.S., *Waste Management and Research*, 9:3–20, January, 1991.

American Forest and Paper Association, Recovered Paper Statistical Highlights 1992, Washington, DC, 1993.

American Paper Institute, Paper Recycling and Its Role in Solid Waste Management, New York, 1990.

American Plastics Council, Plastic Packaging Opportunities and Challenges, prepared by R.F. Testin and P.J. Vergano, Washington DC, 1992.

Andover International Associates/AIA, The Guide to Marketing Recycled Paper in North America: 1991, 1993, and 1995, Danvers, MA, 1992.

Ansheles, C.J., Scrap Tire Management in the NEWMOA States, Portland, ME, 1991.

Apotheker, S., Everyone wants the aluminum can, *Resource Recycling*, 10(10), 60–70, October, 1991.

Apotheker, S., Glass container recycling, *Bottle/Can Recycling Update*, 4(4), 4, April, 1993.

Arcata Community Recycling Center, Recycling Entrepreneurship: Creating Local Markets for Recycled Materials, prepared by Gainer and Associates, Arcata, CA, 1990.

Argent, R.D., Batch charging systems adapt to increased cullet levels, *Glass Industry*, 73(8), 14–17, July 10, 1992.

Central States Glass Recycling Program, Recycling Program Newsletter, Indianapolis, IN, 1991.

Council for Solid Waste Solutions, How to Implement a Plastics Recycling Program, Washington, DC, 1991.

Franklin Associates, Ltd., Comparative Energy and Environmental Impacts for Soft Drink Delivery Systems, prepared for The National Association for Plastic Container Recovery (NAPCOR), Charlotte, NC, March 1989.

Garino, R.J., 1992 commodity wrap-up, *Scrap Processing and Recycling*, 50(3), 53–60, May/June, 1993.

Hurd, D.J., Muchnick, D.M., Schedler, M.F. and Mele, T., Getting a charge out of the wastestream: Final report, The Council of State Governments, Lexington, KY, 1992.

Illinois Department of Energy and Natural Resources, Guide for Recyclers of Plastics Packaging in Illinois, prepared by Bottom Line Consulting, Inc., Report ILENR/RR-90/10, Springfield, IL, 1990.

Institute of Scrap Recycling Industries, Recycling Nonferrous Scrap Metals, Washington DC, 1993a.

Institute of Scrap Recycling Industries, Recycling Plastics, Washington DC, 1993b.

Institute of Scrap Recycling Industries, Recycling Scrap Iron and Steel, Washington, DC, 1993c.

Johnson, R. and Hirth, C., Collecting household batteries, *Waste Age*, 21(6), 48–49, 52, June 1990.

Kohrell, M.G., Business Opportunities in Wisconsin's Post-Consumer Waste Stream, University of Wisconsin-Green Bay, 1993.

Lund, H.F., Ed., *The McGraw-Hill Recycling Handbook* McGraw-Hill, Inc., New York, NY, 1993.

Marsh, K.S., Effective management of food packaging: From production to disposal, *Food Technology*, 45(5), 225–234, May 1991.

Miller, C., Glass containers, *Waste Age*, 23(5), 87–88, May, 1992a.

Miller, C., High-density polyethylene (HDPE) bottles and containers, *Waste Age*, 23(8), 113–114, August, 1992b.

Miller, C., Polyethylene terephthalate, *Waste Age*, 23(9), 73–74, September, 1992c.

Pillsbury, H., Markets for scrap tires: An EPA assessment, *Resource Recycling*, 10(6), 19–24, June, 1991.

Pollock, C., Mining Urban Wastes: The Potential for Recycling, Worldwatch Paper 76, Worldwatch Institute, Washington, DC, 1987.

Powell, J., U.S. Recycling Markets, Turning Recyclables Into New Products Satellite Seminar, Harrisburg, PA, 1993.

Powelson, D.R. and Powelson, M.A., *The Recycler's Manual for Business, Government, and the Environmental Community*, Van Nostrand Reinhold, New York, NY, 1992.

Rattray, T., Source Reduction—An endangered species? *Resource Recycling*, 9(11), 64–65, 1990.

Raymond, M., State Recycling Laws Update: Special Mini-Report, Raymond Communications, Riverdale, MD, 1992.

Smook, G.A., Handbook for Pulp and Paper Technologists Canada, Joint Textbook Committee of the Paper Industry, Montreal, QB, Canada, 1989.

Steel Recycling Institute, Steel Cans: Quality Control and Market Considerations, Pittsburgh, PA, 1992.

Steuteville, R. and Goldstein, N., The state of garbage in America: Part I, *BioCycle*, 34(5), 42–50, May, 1993.

Steuteville, R., Goldstein, N., and Grotz, K., The state of garbage in America: Part II, *BioCycle*, 34(6), 32–37, June, 1993.

U.S. Congress Office of Technology Assessment (USCOTA), Facing America's Trash: What Next for Municipal Solid Waste?, U.S. Office of Technology Assessment, OTA-O-424, Washington, DC, U.S. Printing Office, 1989.

U.S. Department of Transportation, Federal Highway Administration, Highway Statistics 1990, PHWA-PL-91-003, Washington, DC, U.S. Printing Office, 1990.

U.S. Environmental Protection Agency, Decision-Makers Guide to Solid Waste Management, U.S. Environmental Protection Agency, EPA/530-SW-89-072, Washington, DC, U.S. Printing Office, 1989.

U.S. Environmental Protection Agency., Markets for Scrap Tires, U.S. Environmental Protection Agency, EPA/530-SW-90-074A, Washington, DC, U.S. Printing Office, 1991.

U.S. Environmental Protection Agency, Characterization of Municipal Waste in the U.S.: 1992 Update, U.S. Environmental Protection Agency, EPA/530-R-92-019, 1992.

Wisconsin Department of Natural Resources, Waste Glass Container Markets in the Wisconsin Region, prepared by Resource Management Associates, Madison, WI, 1992.

CHAPTER 5

Processing Solid Wastes and Recyclable Materials

5.1 OVERVIEW

Until recently, the majority of MSW processing systems were designed for the purpose of preparing wastes for use as a fuel or for biological processing such as composting. The typical processing steps for these purposes include shredding the raw MSW to obtain more uniform particle sizes, extracting ferrous metals for recycling, and recovering the organic portion that is suitable for thermal or biological processing. The processing sequence used to convert raw waste into a feedstock material is often called the "front-end" system. The processing sequence which occurs when the feedstock is utilized for a specific end-use or purpose is called the "back-end" system.

Many communities have constructed, or are currently constructing, materials recovery facilities (MRFs) to receive and sort recyclable materials. Some facilities receive source-separated recyclable materials that require minimal processing, perhaps no more than the inspection and removal of contaminants, followed by baling of paper and cans, or filling of containers for shipping. But other communities, desiring to save collection costs, have their collection vehicles pick up mixed or commingled recyclable materials and deliver these more heterogeneous materials to a MRF for sorting, baling, and shipping.

Processing systems are comprised of a sequence of unit or single-purpose process components. The basic principles of operation of the unit processes are discussed in this chapter. Many equipment vendors supply equipment based on these principles, but with proprietary or patented refinements in the design to achieve particular results or improvements.

Although this chapter focuses on mechanical processing systems, manual sorting continues to play an important role in processing. Manual source separation prior to collection is commonly an important step in the design of a solid waste management system. Examples of such activities are the separating of recyclable from nonrecyclable materials, combustible from noncombustible materials, and hazardous from nonhazardous materials. An effective education program is needed to create an awareness of the importance of segregating particular waste materials and to teach how to recognize and categorize different materials. The amount of manual sorting at a processing facility varies greatly. Some smaller facilities rely exclusively on manual sorting, but at larger facilities, manual sorting might be limited to screening the incoming materials for unacceptable items, or to performing separations where a mechanical separator is deemed too unreliable or too expensive. Separations of glass and plastics by color, HDPE and PET plastics, and paper and corrugated cardboard are often done manually.

5.2 PROCESS MODELING

One cannot fully understand or effectively design a processing system unless a detailed mass balance can be established to trace the flow of materials through the sequence of unit processes. A mass balance is simply the application of the law of conservation of mass to the system. The mass balance traces the movement of materials through the system; that is, the total mass of the feedstock entering the system must equal the sum of the masses of the output streams. The results of such an analysis may be used for specifying the proper capacities for individual processing units.

Using the concept of mass balance, Diaz et al. (1982) modeled processing systems using a recovery factor transfer function (RFTF) for each unit process. The RFTF can be represented by a diagonal matrix whose elements specify the fraction of each of the waste components—ferrous metal, nonferrous metal, glass, paper, plastics, organic residue (OR) and inorganic residue (IR)—that remain in the primary material stream after unit processing occurs. To illustrate, consider the unit process shown in Figure 5.1. The input to the system is the vector $\mathbf{U}$ whose elements spec-

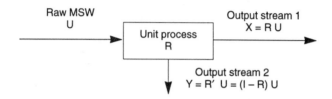

Figure 5.1 A unit separation process.

ify the quantities of each of the waste components listed above. The quantities of the waste components in output stream 1 and output stream 2 are indicated by the elements in the vectors $\mathbf{X}$ and $\mathbf{Y}$, respectively. The $\mathbf{U}$, $\mathbf{X}$, and $\mathbf{Y}$ vectors are written:

$$\mathbf{U} = \begin{bmatrix} u_1 \\ \cdot \\ \cdot \\ \cdot \\ u_n \end{bmatrix} \quad \mathbf{X} = \begin{bmatrix} x_1 \\ \cdot \\ \cdot \\ \cdot \\ x_n \end{bmatrix} \quad \mathbf{Y} = \begin{bmatrix} y_1 \\ \cdot \\ \cdot \\ \cdot \\ y_n \end{bmatrix}$$

The elements in the recovery factor transfer function matrix $\mathbf{R}$ are the fractions of each waste component retained in the main material stream (i.e., r_1 = fraction of ferrous metals, r_2 = fraction of nonferrous metals, etc.). Those in the complementary matrix $\mathbf{R}' = \mathbf{I} - \mathbf{R}$, where I is the identity matrix, are the fractions of each component in the secondary waste stream. Thus,

$$\mathbf{R} = \begin{bmatrix} r_1 & 0 & \cdot & \cdot & 0 \\ 0 & r_2 & 0 & \cdot & \cdot \\ \cdot & 0 & \cdot & \cdot & \cdot \\ \cdot & \cdot & \cdot & \cdot & 0 \\ 0 & \cdot & \cdot & 0 & r_n \end{bmatrix} \quad \mathbf{R}' = \begin{bmatrix} 1-r_1 & 0 & \cdot & \cdot & 0 \\ 0 & 1-r_2 & 0 & \cdot & \cdot \\ \cdot & & 0 & \cdot & \cdot \\ \cdot & \cdot & & \cdot & 0 \\ 0 & \cdot & \cdot & 0 & 1-r_n \end{bmatrix}$$

The amount of component i in output stream 1 is found in the vector $\mathbf{X}$ resulting from the matrix multiplication $\mathbf{X} = \mathbf{R}\,\mathbf{U}$. Since $\mathbf{R}$ is a diagonal matrix, the amount of component I is determined by the simple multiplication

$$x_i = r_i u_i$$

In a similar manner, the amounts in output stream 2 are obtained from the matrix multiplication $\mathbf{Y} = \mathbf{R}'\,\mathbf{U}$. Again because $\mathbf{R}'$ is a diagonal matrix

$$y_i = (1 - r_i)\, u_i$$

A straightforward way to perform the necessary calculations for the RFTF modeling approach is to use a computer spreadsheet. To compute the components of the output stream vector $\mathbf{X}$, the r_i and u_i values can be

Table 5.1. Representative Values of the Diagonal Elements in the RFTF Matrix for Common Unit Processes

	Trommel (oversize)	Shredder		Magnetic Separator	Air Classifier (Light Fraction)	Cyclone	
		Solid	H_2O			Solid	H_2O
Ferrous	0.80	1.00	0.80	0.20	0.10	1.00	0.90
Nonferrous	0.80	1.00	0.80	1.00	0.50	1.00	0.90
Glass	0.20	1.00	0.80	1.00	0.60	1.00	0.90
Paper	0.85	1.00	0.80	0.98	0.98	1.00	0.90
Plastic	0.90	1.00	0.80	0.98	0.98	1.00	0.90
Organic residue	0.25	1.00	0.80	1.00	0.20	1.00	0.90
Inorganic residue	0.25	1.00	0.80	0.95	0.70	1.00	0.90

Source: Diaz, L.F., Savage, G.M., and Golueke, C.G., *Resource Recovery from Municipal Solid Wastes*, Vol. 1, CRC Press, Inc., Boca Raton, FL, 1982.

entered into columns in a spreadsheet. The x_i values are then computed by multiplying the corresponding elements in these two columns.

The entries in the RFTF matrix R must be determined analytically or empirically from field data. Representative values for the RFTF elements for several processing systems are listed in Table 5.1 (Diaz et al., 1982). The RFTF modeling method may also be used to calculate other bulk properties such as moisture content, heating value, and ash content.

Example 5.1: A trommel (i.e., a horizontally mounted rotating cylindrical screen as described in Section 5.4) is to process 100 t of mixed municipal solid waste. Assume that the component categories and the fraction of the components passing through the trommel are those shown in Table 5.1. The MSW consists of 38.0 t of paper, 7.6 t of glass, 7.5 t of ferrous metals, 1.4 t of nonferrous metals, 11.5 t of plastics, 27.7 t of other organic material, and 6.3 t of other inorganic material. (a) What amount of each component will pass through the trommel and remain in the primary stream (i.e., the oversize material)? (b) What amount of material will be diverted (i.e., drop through the holes of the screen)?

Solution to (a): Construct a spreadsheet with the components as the entries in column 1 and their corresponding quantities in column 3. The matrix elements for a trommel—the fractions of the components passing through the unit process and remaining in the main material stream are entered as column 2. These matrix elements are listed in column 2 of Table 5.1. The amount of each component retained in the primary material stream is calculated by multiplying the corresponding row element in column 2 (the fraction retained) by the element in column 3 (the initial quantity of that component). These amounts are the elements in column 4.

Component	Fraction (R element)	Quantity (U element) (t)	Quantity Retained (X element) (t)
Ferrous	0.80	7.5	6.0
Nonferrous	0.80	1.4	1.1
Glass	0.20	7.6	1.5
Paper	0.85	38.0	32.3
Plastics	0.90	11.5	10.4
Organic residue	0.25	27.7	6.9
Inorganic residue	0.25	6.3	1.6

Solution to (b): Construct another spreadsheet like the one above, but re-place column 3 with the elements of the matrix $\mathbf{R'} = (\mathbf{I} - \mathbf{R})$. (That is, each element of column 3 of the new spreadsheet is calculated by sub-tracting the element of the spreadsheet in Part (a) from 1.) This produces the spreadsheet below. The answers are in column 4.

Component	Fraction (R' element)	Quantity (U element) (t)	Quantity Diverted (Y element) (t)
Ferrous	0.20	7.5	1.5
Nonferrous	0.20	1.4	0.3
Glass	0.80	7.6	6.1
Paper	0.15	38.0	5.7
Plastics	0.10	11.5	1.1
Organic residue	0.75	27.7	20.8
Inorganic residue	0.75	6.3	4.7

Of course, the amount of each component diverted could have been cal-culated by simply subtracting the amount of that component retained in the primary stream from Part (a) from the amount initially present.

More complicated processing systems consisting of several unit processes are addressed by successively multiplying by the appropriate RFTF matrices. Such a system is illustrated in Example 5.2.

Example 5.2: A processing system consists of a trommel, shredder, mag-netic separator, and air classifier. The flow of materials through the sys-tem is shown in Figure 5.2. The recovery factor transfer functions for the trommel, shredder, magnetic separator, and air classifier are $\mathbf{Z}$, $\mathbf{A}$, $\mathbf{B}$, and $\mathbf{C}$, respectively. Write the matrix-vector equations to describe the flow of materials in this system.

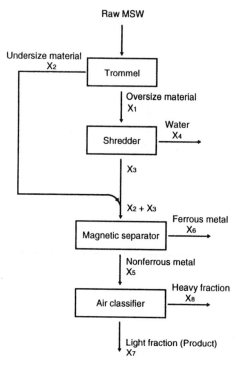

Figure 5.2 A sample processing system consisting of a trommel, shredder, magnetic separator, and air classifier.

Solution: The quantities of the components retained in the main material stream as oversize materials from the trommel are the elements of the vector X_1, whereas the quantities of the components comprising the undersize materials are the elements of X_2. Thus

$$X_1 = Z\,U$$

$$X_2 = Z'U.$$

The vector X_1 is the input to the shredder and the output from the shredder is $A\,X_1$ ($= A\,Z\,U$). Similar equations can be written to find the quantities entering and leaving each of the unit processes. The complete set of equations is the following:

$$X_1 = Z\ U$$
$$X_2 = Z'\ U$$
$$X_3 = A\ X_1 = A\ Z\ U$$
$$X_4 = A'X_1 = A'\ Z\ U$$
$$X_5 = B\ (X_3 + X_2) = B\ (A\ Z\ U + Z'\ U) = B\ (A\ Z + Z')\ U$$
$$X_6 = B'\ (X_3 + X_2) = B'\ (A\ Z\ U + Z'\ U) = B'(A\ Z + Z')\ U$$
$$X_7 = C\ X_5 = C\ B\ (A\ Z + Z')\ U \text{ (the desired product)}$$
$$X_8 = C'\ X_5 = C'\ B\ (A\ Z + Z')\ U$$

It is left as an exercise for the reader to show how these various material streams can be calculated with a computer spreadsheet (Problem 8).

5.3 SHREDDING

Shredding is a generic term used to describe the process by which mechanical forces are used to break materials into smaller, uniform particle sizes. Municipal waste has a low, but nonuniform bulk density. Size reduction increases the homogeneity, the bulk density, and the surface area-to-volume ratio. The increased density results in more efficient transporting of wastes and in the need for less landfill space. When shredded wastes are placed in a landfill and compacted in the same manner as unshredded wastes, the effective density (i.e., refuse weight divided by the sum of the volumes of the refuse and cover material) is 25% to 60% greater than for the unshredded waste. Furthermore, the landfill operator may be permitted to cover the shredded wastes less frequently than is the case for unprocessed waste, thereby conserving landfill space and cover material as well as improving the operating efficiency.

Size reduction is also incorporated into other solid waste management alternatives. Small uniform particle sizes are required for many separation techniques used in resource recovery processes. Shredded MSW is more easily handled in separation and sorting operations, such as magnetic separation and air classification. Also, the greater surface area-to-volume ratio of the shredded particles improves the efficiency in both composting and incineration processes.

TYPES OF SHREDDERS

Several types of shredders that have been used for processing solid wastes are flailmills, hammermills, grinders, shears, and wet pulpers. When MSW is processed for use as fuel, flailmills and hammermills are most commonly employed. Frequently, a flailmill provides the first step in primary shredding. Size separation techniques are then used to remove rocks, glass, or grit, and magnetic separation is employed to remove ferrous metals. Finally, paper, plastics, and other combustible components are shredded by a hammermill to a smaller size.

Flailmills

A flailmill is a coarse shredder. It typically consists of two sets of hammers or flails mounted on parallel shafts that rotate in opposite directions. The flails tear open refuse bags and provide a coarse shredding. Flailmills are able to process large quantities of refuse with low energy requirements. Items that are difficult to shred pass through the flailmill with little size reduction.

Hammermills

A hammermill consists of a central rotor or shaft with radial arms (hammers) protruding from the rotor circumference, all mounted in a heavy duty cylindrical housing. The hammers may either be fixed or free swinging. A hammermill is classified as horizontal or vertical, based on the orientation of its rotor.

Waste is fed into the hammermill through the top of the housing. Upon entering the hammermill, the waste is struck by the hammers. The hammers are moving with such speed that most of the size reduction is accomplished under the initial impact. Breaker bars are set along the inside of the housing. The purpose of the breaker bars is to reduce the clearance between the hammers and the walls of the housing. By reducing this clearance, waste gets pinched between the breaker bars and the hammers. This pinching action crushes and shears the waste into smaller particles. A grate at the bottom of the housing permits particles which have been sufficiently reduced in size to fall through the grate and onto a conveyor. Particles which are too large to pass through the grate remain in the hammermill until they are sufficiently reduced or are ejected though a rejection chute. A rejection chute at the top of the housing provides a means for hard-to-shred items to escape.

A vertical hammermill is similar to a horizontal hammermill in design and operation, except that the rotor is mounted vertically in a conical housing as shown in Figure 5.3. The housing is wider at the top. The rotor is also unique in that it has fewer hammers at the top than at the bottom. As in the horizontal hammermill, waste is fed into the top of the vertical hammermill. The waste is initially impacted by the high speed hammer and the reduction in size begins. As the waste is reduced in size, it falls deeper into the hammermill where the hammer density is higher and the spacing between the housing is narrower. It is in this lower part of the

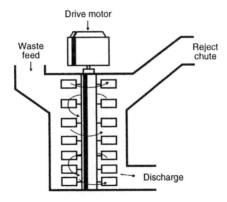

Figure 5.3 Vertical hammermill.

mill that most of the size reduction occurs. The action of the hammers and the tapered walls prevent a piece of waste from dropping deeper into the hammermill if it has not been sufficiently reduced in size. This property enables vertical hammermills to reject unshreddable objects by propelling them with a hammer to a rejection chute near the top of the housing. When the waste finally reaches the bottom of the housing it has been sufficiently reduced in size so that a sizing grate is not needed. The waste is ejected through a hole in the side of the housing.

The hammers in both types of hammermills require regularly scheduled maintenance to minimize wear. In a routine maintenance procedure, the hammers are removed from the housing and hardened steel is welded to their edges.

Grinders

A grinder is similar to the vertical hammermill. It has a vertical rotor and a conical housing, but instead of hammers the grinder has star wheels mounted on the rotor. These star wheels resemble massive gears. Reduction occurs when the waste is crushed between the star wheels and the walls of the housing. The operation of a grinder is similar to that of a vertical hammermill. The waste enters at the top of the housing and as it is reduced in size it drops deeper into the housing until it reaches the bottom where it is discharged.

Wet Pulping

Wet pulping is a very different approach to shredding. The wet pulping process reduces the size of wastes in a water slurry using a hydropulper. A hydropulper, or wet pulper, consists of two disks, mounted with a small space between their faces, that rotate in opposite directions in a vat. The action of the disks grinds the waste into small particles. The waste and water are introduced between the disks near the center and the processed waste is discharged along the edge of the disks in the form of a slurry with about 10% solids. These solids consist primarily of low grade fibers.

The wet pulping process has not been widely used. It was first used in a demonstration project in Franklin, Ohio, in the early 1970s by Black-Clawson, a company which manufactures equipment for the paper industry. In the Franklin project, the fibers were used as ingredients in the manufacture of roofing materials. An 1800 tonne per day wet process plant was built in Hempstead, New York, and operated for a year at the rated capacity; however, the plant was closed, not as a result of technical problems, but because of political problems. Another plant in Dade County, Florida, was designed to use both dry and wet processing, but the wet processing component was discontinued.

OPERATIONAL PROBLEMS WITH SHREDDERS

Several operational problems may arise in the shredding process, such as material jams, fires, and explosions. In addition, workers may be exposed to high dust levels in the surrounding air. Potentially explosive items, including propane bottles, gas cans, and ammunition, often make their way into the waste stream. A common practice in combating explosions is to provide an escape channel, running from the inside of the shredder to the outside of the building, to act as a conduit for the expanding gases of an explosion. The escape channel diverts expanding gases away from employees and equipment.

Jamming occurs when a shredder is unable to break up a material, such as heavy wire, while at the same time failing to expel it through the rejection chute. Jammed material may overload and cause serious damage to the motor if it is not stopped.

Fires started by friction in a shredder are also a problem. During normal operation the waste materials pass through a shredder quickly enough so that a kindling temperature is not reached. However, if a shredder is stopped or becomes jammed, a fire may start. Fires are extinguished manually with the aid of carbon dioxide and water extinguishers, or through the use of heat detectors and automatic extinguishers.

Dust can be controlled by spraying a mist of water on the waste. The water reduces the dust but increases the moisture content of the discharged waste. The additional water may increase leaching at landfills, and reduce the effectiveness of the waste as a fuel.

ENERGY AND POWER REQUIREMENTS

The factors that affect the energy consumption of a shredder are the type of feed material to be shredded, the moisture content, the desired particle size of the product, the design of the shredder, and the state-of-repair of the machine.

According to Drobney et al. (1971), the energy required for dry shredding MSW to a nominal 15 cm (6 in.) size is 8.2 to 16.4 kW h/t. Power is the rate at which this energy is expended and, as such, depends upon the feed rate of the raw material. The energy (and power) requirements are greater for smaller product sizes and vary according to material type. Table 5.2 provides multipliers for determining the energy and power requirements for different product sizes and material strengths. The adjusted power requirement P_a, using 12 kW h/t as the average shredding energy, is calculated with the equation:

$$P_a = 12 \text{ kW h/t} \times Q \times \text{size multiplier} \times \text{material multiplier},$$

where Q is the feed rate of the raw material in t/h.

Example 5.3: Using the assumption that the energy for shredding MSW is 12 kW h/t, determine (a) the power required to shred 200 t/d (assuming an 8 hour shift) of MSW to a nominal size of 0.10 m (4 in.); (b) the energy required to shred the 200 t of waste; and (c) the electrical energy cost of shredding the waste if the cost of electricity is $0.05 /kW h.

Table 5.2 Multipliers to Adjust Energy and Power Requirements for Dry Shredding for Product Size and Material Type

Product size	Multiplier
.15 m	1.00
.10 m	1.39
.05 m	1.64
.025 m	2.38
Material Type	Multiplier
Muncipal solid waste	1.00
Picked municipal solid waste	0.65
Wood and fibers only	0.45
Automobile bodies	2.82

Source: Drobney, N.L., Hull, H.E., and Testin, R.F., Recovery and Utilization of Municipal Solid Waste, U.S. Environmental Protection Agency, SW-10c, 1971.

Solution: (a) The feed rate, Q, is calculated as 200 t/8 h = 25 t/h. The size multiplier (from Table 5.2) for a 0.10 m product is 1.39 and the material type multiplier is 1.00. The required power is

$$P_a = 12 \text{ kW h/t} \times Q \times \text{size multiplier} \times \text{material multiplier.}$$

$$P_a = 12 \text{ kW h/t} \times 25 \text{ t/h} \times 1.39 \times 1.00 = 417 \text{ kW.}$$

(b) The energy, E, required is

$$E = 417 \text{ kW} \times 8 \cong 3330 \text{ kW h}$$

(c) At $0.05 /kW h, the cost of the energy is

$$\text{Cost} = 3330 \text{ kW h} \times \$0.05 = \$166, \text{ or about } \$0.83 /t.$$

In another approach, Diaz et al. (1982) analyzed data from seven re-source recovery facilities in the United States and developed the follow-ing empirical relationship between energy requirements and product size:

$$E_o = 35.55 \, X_{90}^{-0.81}$$

where E_o is the average net specific energy (kW h/t) and X_{90} is the nomi-nal product size in centimeters (90% of the product particle sizes are smaller than this value). Their expression for the gross power (power to shred the wastes + freewheeling or no-load power) to shred MSW to a nominal particle size X_{90} (in cm) is

$$P_G = 35.55 \, X_{90}^{-0.81} \, Q / (1 - \alpha)$$

where P_G is the power in kW, Q is the feed rate in t/h, and α is the ratio of the freewheeling power to the gross power (typically in the range of 0.2 to 0.5). For MSW, Diaz et al. (1982) suggest that the average design power P_D should be larger than the gross power by a factor of 1.2 to 1.7.

5.4 SEPARATION BY PARTICLE SIZE

Separation by particle size is accomplished by passing materials over a screening surface, where objects of a size smaller than the holes in the screen drop through while objects of larger size are retained on the screen. Screening is used in processing both raw and shredded refuse. A rotary screen, or trommel, has been effectively used to process raw refuse prior to shredding, by removing glass, rocks, dirt, and metal from oversized paper and plastics materials. The paper and plastics materials are then shredded in order to produce refuse-derived fuel (RDF). The initial screening reduces the quantity of waste to be shredded and removes components that cause the greatest wear on the shredder equipment. Screening may also be used after shredding to separate those shredded materials that have attained a desired size from those that require further shredding.

TYPES OF SCREENS

There are three main types of size-separating devices: vibrating screens, trommels, and disk screens. Each contains a screen for separating particles, but the manner in which the screens move varies. The efficiency of each depends upon the sizes of the holes, the speeds of the rotation or vibration, the characteristics of the feed materials, and the feed rates.

Vibrating Screens

A vibrating screen consists of a mounted flat screen which undergoes a reciprocating or gyrating motion. The screens are often mounted at a slight incline to encourage the material to move in a particular direction. Depending on the objective of the process, several screens with varying mesh size may be used. The feed material (e.g., shredded waste, composted material, recyclables) is placed on top of the screen. As it vibrates, the particles which are smaller than the apertures in the screen drop through the openings and are collected below, while the larger particles remain on top. Flat screens are generally not used in processing mixed solid waste. They are most successfully applied to concentrated fractions of wastes which have been previously processed into a relatively fine particle size. For example, flat screening may be used to remove impurities (glass, rocks, etc.) from composted materials.

Trommels

A trommel (Figure 5.4) is a horizontally mounted rotary cylindrical screen. The rotational axis is on a slight incline so that waste fed into the higher end will move along the trommel as it rotates. Along the way, the particles that are smaller than the holes in the trommel fall through and are collected below. The aperture size is chosen so that the desired concentrations of certain refuse components in the oversize or undersize fraction can be achieved. Using information such as that shown in Figure 5.5, the separation efficiency resulting from using a particular aperture size can be predicted. For example, a trommel with an opening size of 10 cm will permit about 90% or more of the metals and glass, and about 30% of the paper and plastics, to fall through as undersize material. About 10%

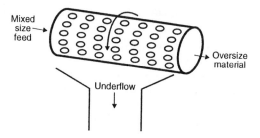

Figure 5.4 Trommel.

of the metals and glass and 70% of the paper and plastics would be retained in the oversize portion.

In addition to the size of the screen openings, other parameters that affect the operation are the diameter and length of the trommel, the incline angle, the rate of rotation, and the feed rate. The optimal operating condition is achieved when the speed of rotation is such that the load rides about two-thirds of the way along the circumference toward the top of the trommel before it tumbles (Hecht, 1983). The rotation rate is often expressed as some fraction of the "critical frequency." The latter term refers to the frequency of rotation at which the centripetal force of the trommel on the materials prevents a tumbling action and holds the material against the trommel walls throughout a complete revolution. The critical frequency is related inversely to the square root of the radius of the trommel according to the equation

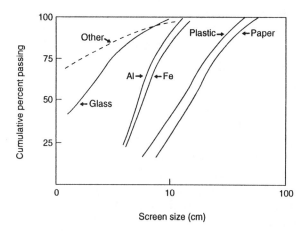

Figure 5.5 Hypothetical particle size distribution (Diaz et al., *Resource Recovery from Municipal Solid Wastes*, Vols. 1 & 2, CRC Press, Inc., Boca Raton, FL, 1982). Reprinted with permission.

$$f_c = 29.9 \, r^{-0.5}$$

where f_c is the critical frequency in revolutions per minute and r is the radius of the trommel in meters. The rotational frequency of a trommel is typically chosen to be 30% to 50% of the critical frequency.

Effective screening requires a shallow depth of material on the screen surface. Otherwise, much of the material is shielded from the screen openings and fails to fall through the trommel. The depth of material on the screen surface depends upon the loading rate.

Stessel (1991) developed a computer model to provide insights into trommel screen design, operation, and performance. The results of the model were consistent with laboratory investigations as well as intuitive insights. The model indicates that a trommel designer should seek the longest affordable trommel, operating it at the greatest practical angle of incline with respect to the horizontal to maximize throughput. The average residence times for waste in real trommels range from 25 to 60 seconds for raw refuse prior to shredding to about 10 seconds for shredded air-classified "light" materials. Long trommels are also desirable because they achieve a more complete screening. The study also relates recovery rates to the rotational speed of the trommel and aperture size.

Disk Screens

A disk screen consists of lobed or star-shaped disks mounted on rotating shafts perpendicular to the direction of material flow. The undersized materials fall through the spaces between the disks. If a blockage occurs, an electronic controller reverses the rotation of a shaft to clear the jammed materials. Disk screens are used in many of the same applications as trommels.

SEPARATION EFFICIENCY

The degree to which a particular separation process extracts a material from a mixture is partially described quantitatively by the efficiency of the operation. Tchobanoglous et al. (1977) define the efficiency of a screening operation as the fraction of the desired material that is recovered, expressed as a percentage. Thus,

$$\text{efficiency } (\%) = \frac{R_d \; (\text{t/h})}{T_d \; (\text{t/h})} \times 100\%$$

where
R_d = weight of desired material recovered, and
T_d = total weight of desired material in input stream.

For separation processes in which contamination is an important consideration, it may be useful to define another entity to describe how effectively contamination is controlled. A rejection factor may be defined as follows:

$$\text{rejection} = 1 - \text{recovery of the undesired material}$$

or mathematically,

$$\text{rejection} = 1 - \frac{R_u \ (t/h)}{T_u \ (t/h)}$$

where
   R_u = weight of undesirable material recovered, and
   T_u = total weight of undesirable material in input stream.

Tchobanoglous et al. (1977) then define **effectiveness** as the product of the recovery and rejection fractions. Effectiveness, with a value in the range from 0 to 1, provides an overall measure of the screening process. A value of one is associated with a system which recovers all of the desired material with no contamination.

Although these concepts of efficiency and effectiveness have been defined for screening, they may be useful in describing the operation of other separation processes as well. Example 5.4 illustrates the application of these concepts to a magnetic separation of ferrous materials from a waste stream.

Example 5.4: A mixed municipal waste stream is processed at an average rate of 100 t/h. The refuse contains 6.0% ferrous materials (including tin-coated steel cans) that are to be extracted with a magnetic separator. The separator extracts materials at an average rate of 5.4 t/h with an average contamination of 8.0%. What is the efficiency, rejection, and effectiveness of the magnetic separation process?

Solution: The total mass of the ferrous materials in the input stream is 100 t/h × 0.060 = 6.0 t/h. Of the 5.4 t of material extracted each hour, 92%, or 5.0 t, is ferrous metal. Therefore the efficiency is

$$\text{efficiency} \, (\%) = \frac{R_d \ (t/h)}{T_d \ (t/h)} \times 100\% = \frac{5.0 \ (t/h)}{6.0 \ (t/h)} \times 100\%$$

or

$$\text{efficiency} \, (\%) = 83.3\%.$$

To calculate the rejection, we first calculate the recovery of the undesirable material, which is everything other than the ferrous, in both the incoming stream (100% − 6.0% = 94% of the other materials or 94 t/h) and the extracted portion (5.4 t/h − 5.0 t/h = 0.4 t/h). Thus

$$\text{rejection} = 1 - \frac{R_u \ (t/h)}{T_u \ (t/h)} = 1 - \frac{0.4 \ (t/h)}{94 \ (t/h)}$$

or

$$\text{rejection} = 1 - 0.004 = 0.996.$$

The effectiveness is then calculated to be:

$$\text{effectiveness} = \text{recovery} \times \text{rejection}$$
$$= 0.833 \times 0.996 = 0.830.$$

5.5 SEPARATION BY DENSITY

Materials are readily characterized by their densities. Density, the weight or mass per unit volume, is a physical property which can be exploited to inexpensively separate different types of materials in a mixture. Separation by density may be accomplished in several ways: in a liquid or dry medium, through air classification, or separation by inertia. The basic principle of operation is that in a mixture of materials of different densities, materials with the higher densities tend to sink while those with lower densities tend to float.

METHODS OF SEPARATION BY DENSITY

Common methods of separation that use a liquid or dry density medium are liquid flotation, dry fluidized bed, and pinch sluice. In the latter two methods a dry medium is employed. Any component of the waste with a density that differs by more than 0.2 g/cm^3 from that of another component can be separated by either of these dry methods.

Liquid Flotation

Liquid flotation is used to separate materials with a density lower than the liquid (floating fraction) from those materials with a density greater than the liquid (sinking fraction). A common application is the separation of materials obtained from recycled plastic soft drink beverage containers. The entire container, including base cup, and cap is fed into a grinder. The constituents are then separated by water flotation. The polyethylene terephthalate (PET) from the body of the bottle and cap sink, while the less dense base cup made of high-density polyethylene (HDPE) floats. The aluminum caps and PET material can then be separated by a similar liquid flotation operation.

Dry Fluidized Bed

The equipment for the dry fluidized bed process consists of a rectangular basin holding a powder. The powder is selected so that its specific gravity falls between the specific gravities of the components which are to be separated. The powder is then fluidized by forcing air up from the bottom of the basin. The waste is placed on top of the fluidized powder. The components of the waste which are more dense than the powder medium sink to the bottom of the basin while those that are less dense remain on top. The separated components are then collected at the top and bottom.

Pinch Sluice

The process involving a pinch sluice is similar to the fluidized bed process in that it also uses a fluidized powder medium with a density which falls between the densities of the waste components to be separated. The difference is in the design and operation of the pinch sluice. The pinch sluice uses a trough which is wide and shallow at one end, and narrow and deep at the other. The wide end of the trough is elevated slightly above the narrow end. A mixture of the waste and the powder medium is fed into the wide end of the trough. Air is forced up through many tiny holes in the bottom of the trough fluidizing the powder. Because of the slope of the trough, the mixture flows toward the narrow end. As the fluidized mixture flows, the denser components of the waste sink to the bottom and the lighter components ride on top. As the trough narrows and deepens the thickness of the powder medium increases and the vertical distance between the heavy and light components becomes greater. At the narrow end, the powder medium is split into the heavy and light components by a horizontal divider and removed from the sluice. The powder is then separated from the waste, usually by a vibrating screen, and reused.

Air Classification

Air classifiers are sometimes used in MSW processing systems for separating components of low density and high air resistance, called the light fraction, from components of high density and low air resistance, called the heavy fraction. The light fraction consists largely of paper, cloth, and cardboard, but it also contains light plastics. The heavy fraction consists of metal, glass, stones and heavy plastics. The light fraction consists almost entirely of combustible components, whereas the heavy fraction contains almost no combustible products. In most municipal solid waste, the light fraction constitutes 65% to 75% of the total. The basic principle of operation for air classification systems is that low density materials tend to be carried along by the moving air while the heavy fraction is not, or only slightly, affected and tends to remain stationary.

In the 1970s many experimental projects were conducted with vertical air classifiers. In a vertical air classifier, shredded waste is fed into a vertical shaft while air is blown upward as shown in Figure 5.6. The air lifts the light fraction upward while the heavy fraction settles to the bottom of the shaft from where it is collected. Some designs of the vertical air classifier include a zigzag shaft or baffles.

As noted in Robinson (1986), most air classification systems are now air knife or horizontal air classifier systems. A diagram of an air knife system is shown in Figure 5.7. The shredded waste is subjected to a moving air stream, which causes the light materials to be blown farther than the heavy materials.

In a horizontal air classifier system the waste components are mixed with air and both the light and heavy fractions are carried by the air in the

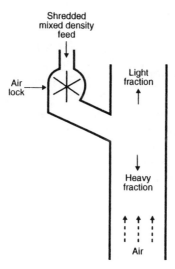

Figure 5.6 Vertical air classifier.

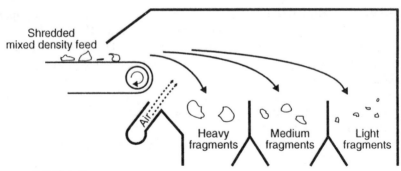

Figure 5.7 Air knife.

same direction. The horizontal air classifier uses a horizontal air shaft. The waste and air enter at one end of the shaft and move toward the other end. The separation occurs when the waste reaches a hole in the bottom of the shaft. The heavy fraction, which has hugged the bottom of the shaft, falls through the hole and is collected, while the light fraction is blown past the hole to a separate collection area.

Some recycling operations may be able to use air separation as an inexpensive technique for separating mixtures of heavy and light materials, such as glass and aluminum cans or glass and plastics, as the mixture moves along a conveyor. An air stream oriented perpendicular to the conveyor blows the light materials from the conveyor.

Incline and Moving Chain Curtains

A process developed by New England CRInc. to separate recyclable glass, plastic, and aluminum cans uses a curtain consisting of hanging chains moving perpendicular to the mixture of materials as they slide down an incline. Glass objects are heavy enough to push through the chains, but plastic bottles and aluminum cans are not. The plastic and aluminum objects are swept to the side for further separation.

5.6 SEPARATION BY INERTIA

Inertia may also be used to separate solid waste components. Inertia, which is the resistance of an object to changing its motion, is dependent on density; that is, for objects of similar size, the objects with the greatest density have the greatest inertia. Common inertial separators are the cyclone, vibrating table (stoner), ballistic separator, and inclined conveyor.

Cyclone Separators

A cyclone is an air classifier that uses a cylindrical chamber with an inlet that allows the shredded mixture to be entered tangentially. As the mixture swirls around the cylinder while suspended in the air, inertia carries the more dense materials to the wall, where they are aggregated. Eventually they drop through the conical section at the bottom.

Vibrating Tables

A vibrating table, or stoner, was originally developed for use in the mining industry. The unit consists of a long narrow inclined table as shown in Figure 5.8. A unique transverse vibrational motion consisting of a relatively slow movement in one direction is followed by an action which causes the table to return rapidly to its original position. As waste enters at one end, the vibration and the incline cause it to move down the table. The nature of the vibration causes the particles to separate according to their densities. The light particles collect on the side that is in the direction of the fast movement, and the heavy particles gather in the direction of the slow movement. The particles of intermediate density gather in the middle of the table. At the low end of the table a divider is used to direct the components to collection bins.

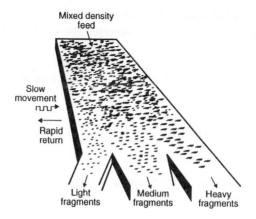

Figure 5.8 Vibrating table.

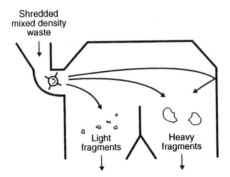

Figure 5.9 Ballistic separator.

Ballistic Separator

A ballistic separator is a unit process that uses a rotor to propel shredded waste particles through the air. The particles with a high density, and therefore high inertia, are propelled farther than particles with a low density and inertia. As shown in Figure 5.9, collection bins are set to catch the particles as they fall. Separation is determined by the distance the particles travel.

5.7 MAGNETIC SEPARATION

Magnetic forces of attraction or repulsion can readily separate metal objects from other materials. The most common type of magnetic separation relies on the strong attraction of ferromagnetic metal objects (i.e., steel and steel-based objects) toward a magnet. Magnetic separators to extract steel cans from the waste stream have been in service at MSW processing plants for many years because they are efficient, reliable, and relatively inexpensive.

A less familiar technology, eddy current separators, are now being used by some vendors to separate aluminum cans from plastic containers in MRFs.

TYPES OF MAGNETIC SEPARATORS

Ferromagnetic Separators

There are several types of magnetic separators. Most make use of a magnetic end pulley on a conveyor belt or a drum located at the end of the conveyor belt as shown in Figure 5.10. As the refuse passes down the conveyor belt and over the magnetic portion of the pulley or drum, the

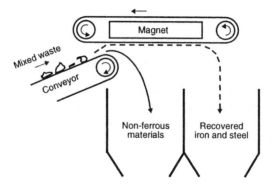

Figure 5.10 Magnetic belt separator.

ferrous metal objects adhere to the belt or drum. The nonferrous objects drop into a collection bin. The ferrous objects remain attached and are carried away from the remaining waste and dropped into a collection bin when they are no longer affected by the magnetic field. Another design uses a conveyor belt suspended over and perpendicular to the conveyor belt carrying the waste. Ferrous metal objects are picked up and held to the suspended conveyor belt by a magnetic field. The conveyor belt holding the ferrous metal moves away from the remaining waste and drops the metal into a collection bin when the magnetic field is removed.

Eddy Current Separator

Nonferrous metals may be separated from nonmetals by inducing currents in the metals with an alternating magnetic field. These induced currents, called eddy currents, produce a magnetic field in the opposite direction to the applied magnetic field. The eddy currents have a negligible effect on the nonmetallic materials. The result is that metal is repelled from other materials of the waste stream. Different nonferrous metals and even different forms (e.g., aluminum foil, aluminum pie tins, and similar objects) can also be separated from each other with eddy currents.

5.9 OPTICAL/INFRARED/X-RAY SORTING

Laboratory spectroscopy techniques coupled with modern computer technology can be used to separate glass by color, and plastics by color and resin type. The color of an object, such as a bottle, is determined by passing light through it and detecting the transmitted light with one or more photodetectors. The absorption characteristics registered by the detectors permit the instrument to determine the color and activate air jets to remove the object from the processing line. (Removing objects from a processing line is sometimes called positive sorting. Negative sorting means that the identified objects are left on the processing line.) The process for sorting glass pieces by color has been described in the literature since the 1970s, but its use has never been widespread.

Modern uses for the optical sorting technologies are directed toward plastic bottle separation. Here, absorption of near-infrared or x-ray radiation can be used to identify plastics by resin type. Visible light sensors enable further sorting by color.

5.8 VOLUME REDUCTION

The compaction of materials into bales reduces their volume and increases their density. Many recyclable materials (e.g., aluminum cans, paper, and cardboard) are baled to provide more convenient handling and to reduce transportation costs. A forklift vehicle can be used to load and stack the bales on a truck or railroad car. Unprocessed MSW is sometimes baled prior to landfilling. Baling reduces litter and, because of the higher density, increases the working lifetime of a landfill. Landfills in which MSW is deposited in baled form are sometimes called **balefills**.

Baling equipment is designed so that the materials are compressed in one or more directions using hydraulically operated pistons at pressures of 1.4×10^4 to 2.4×10^4 kPa (2000 to 3500 lb/in.^{2}). Figure 5.11 depicts a

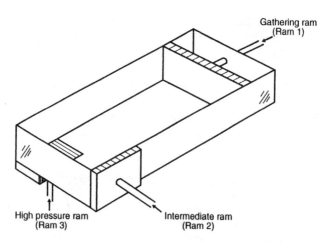

Figure 5.11 Baler.

three stroke baler. Medium density balers usually employ a single stroke (one direction) design to produce bales that are tied or banded to avoid expansion when the pressure is removed. The density of a bale of MSW is about 1200 kg/m³ (2000 lb/yd³), which is about twice that of compacted wastes in a landfill. High pressure balers typically use a three stroke design and are capable of producing bales that do not require wire ties. The factors that affect bale quality are the type of material, pressure applied by the pistons, the time that the pressure is applied, the volume of the bale, and the moisture of the materials.

DISCUSSION TOPICS AND PROBLEMS

1. Does your community process MSW? If so, what is the purpose of the processing, what unit processes are employed, and what products are obtained?
2. Assuming that the energy required to shred MSW is 12 kW h/t, determine (a) the power required to shred 25 t of MSW in an 8 hr period to a nominal size of 0.05 m; and (b) the energy required to accomplish the shredding.
3. A mixture of HDPE plastic milk jugs and PET plastic beverage bottles is to be separated by passing the mixture over sets of suitably-spaced bars that permit the PET bottles and smaller milk jugs to fall through, while retaining the large milk jugs. The mixture is 75% HDPE jugs by weight. Most of the jugs are large and will pass over the screening bars. The mixture is screened at an average rate of 3.6 t/h with an average of 2.4 t/h of HDPE and 0.2 t/h of PET bottles passing over the screen. What is the efficiency, rejection, and effectiveness of this process for the separation of the large milk jugs?
4. A shredder is designed to process 10 t of MSW per hour. The desired nominal size of the product is 5 cm. (a) Calculate the appropriate design power (in kW) for the shredder using Diaz's expression

$$P_G = 35.55 \ X_{90}^{-0.81} \ Q \ /(1 - \alpha).$$

Assume $\alpha = 0.5$ and a design multiplier of 1.5.
5. Justify the formula that relates the critical frequency of rotation of a trommel in revolutions per minute to the radius of the trommel r in meters:

$$f_c = 29.9 \ r^{-0.5}.$$

(Hint: At the slowest rotational speed for which a particle travels in a circular path, the gravitational force on a particle provides the centripetal force at the top of the circular path. That is,

$$mg = \frac{mv^2}{r}$$

where m is the mass of the particle, v = linear velocity of the particle, r is the radius of the circular path, and g is the acceleration due to gravity (g = 9.8 m/s^2). Furthermore, the rotational frequency f is related to the linear velocity by the equation:

$$f = \frac{v}{2\pi r}$$

6. The densities of common plastics are given in Table 4.8. (a) Evaluate the liquid flotation technique as a means of separating mixtures of shredded plastics by type. (b) What are the limitations of this technique?

7. Through a search of the literature pertaining to recycling of plastics, prepare a list of techniques for separating plastics by type.

8. (a) Prepare a spreadsheet to calculate the values of the elements of the vectors X_1, ... ,X_8 for the processing system described in Example 5.2 with the assumption that one metric tonne (1000 kg) of MSW is to be processed with the composition:

	Composition	
	Solids (kg)	Moisture (kg)
Ferrous metals	70	5
Nonferrous metals	20	1
Glass	80	5
Paper	420	80
Plastics	95	10
Organic residue	120	50
Inorganic residue	35	9

Use the RFTF values provided in Table 5.1. (Note: Calculate the masses of dry material and moisture for each component separately, then add at any stage to get the total mass at that point in the process.) (b) Verify that the sum of the masses calculated for the process streams X_1, ... , X_8 equals 1000 kg.

REFERENCES

Diaz, L.F., Savage, G.M., and Golueke, C.G., *Resource Recovery from Municipal Solid Wastes* Vols. 1 & 2, CRC Press, Inc., Boca Raton, FL, 1982.

Drobney, N.L., Hull, H.E., and Testin, R.F., Recovery and Utilization of Municipal Solid Waste, U.S. Environmental Protection Agency, SW-10c, 1971, p. 42.

Hecht, N.L., *Design Principles in Resource Recovery Engineering*, Butterworth Publishers, Woburn, MA, 1983.

Robinson, W.D., *The Solid Waste Handbook: A Practical Guide*, John Wiley & Sons, New York, NY, 1986.

Stessel, R.I., A new trommel model, *Resources, Conservation, and Recycling*, 6, 1–22, 1991.

Tchobanoglous, G., Theisen, H., and Eliassen, R., *Solid Wastes: Engineering Principles and Management Issues*, McGraw-Hill Book Company, New York, NY, 1977.

CHAPTER 6

Wastewater Treatment

6.1 OVERVIEW

The purpose of wastewater treatment is to remove the suspended solids, dissolved chemicals, pathogens, and other biological organisms from municipal sewage and contaminated industrial process waters, to yield an effluent that can legally and safely be discharged into the environment. To achieve this removal, large-scale wastewater treatment employs a variety of physical, chemical, and biological processes. The typical processing stages in a municipal wastewater treatment system are shown in Figure 6.1.

The preliminary and primary treatment stages rely primarily on physical processes such as screening and gravity settling. Chemicals may be added to promote the settling of suspended solids.

Secondary treatment commonly employs biological processes which are especially effective in the removal of soluble organic material (BOD) from a wastewater. With an appropriate process design it is also possible to remove much of the dissolved nitrogen and phosphorus as well.

Although secondary treatment can produce a relatively good quality effluent with low BOD, water quality standards may require tertiary treatment to produce still greater reductions in BOD and lower concentrations of nitrogen and phosphorus. Tertiary treatment relies on chemical and physical processing.

This chapter focuses on processes for treating municipal wastewaters, but it should be noted that similar processes can be applied to treat many industrial wastewaters. There are, however, some industrial waste streams with little organic matter but large amounts of dissolved inorganic

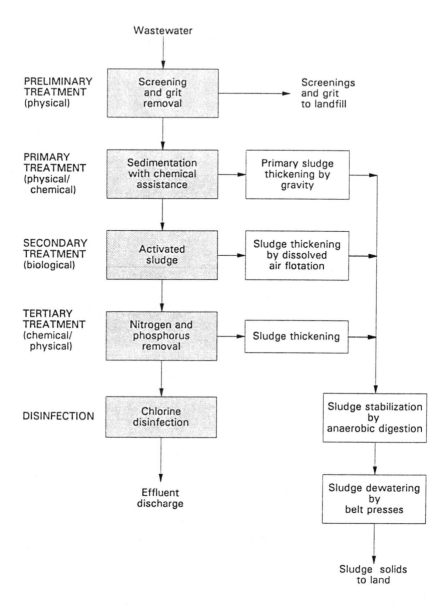

Figure 6.1 The stages of a typical municipal wastewater treatment system incorporating specific physical, chemical, and biological process stages.

compounds and metal ions. The removal of these contaminants may require specialized treatment processes such as ion exchange or membrane filtration.

6.2 PRELIMINARY TREATMENT (SCREENING AND GRIT REMOVAL)

In the preliminary treatment stage, coarse solids and grit are removed through a combination of screening and settling by gravity. A common screening device, the bar screen, is constructed with a set of bars spaced about 1.5 to 2 cm apart and usually oriented at an incline with the bottom of the screen toward the incoming flow. The screen is equipped with mechanical scrapers which remove accumulated solids. The coarse materials that pass through the screen but settle quickly (i.e., the grit) are captured in a grit chamber. Screening and grit removal is an important step because these coarse materials may damage pumps, centrifuges or other equipment used in the later stages of the treatment process. The solids removed in the screening and grit removal processes are usually landfilled. The wastewater next moves to the primary treatment stage.

6.3 PRIMARY TREATMENT (SEDIMENTATION)

One of the oldest and simplest means of achieving at least some degree of primary treatment is to retain wastewater in a basin or lagoon so that suspended solids can settle out under the influence of gravity. Settling (sedimentation) systems are as widely used as ever, but with an enhanced understanding of the physical principles involved, they are now engineered for increased efficiency and are capable of handling much larger quantities of wastewater per unit volume of basin size. The wastewater is retained for 1-1/2 to 2-1/2 hours in a circular or rectangular sedimentation basin, also called a primary clarifier, with a low flow speed and minimum turbulence. The suspended solids with a density greater than water settle, while those with a density less than water rise. The settled solids and the solids skimmed from the surface are removed and constitute the primary sludge. A well designed and operated primary sedimentation tank in a municipal wastewater treatment plant can remove 50% to 70% of the suspended solids and reduce the BOD by 25% to 40%.

Figure 6.2 depicts a radial flow tank, a common sedimentation basin designed with a circular configuration. Wastewater enters at the center of the tank and flows slowly toward the edges where it is removed. The tank has a cone-shaped bottom and, as the solids settle, they form a layer which tends to slide toward the center of the tank where the sludge is continuously removed. Floating materials are skimmed from the surface.

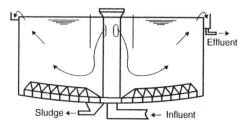

Figure 6.2 The radial flow sedimentation tank used in the primary stage of waste-water treatment.

Rectangular sedimentation tanks operate on essentially the same principle as circular tanks except that wastewater enters at one end and flows horizontally along the length of the tank. Flow distribution is critical because efficient sedimentation requires uniform distribution of the incoming wastewater across the full width of the tank. Baffles are typically engineered into rectangular tanks to maintain flow velocities in the range of 3–9 m/min (10–30 ft/min). The settled sludge is removed from the bottom of the rectangular tank with scrapers and, once again, floating material is skimmed from the surface.

The design details for sedimentation tanks and other processes are beyond the scope of this book. The interested reader is referred to Metcalf and Eddy (1991) and other references where extensive discussions on the design and operation of wastewater treatment processes are provided.

CHEMICAL ADDITION IN PRIMARY TREATMENT

An important adjunct in primary treatment is the addition of chemical agents which enhance the removal of suspended solids, thereby increasing the sedimentation efficiency of a primary treatment basin. The basic principle involved in chemical precipitation is that a change is induced in the physical state of dissolved and suspended solids in order to facilitate their removal in sedimentation basins. Without chemical treatment the suspended particles with the greatest density settle readily in sedimentation basins, but a significant amount of suspended matter remains as colloids and fines. Colloids and fines are particles in the 1–10 μm size range which settle slowly, if at all. These materials, as well as some dissolved ions, can be removed by treating the water with appropriate chemical reagents which promote the formation of particulates that will precipitate or flocculate and settle. The effectiveness of chemical addition during primary

Table 6.1 Enhanced Removal of Suspended Solids During Primary Treatment with the Use of Chemical Precipitation

Process	Suspended Solids Removal	BOD Removal
Primary sedimentation without chemical addition	50–70%	25–40%
Primary sedimentation with chemical addition	80–90%	50–80%

treatment is evident from Table 6.1 which compares suspended solids and BOD removal with and without the use of chemical treatment.

The compounds used for chemical precipitation are often one of the inorganic compounds: alum ($Al_2(SO_4)_3 \cdot 18H_2O$), ferric chloride ($FeCl_3$), ferric sulfate ($Fe_2(SO_4)_3$), ferrous sulfate ($FeSO_4$), or lime. They are relatively low in cost, but considerable quantities are required to achieve a given level of chemical precipitation; hence they add to the total volume of sludge generated and the cost of sludge disposal. Table 6.2 shows the impact of chemical precipitation on sludge mass and volume when ferrous sulfate and lime are used as chemical precipitants.

Polyelectrolytes (polymers), another category of chemical additives, are organic molecules composed of numerous long chains whose subunits are ionizable. As an example, it has been estimated that wastewater treated with a concentration of 0.2 mg/L of a polyelectrolyte with a molecular weight of 100,000 might provide 120 trillion active chains per liter (USEPA, 1979). The polyelectrolytes enhance removal of minute particles through the processes of adsorption and bridging which occur as a result of the charge attraction between the active sites of the polymer and the wastewater solids. Cationic (positively charged) polymers are widely used for wastewater and sludge treatments since many of the wastewater solids carry a negative charge. A single polymer molecule may attract numerous solids particles, resulting in the formation of a large floc particle which will readily settle in primary sedimentation basins and during sludge thickening and dewatering processes. Sludge thickening and dewatering will be described later in this chapter. Relative to the inorganic chemical additives mentioned above, the polyelectrolytes are quite costly, but only a comparably small amount is needed to treat a given volume of wastewater. Since they are used in such small dosages their contribution to increased sludge volume is negligible.

A combination of gravity settling, skimming, and chemical addition results in the removal of much of the wastewater suspended solids in a primary sedimentation basin. The primary sludges generated require

Table 6.2 Comparing Sludge Quantities from Primary Treatment With and Without the Use of Chemical Precipitation[a]

	Sludge Generated[b]	
Treatment	Dry Mass (kg/d)	Volume (L/d)
Without chemical precipitation	133	2,150
With chemical precipitation		
Suspended solids removed	188	—
Ferric hydroxide formed	3	—
Calcium carbonate formed	175	—
Total	366	4,663

[a] The calculations on which this example is based are given in detail in Metcalf and Eddy, 1991, beginning on page 489.

[b] Based on 10^6 liters of wastewater per day with a suspended solids concentration of 220 mg/L. The chemical precipitants are ferrous sulfate (8.43 kg) and lime (72 kg).

further processing, a topic which is discussed later in this chapter and in Chapter 7. The effluent derived from primary treatment is discharged to the secondary treatment stage.

6.4 SECONDARY TREATMENT (BIOLOGICAL TREATMENT)

The effluent which leaves a primary treatment settling basin is greatly reduced in suspended solids but it still contains significant amounts of dissolved organic and inorganic matter which require further treatment. While some physical and chemical processing may be employed, the secondary treatment stage is largely based on the biological activity of a variety of microorganisms.

A major objective of the secondary stage of wastewater treatment is the biological removal of dissolved organic matter. However, there are inorganic compounds of environmental interest in wastewater, especially phosphate and the nitrogen compounds, ammonia and nitrate, which, as noted earlier in Chapter 2, can degrade the quality of receiving waters. Some of these compounds can also be removed in the secondary treatment stage of an appropriately designed wastewater treatment system.

AEROBIC DEGRADATION OF ORGANIC MATTER

Most secondary wastewater treatment processes are based on aerobic designs where, in the presence of oxygen, microbes metabolize (degrade) organic molecules, converting them to carbon dioxide and water and cell material. A common method of depicting this process is shown in Figure 6.3, where the oxidation of glucose is used as an example.

The microbial oxidation of glucose releases a substantial amount of energy. This energy is indicated by the term $\Delta G^{o'}$, which is the change in the Gibbs function. (In the older literature, the Gibbs function is called the Gibbs free energy or sometimes, the free energy.) A negative value of $\Delta G^{o'}$ signifies that the reaction will occur spontaneously and energy is released to do useful work. In a biological system, useful work commonly refers to the growth and maintenance of living cells or tissue. The superscript "o'" means that the value of the Gibbs function was obtained under the standard conditions of pH = 7 and all reactants and products are initially at 1 molar concentration.

Many cellular reactions, such as synthesis of proteins, membrane lipids and other molecules, require energy and do not occur spontaneously. These reactions are characterized by a positive $\Delta G^{o'}$. The necessary energy for these cellular reactions to proceed is obtained from the energy released by the oxidation of glucose or other organic substances. In the example shown in Figure 6.3, a cell, such as a bacterium, having access to a given amount of glucose and oxygen oxidizes about 50% of the glucose to carbon dioxide and water. This releases energy that can then be used by the organism along with the other 50% of glucose for cell growth. Thus, one would expect that substantial amounts of microbial cell mass will be generated during the aerobic degradation of the organic matter in wastewater. The generation of microbial cell mass is a significant factor in the design of wastewater treatment systems because this cell

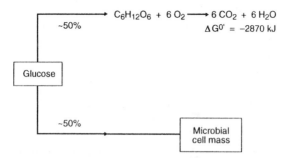

Figure 6.3 Microbial utilization of glucose for energy production and carbon for cell growth and maintenance.

mass becomes part of the sewage sludge solids which require appropriate handling and disposal. Sludge processing will be discussed later in this chapter and in Chapter 7. Final disposal of sludges through land application will be described in Chapter 11.

It is difficult to precisely determine the amount of cell yield from waste because it varies with the organic content of the wastewater and the method of biological treatment. Nevertheless, methods are available for estimating approximate cell yields. One such method is based on the COD test which was introduced in Chapter 2. Numerous studies have shown that for every gram of organic matter expressed as COD and which is aerobically degraded, approximately 0.4 gram of cell mass is generated in the activated sludge process, a common form of aerobic biological wastewater treatment (Grady, 1980).

NITRIFICATION AND DENITRIFICATION

The significance of ammonia (NH_3), and nitrate (NO_3^-), was discussed briefly in Chapter 2. While these nitrogen compounds may not be a significant problem in solid wastes, their presence in wastewater effluent may have serious environmental consequences because of potentially toxic effects on some aquatic organisms and their contribution to eutrophication of receiving waters. Hence, there is now an increasing emphasis on the removal of ammonia and nitrate from wastewater prior to its discharge to surface waters. While the secondary treatment stage of wastewater treatment is often viewed primarily in terms of removal of BOD, this stage of wastewater treatment can also be designed to remove ammonia and nitrate.

Nitrification

Ammonia can be removed effectively through a biological conversion to nitrate, a process called **nitrification**. This conversion is summarized below as a two-stage process involving certain groups of aerobic bacteria. Species in the bacterial genus *Nitrosomonas* convert ammonia, present as ammonium ions (NH_4^+) in aqueous solution, to nitrite ions (NO_2^-) in the first stage of the following process:

$$NH_4^+ + O_2 \xrightarrow{\text{\textit{Nitrosomonas}}} NO_2^- + O_2 \xrightarrow{\text{\textit{Nitrobacter}}} NO_3^-$$
$$\text{ammonium} \qquad\qquad \text{nitrite} \qquad\qquad \text{nitrate}$$

In the second stage, nitrite ions are oxidized to nitrate ions by bacteria in the genus *Nitrobacter*. It is evident from these reactions that nitrification is a highly aerobic process; hence, engineering design for nitrification in wastewater treatment requires efficient delivery of oxygen for the nitrifying bacteria.

Denitrification

After the aerobic treatment to reduce the BOD and the nitrification stages in a treatment process, the concentration of nitrate ions may be unacceptably high. Nitrate ions can be removed from a wastewater through a conversion to nitrogen gas by a biological process called **denitrification**. Denitrification occurs via a complex multistep reaction sequence shown below:

$$NO_3^- \longrightarrow NO_2^- \longrightarrow NO \longrightarrow N_2O \longrightarrow N_2$$

| nitrate | nitrite | nitric oxide | nitrous oxide | nitrogen |

In contrast to nitrification which requires aerobic conditions, denitrification reactions are favored in an environment lacking oxygen but which contains nitrate and some organic matter (anoxic conditions). The bacteria which carry out denitrification reactions are among the same ones which are involved in aerobic degradation. In the presence of oxygen these bacteria use the oxygen as an electron acceptor in their cellular respiration. When oxygen is lacking many of these bacteria are able to use an alternate electron acceptor such as nitrate. Hence it is possible to engineer a wastewater treatment system to incorporate a nitrate removal step by creating an environment which favors those bacteria which have the capacity to carry out the steps in denitrification, thereby converting the nitrate to the environmentally innocuous nitrogen gas.

BIOLOGICAL REMOVAL OF PHOSPHORUS

Many wastewater treatment plants remove phosphate ions (PO_4^{3-}) from wastewater by chemical treatment. Certain aluminum and iron salts react with phosphate, forming precipitates that can then be removed in a sedimentation basin. This approach to phosphorus removal is somewhat expensive due to the cost of the chemical additives and the disposal costs incurred from the generation of additional sludge solids.

Operators at some wastewater treatment plants in the 1960s and 1970s observed that some phosphate can be removed without addition of chemicals. Further study of this phenomenon has provided evidence that large amounts of phosphate can be removed from a wastewater by appropriate manipulation of the biological environment. Under a sequence of anaerobic/aerobic conditions the microbes in activated sludge systems may accumulate more than the normal 2 to 3% phosphorus. A wastewater treatment system designed for biological removal of phosphorus consists of an anaerobic stage (no oxygen) without the presence of nitrate in which many bacteria transport organic substrates (BOD) into their cells and store much of this material as an energy reserve called poly-beta-hydroxybutyrate (PHB). If the anaerobic stage is followed by an aerobic stage the same microbes will revert to aerobic respiration. Under these conditions the bacteria (bio-P organisms) will oxidize the stored PHB and also take up large amounts of phosphate from the wastewater (Zitomer and Speece, 1993). The cells with their accumulated phosphorus will later be removed in the clarification stages of the wastewater treatment process.

ENGINEERED BIOLOGICAL PROCESSES IN SECONDARY TREATMENT

A variety of approaches are used to achieve biological removal of organic matter and nitrogen and phosphorus compounds during wastewater treatment. Several examples are the activated sludge process, the trickling filter, and the rotating biological contactor.

The Activated Sludge Process

The activated sludge process is a commonly used method of secondary treatment for the large volumes of wastewater generated by urban communities. It is a suspended growth system in which dense populations of microorganisms are kept continuously supplied with oxygen and organic matter (from incoming primary effluent). Under these aerobic conditions the microbes metabolize the organic matter in wastewater, converting it into carbon dioxide, water, and added cell growth similar to the glucose example shown in Figure 6.3.

In order to maintain aerobic conditions the microbial biomass in an activated sludge basin must be constantly supplied with oxygen to promote the efficient breakdown of the organic matter. An activated sludge basin may have a depth of 4–5 m (13–16.5 ft) and a total volume of hundreds of cubic meters. Supplying sufficient oxygen to wastewater in large activated sludge basins is complicated by the fact that the solubility of oxygen in water is low and hence the rate of oxygen transfer into water is slow. The normal air-water surface is insufficient, so some type of aeration system must be designed. Figure 6.4 shows two examples of aeration devices used in wastewater treatment. One device is based on the injection (sparging) of air or oxygen into the wastewater in the vicinity of a turbine. The second passes air or oxygen through a finely porous matrix to produce minute bubbles of gas.

The wastewater leaving an activated sludge basin has greatly reduced dissolved organic matter (BOD) but does contain large numbers of suspended microbial cells often aggregated into flocs. Some of these micro-

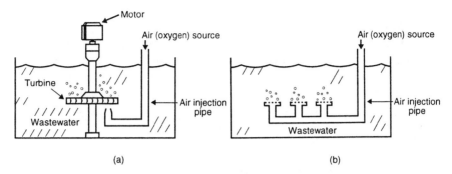

Figure 6.4 Examples of aeration devices used in aerobic biological treatment basins: (a) aeration using air injection with a turbine to provide mixing, and (b) aeration by injecting air through porous plates, domes or tubes.

bial solids are recycled back to the activated sludge tank in order to maintain the high population of microbes needed to remove BOD most efficiently. Figure 6.5 illustrates the recycling of microbial matter and also shows the stream of excess solids called waste activated sludge.

The solids are sometimes difficult to separate from the wastewater stream by sedimentation techniques alone because their densities are near that of water. They do not settle as efficiently as the primary sludge solids described earlier. For this reason waste activated sludges are sometimes removed in a process called dissolved air flotation (DAF). Air under a pressure of several atmospheres is introduced into wastewater from an activated sludge basin. The wastewater is then passed through a release valve in a flotation tank where the pressure drop causes air to leave the solution as minute bubbles. The bubbles attach to the small suspended microbial flocs in the liquid and carry them to the surface where they are removed by skimming. The subsequent handling and disposal of the waste activated sludge solids is described later in this chapter and in Chapters 7 and 11.

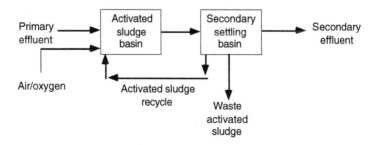

Figure 6.5 Diagrammatic depiction of the activated sludge process in secondary treatment.

The activated sludge process has the advantage of removing BOD from large volumes of wastewater in a relatively compact reactor. The combination of primary treatment followed by secondary treatment using the activated sludge process is capable of removing over 90% of the BOD from a given wastewater.

Sequential Reactor Designs for Biological Removal of BOD, Nitrogen, and Phosphorus

The activated sludge process, as described above, is effective in the removal of soluble BOD from wastewater but it can also be designed and operated in a manner which permits nitrogen removal (nitrification and denitrification) in selected parts of the system (Horan, 1990). With an improved understanding of how phosphorus can be removed by the biological activity of the microbes in the activated sludge process, it has become possible to integrate treatment objectives still further by designing wastewater treatment plants which incorporate BOD, nitrogen, and phosphorus removal in sequentially arranged reactors (Zitomer and Speece, 1993). A typical arrangement consists of an anaerobic reactor, followed by an anoxic tank (optimized for denitrification), followed by an aerobic tank as shown in Figure 6.6. The operation of such a system is summarized as follows:

- **Anaerobic stage**: Wastewater from primary treatment is mixed with return activated sludge in the anaerobic stage where bio-P organisms (microbes able to accumulate phosphorus under certain conditions) store BOD and release phosphorus.

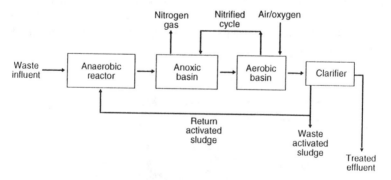

Figure 6.6 Sequential reactor system for removal of BOD, phosphorus, and nitrogen during secondary treatment. Adapted with permission from Zitomer, D.H. and Speece, R.E., Sequential environments for enhanced biotransformation of aqueous contaminants, *Environmental Science and Technology*, 27(2), 226–244, 1993.

- **Anoxic stage**: Mixed liquor from the anaerobic stage is pumped to the anoxic stage where nitrate (recycled from the aerobic stage), serving as an electron acceptor for microbial respiration, is reduced to nitrogen gas. This respiration also removes more BOD from the wastewater.
- **Aerobic stage:** The mixed liquor next moves to the aerobic stage where much of the remaining soluble BOD is oxidized. Ammonia is oxidized to nitrate and the bio-P bacteria oxidize stored substrate while removing much of the phosphate from the wastewater.

The wastewater which leaves the final stage of this sequential treatment system has greatly reduced concentrations of dissolved organic matter, nitrogen, and phosphorus. However, before the secondary effluent is discharged from the treatment plant, it contains high concentrations of suspended microbial flocs which must be removed using a secondary sedimentation basin or the DAF process.

Trickling Filters

Another method of secondary wastewater treatment uses a trickling filter. A trickling filter is a fixed film process in which the reactor is filled with a medium, such as small rocks. Primary effluent is distributed over the medium with a moving overhead distribution arm. The large surface area of the rock, or other type of medium, serves as a support for microbial growth and provides good aeration. As the wastewater flows over the medium it comes into close contact with the microbial film, resulting in aerobic degradation of the BOD in the wastewater. After passage through the trickling filter, the wastewater is collected in an underdrain system at the base of the filter bed.

The word filter implies filtration but the trickling filter system is really a biological treatment process. Its efficiency depends upon a uniform distribution of wastewater across the medium and on circulation of air through it. The successful operation of a trickling filter system requires a suitably sized medium which provides an optimal combination of large surface area, large void space for good air circulation, and minimal tendency to plug. Traditionally, crushed rock of diameter 2.5 to 10 cm (1–4 in.) has been used with a depth limited to about 3 m (10 ft). Recently, lightweight plastic media have been developed which allow depths of up to 12 m (Horan, 1990).

Trickling filters are simpler to operate than the activated sludge process and have lower operational costs. This makes them attractive for wastewater treatment in smaller communities and in certain industries. They are not suited to large flow volumes because land requirements are high and they are not able to achieve quite the high level of BOD reduction that is possible with a well designed activated sludge system. They may also be a source of odors and fly nuisance, especially during periods of warm weather.

Rotating Biological Contactors

Another reactor suitable for small wastewater flows is the rotating biological contactor (RBC). It is similar in principle to the trickling filter (fixed film); however, its operation is based on thin films of microbes attached to rotating disks immersed in a tank of wastewater. The rotating disks are arranged so that approximately half of each disk is immersed while the remainder is exposed to the atmosphere. As the disks rotate, the microbes on them are alternately wetted with wastewater and exposed to air, thus providing an excellent aerobic environment for BOD removal. With time a substantial microbial community develops to a thickness of up to 3 mm (Horan, 1990). Eventually the film will slough off in a manner similar to that in trickling filter systems. The rotation of the disks provides aeration, keeps the tank contents mixed, and introduces shear forces so that the microbial films will slough off before they become too thick.

A major advantage of the RBC system is its ease of operation. The system can be designed as a package plant to fit the wastewater flow needs of a small community or an industry. Other advantages of the RBC are stability against hydraulic shock loadings and elimination of clogging of filter nozzles, a problem which is associated with the trickling filter process.

6.5 TERTIARY TREATMENT (ADVANCED TREATMENT)

The secondary treatment processes in combination with primary treatment can achieve about 90% removal of BOD as well as significant reductions in nitrogen and phosphorus. However, further treatment may be required to meet current regulatory discharge standards. The objective of tertiary treatment, also called advanced treatment, is to produce a high quality effluent using the best available treatment technology. These processes are used to further lower BOD, and reduce concentrations of nitrogen, phosphorus, and any other chemical species present at unacceptable concentrations and capable of degrading the quality of receiving waters.

CHEMICAL TREATMENT

Chemicals are frequently added to wastewater to promote the removal of suspended solids, to remove dissolved ions, and to disinfect effluent before discharging it into the environment. Some common chemical treatment processes involve precipitation, flocculation, ion exchange, reduction and precipitation, phosphorous removal, and nitrogen removal. Many of the same chemicals; lime, ferric or ferrous sulfate, alum, ferric chloride, and polyelectrolytes (polymers) discussed earlier in this chapter in the description of primary treatment processes can also be used to improve the quality of secondary effluent prior to its discharge.

One example of chemical treatment involves the precipitation of phosphorus by adding salts of multivalent metal ions to form phosphate precipitates of very low solubility. Here, the addition of calcium, usually in the form of lime, reacts with phosphate to form hydroxylapatite:

$$10Ca^{2+} + 6PO_4^{3-} + 2OH^- \rightarrow Ca_{10}(PO_4)_6(OH)_2$$
$$\text{Hydroxylapatite}$$

Hydroxylapatite is very insoluble in water and forms a precipitate which can be readily removed in a sedimentation basin or by filtration.

ACTIVATED CARBON

Adsorption refers to the assimilation of gas, vapor, or dissolved matter onto the surface of a solid (the adsorbent). Activated carbon is an adsorbent that has come into wide use in advanced wastewater treatment and the treatment of certain types of industrial wastewater. It adsorbs a wide variety of organic compounds, and residual amounts of inorganic compounds such as sulfides and heavy metals. Its ready production and regeneration makes it economically feasible for use with large volume waste streams.

Activated carbon is manufactured from various plant tissues such as wood, sawdust, coconut shells, walnut hulls, or from coal, lignite, or petroleum residues. These materials are carbonized at high temperatures and then activated using hot air or steam to make the final activated carbon product. Activated carbon in granular or powdered form has an extremely high adsorbing surface area.

In wastewater treatment, filtered secondary effluent is passed through the activated carbon units, either columns or beds, which are sized according to the effluent volume, contact time, and carbon depth. Contact time is a very important factor in activated carbon treatment. If COD levels of 10 mg/L or less are desired, the contact time may need to be 30 minutes or more.

FILTRATION

Filtration, usually with sand filters, has been widely used for many years in the treatment of potable water. It is a recent practice to use filtration in wastewater treatment where this process can achieve the removal of suspended solids beyond the level attained with sedimentation, flotation, or chemical treatment.

There are many filter designs using granular media such as sand, gravel, or anthracite. The water flows through them by gravity or applied pressure. While granular media can effectively remove much finely suspended matter, these filters are not suitable for the removal of dissolved organic or inorganic matter.

Filter performance deteriorates with use as suspended matter is trapped in the medium and clogs pores. Therefore the filters must be designed with a provision for cleaning. One method of cleaning is to take a filter out of operation briefly and backwash it with treated effluent. The backwashing process loosens the sand or other medium in the filter carrying accumulated solids from the system.

Membrane filtration

A more recently developed filtration technology based on the use of membrane filters is effective in removing colloidal-sized material as well as dissolved molecules. Membrane filters are molecular screens with pore sizes that determine which molecules will pass through. Two examples of membrane-based filtration systems are ultrafiltration and reverse osmosis.

Ultrafiltration membrane systems are typically designed to remove colloidal matter (5–100 nm or more) and large molecules with molecular weights in excess of 5000 amu. Given the small pore size in ultrafiltration membranes, waste streams that are to be treated with ultrafiltration require extensive preprocessing. For example, granular filtration is recommended in order to avoid rapid fouling of the membrane surfaces. Ultrafiltration can produce a good quality effluent. If still greater effluent quality is desired, the output from an ultrafiltration process can be used as the feedstock to a reverse osmosis unit.

Reverse osmosis can effectively remove particles in the size range of 0.1 to 15 nm. The term osmosis describes the natural movement of water from one solution through a selective membrane into a more concentrated solution. But, by applying pressure to the more concentrated solution, the flow direction can be reversed so that water passes through the membrane, leaving the dissolved materials behind in the concentrated solution; hence the name, reverse osmosis. The pressures required for this process are typically 2000–7000 kPa (300–1000 lb/in.^{2}).

The membranes in a reverse osmosis unit are easily fouled by colloidal matter in a feed waste stream; therefore, a feed stream free of these materials is required for efficient performance of a reverse osmosis unit. It may also be necessary to remove iron to minimize scaling.

Reverse osmosis has been used for over two decades to produce fresh water from seawater or brackish waters. The application of reverse osmosis in the treatment of secondary-treated wastewaters has been limited because the operating and maintenance costs are relatively high. However, this process may prove to be a suitable and cost-effective choice for purifying certain industrial streams.

6.6 DISINFECTION

Disinfection is commonly defined as destruction of disease causing organisms. Disinfection is a concern in municipal wastewater treatment because municipal wastewater effluent contains high populations of microbes and viruses, some of which are capable of causing disease. Many chemical agents have been used as disinfectants, including chlorine, bromine, iodine, ozone, phenols, alcohols, hydrogen peroxide, quaternary ammonium compounds, heavy metal based compounds and various alkalis and acids. In addition, the disinfectant properties of physical agents such as ultraviolet radiation (UV) and ionizing radiation have also been investigated. Only three of the above, chlorine, ozone, and ultraviolet (UV) radiation, will be discussed here. Some of the properties of chlorine, ozone, and UV radiation are compared in Table 6.3.

CHLORINE

Chlorine in its various forms (chlorine gas, calcium and sodium hypochlorite, and chlorine dioxide) has been the most widely used disinfectant in wastewater treatment. This is because of its effectiveness in destroying bacterial cells and low cost.

Chlorine is very effective in destroying young actively growing bacterial cells, but less effective in killing older cells and much less effective in killing bacterial spores. The oocyst of the protozoan parasite *Cryptosporidium*, which causes an intestinal disease known as Cryptosporidiosis, is very resistant to chlorine. Also, some viruses may be resistant to the levels of chlorine that are used to kill many bacteria in wastewater effluent.

In recent years the use of chlorine has been criticized because of its potential for producing toxic compounds. Chlorine is highly reactive and in the presence of organic compounds may produce a variety of halogenated organic molecules. Some of these, for example, chloroform, have

Table 6.3   Comparing the Characteristics of Chlorine, Ozone and Ultraviolet Radiation in Wastewater Effluent Disinfection

Characteristic	Chlorine	Ozone	Ultraviolet Radiation
Effectiveness as a disinfectant	Very effective against bacterial cells, much less so against bacterial spores or the oocysts of a parasite such as *Cryptosporidium*	Effective against bacteria, thought to be a better viricide than chlorine	Effective against microbial cells, Dosage must allow for the UV damage-repair systems of microbes
Toxic residuals or products	Chlorine residuals possible unless dechlorination used, potentially toxic halogenated organics may be formed	Dissipates rapidly, little or no residual formation	No toxic residuals formed
Penetration	High	High	Moderate, suspended solids interfere, UV lamps require frequent cleaning
Cost	Low if no dechlorination is required	Moderately high, but becoming competitive with chlorination/ dechlorination	Moderately high, but becoming competitive with chlorination/ dechlorination

been demonstrated to be carcinogenic. In addition, the residual chlorine may be toxic to aquatic life in receiving waters. This has resulted in environmental regulations requiring dechlorination of wastewater effluent before discharge. The dechlorination step adds to the cost of chlorination so the advantage of chlorination as a low cost disinfectant is diminished. Dechlorination does not eliminate the problem of generation of halogenated organic compounds. The concerns about the detrimental impacts of chlorine has driven a search for alternative disinfectants such as ozone and UV radiation.

OZONE

Ozone (O_3) is a potentially good substitute for chlorine as a disinfecting agent in wastewater treatment. It is very reactive and a good microbial disinfectant, perhaps even a better viricide (capable of destroying viruses) than chlorine. Ozone is unstable, so it must be generated onsite, a factor which increases the cost and complexity of the disinfection stage of wastewater treatment. The instability of ozone is actually an advantage because it dissipates rapidly and little or no ozone residual remains in the effluent by the time it is discharged. Furthermore, ozone decomposition produces oxygen which is beneficial to the waters receiving the effluent. The cost of ozone disinfection is greater than chlorine disinfection; however, recent advances in ozone generation technology have made it more competitive.

ULTRAVIOLET (UV) RADIATION

Ultraviolet radiation (UV) has long been recognized as an effective germicidal agent in laboratories and health care facilities. It has also been used to a limited extent for disinfection of water supplies.

Ultraviolet radiation is produced by low pressure mercury lamps. These lamps generate about 85% of their output radiation at a wavelength of 254 nm, which is very effective in disrupting the nucleic acids (DNA and RNA) of cells or viruses. Because UV radiation is a physical disinfecting agent it is thought that no toxic residuals are produced, as occurs in the use of chlorine disinfection.

Until recently, UV radiation has not received much consideration as a means of the disinfection of wastewater effluent because it is easily absorbed by suspended solids and other UV-absorbing materials often found in wastewater effluent streams. However, with improved tertiary treatment, many wastewater effluents now have much lower concentrations of solids and dissolved chemicals so that UV treatment for disinfection becomes much more efficient and cost-effective. Increasing regulatory pressures to eliminate chlorination as a disinfectant process have stimulated much research into full-scale application of UV as a disinfectant for wastewater effluent.

6.7 SLUDGE THICKENING AND DEWATERING

The best modern methods of wastewater treatment can generate a high quality effluent that is low in organic matter (BOD), nitrogen and phosphorus, and hence suitable for discharge to receiving waters with minimal environmental impact. However, this treatment efficiency results in the generation of large quantities of sludges which require processing and ultimate disposal (see Table 2.16 in Chapter 2).

The name sludge implies solids, but raw sludges are still largely water with a solids content. The activated sludges produced during secondary treatment often contain less than 1% settled solids. This means, in effect, that large sludge volumes need to be reduced in order to obtain a final sludge product which can be economically and safely disposed or used in a beneficial manner. Application of sludges to land is an example of a beneficial use. Therefore, it is common for wastewater treatment plants handling large volumes of wastewater to invest heavily in capital equipment and operational activities designed to condition, thicken, and dewater sludges. The physical and chemical processes used to achieve these objectives are summarized in Table 6.4. The biological processes known as anaerobic treatment and composting, which will be discussed in Chapter 7, may also play a role in the reduction of sludge volume.

SLUDGE CONDITIONING

Sludge conditioning refers to biological, chemical, or physical processes which enhance removal of water from the sludge solids. These treatments may not actually remove water but instead change the characteristics of the sludge stream so that the solids are more effectively separated from the aqueous phase (e.g., by flocculation and precipitation) during the sludge thickening and dewatering steps which come later. Inorganic chemical precipitants such as lime and ferric chloride mentioned earlier are examples of chemical conditioning agents. Other widely used conditioning agents are the polyelectrolytes, also described earlier.

Another type of sludge conditioning process, thermal conditioning, utilizes high temperatures and pressure. In this process sludges are heated

Table 6.4 Comparison of Sludge Conditioning, Thickening, and Dewatering

Process	General Definition
Sludge conditioning	Biological, chemical or physical treatment to enhance water removal from a sludge. Often used prior to sludge thickening and dewatering techniques.
Sludge thickening	Removal of water from a sludge to achieve a reduction in volume. Thickened sludges still have a liquid consistency.
Sludge dewatering	Dewatering is removal of water from sludge beyond that achieved by thickening. A dewatered sludge is no longer a liquid. For example a 25% solids sludge, although still containing 75% water, would have a cake–like consistency.

to 177–240°C under pressures of 1700–2800 kPa (250–400 lb/in.^{2}) for periods of 15–40 min. In some systems air may also be injected to create an oxidizing environment which reduces the solids volume somewhat.

Thermal conditioning produces a sludge with excellent dewatering characteristics. During the final dewatering step it may be possible to achieve sludge cake solids as high as 50%. The high temperatures used in thermal conditioning kill pathogens very effectively. Unless it is recontaminated, the final sludge cake is effectively sterile.

Thermal treatment has some major disadvantages. Capital costs and maintenance costs are very high. Because of the high temperatures and pressure, only high-quality materials can be used for piping, pumps, and other equipment. Scale formation and acid induced corrosion on even high quality stainless steel materials has been a problem. The process also produces a liquid side stream with high BOD as a result of solubilization of some of the organic solids in the sludge being treated. This side stream must be returned to the secondary step in a wastewater treatment plant, thus increasing the load to be handled by this step. The disadvantages of thermal conditioning appear to have inhibited the construction of new facilities with this technology in recent years (Metcalf & Eddy, 1991).

SLUDGE THICKENING

Sludge thickening is any process which removes water to reduce the volume of a sludge stream. For example, a sludge consisting of 3% solids that is thickened to 6% solids will occupy one-half as much volume. Typically, sludges which have been thickened still have enough water to give the sludge the consistency of a liquid. Thickening is accomplished in sedimentation basins, by centrifugation, and by the use of flotation (DAF) or gravity thickening basins. Primary sludge can be thickened with an appropriate design and operation of a primary sedimentation basin. On the other hand, the biological sludges which come from activated sludge basins (secondary treatment) are much harder to thicken by sedimentation. Thickening these sludges may require the DAF processing technique discussed earlier or centrifugation in order to achieve the desired level of thickening.

Centrifugation

The application of centrifugation to wastewater treatment dates back more than 50 years. Early centrifuges were primarily designed for industrial processing and were not well suited for the variable characteristics of wastewater sludges. Hence their early use in wastewater treatment plants was associated with significant operational and maintenance problems.

Since the 1960s, centrifuge designs have appeared that are more trouble-free in dealing with the unique and variable properties of sewage sludges. Two examples of centrifuges used for sludge thickening are the imperforate basket centrifuge and the solid bowl decanter centrifuge. In Figures 6.7 and 6.8 diagrams of their basic designs are shown. The advantages and disadvantages of each are compared in Table 6.5.

The imperforate basket centrifuge operates in a batch mode. Liquid sludge is pumped into a vertically spinning bowl. Solids build up on the wall of the bowl while the liquid (centrate) is decanted. After the solids accumulate to a given depth, the bowl is slowed and a scraper moves through the bowl to remove the accumulated solids. This centrifuge design is well suited for difficult-to-handle sludges such as waste activated sludges or any sludge with soft or fine particulates. The imperforate bas-

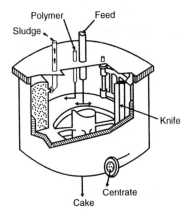

Figure 6.7 Schematic of the imperforate basket centrifuge (Adapted from USEPA, 1979).

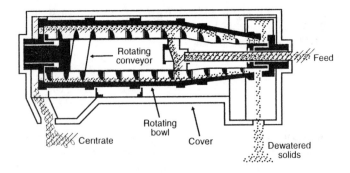

Figure 6.8 Schematic of the solid bowl decanter centrifuge (Adapted from USEPA, 1979).

Table 6.5 Comparing the Imperforate Bowl and Solid Bowl Decanter Centrifuges for Sludge Processing

Centrifuge	Advantages	Disadvantages
Imperforate bowl	Same machine can be used for thickening and dewatering	Unit is not continuous feed and discharge
	Very flexible in meeting process requirements	Requires special structural support
	Tolerates grit	Highest ratio of capital cost to capacity
	Lower operation and maintenance costs	
	Excellent thickener for hard-to-handle sludges, 8–10% solids possible from waste activated sludge	
Solid bowl decanter	High throughput in a small area	Potentially a high maintenance item
	Easy to install	May require polyelectrolytes for good performance
	Quiet	Grit removal important
	No odor problems	Skilled maintenance personnel needed
	Consistently able to achieve 4–6 % solids in thickened sludge	

Sources: U.S. Environmental Protection Agency, Process Design Manual for Sludge Treatment and Disposal, Municipal Environmental Research Laboratory, EPA 625/1–79–011, 1979.

ket centrifuge has the disadvantage of not being able to operate continuously and has the highest ratio of capital cost-to-capacity.

The solid bowl decanter centrifuge is a long horizontal bowl that is tapered at one end. Sludge is fed into the centrifuge continuously and solids concentrate on the sides of the bowl. A helical scroll spins at a slightly different speed, moving accumulated sludge toward the tapered end where it is discharged. The solid bowl centrifuge is capable of very high throughput for a given size and is able to achieve thickened solids concentrations of 4 to 6% on waste activated sludge. This centrifuge design may require extensive maintenance because the scroll is subject to wear from abrasive materials in a sludge stream.

SLUDGE DEWATERING

Dewatering is typically the last sludge processing step prior to the ultimate disposal or use of a sludge. The primary objective of dewatering is to greatly reduce sludge volumes and the consequent cost of sludge handling in final disposal or use. For example, a sludge is thickened from a 2% solids concentration to 6% solids. The 6% solids sludge stream is then dewatered, using one of a variety of processes, to a sludge cake that may consist of 20% solids. The final volume of the 20% sludge cake will be much less than that of the original 2% sludge. One or more of the following factors are significant in deciding whether to dewater a given sludge:

- The cost of transporting the sludge to the ultimate disposal site is highly dependent on volume.
- Removing excess moisture aids in odor control.
- Sludge dewatering is required for sludges to be placed in landfills.
- Dewatered sludge in cake form is usually easier to handle than liquid sludge.
- Dewatering is necessary to reduce the cost of incinerating a sludge.
- If sludge is to be composted it is helpful to dewater it in order to reduce the amount of bulking agents.

The advantages and disadvantages of a variety of sludge dewatering methods are compared in Table 6.6. Among the most significant factors that determine the choice of dewatering methods are the types of sludges produced, the size of the wastewater treatment plant, and the method of final sludge use or disposal. For example, drying beds may be a very effective and low cost method of dewatering sludge in a small community's wastewater treatment plant in a dry climate. On the other hand, drying beds demand considerable land space per unit volume of sludge processed and therefore are often not appropriate for large treatment plants where space and/or climate may be limiting factors.

The first four types of sludge dewatering shown in Table 6.6 are sometimes called mechanical methods, in contrast to the two nonmechanical methods, drying beds and sludge lagoons. The mechanical methods are characterized by relatively elaborate mechanical systems which entail a high capital investment. They may also require high energy inputs and

Table 6.6 Comparison of Sludge Dewatering Methods

Method	Advantages	Disadvantages
Vacuum filter	Fairly low maintenance	Largest amount of energy consumed per unit sludge
	Specialized operator skills not required	Operators must be present during operation
	Filtrate with low suspended solids with most sludges	Equipment noise
Belt filter press	Very high solids, upwards of 50%, sludge cake is possible	Very sensitive to sludge feed characteristics
	Low energy requirements	Hydraulic throughput limited
	Easy shutdown	Short filter belt life
Imperforate basket centrifuge	Same machine can be used for thickening and dewatering	Limited size capacity
	Grit not a problem	Second only to vacuum filter in energy consumption
	Works well on "difficult" sludges	Highest capital cost-to-capacity
	Flexible in meeting process needs, fast startup/shut down, few odor problems	Produces lowest cake solids with most sludges
	May not need chemical conditioning	
Solid bowl centrifuge	Fast startup/shut down, little odor, clean appearance	Scroll wear a possible high maintenance item, grit removal important
	Installation is easy, relatively low capital cost–to–capacity ratio	Skilled maintenance personnel needed
	High cake solids concentration is possible	Possible problems with high suspended solids in centrate

Table 6.6 (Continued)

Method	Advantages	Disadvantages
	High throughput/unit surface area	
	Does not require continuous operator attention	
Sludge drying beds	Lowest capital cost where land is available	Large land area required
	Little energy and chemical consumption	Sludge must be stabilized
	Little operator attention or skill needed	Climate a design variable
	Sludge variability not a major problem	Labor cost of sludge removal
	Higher cake solids than with mechanical methods above	
Sludge lagoons	Low energy and chemical consumption	Potential for odor and vector problems
	Least amount of operational skill	Public perception
	Insensitive to sludge variability	Land intensive
	Further stabilization of organic matter	Climate effects, ground water contamination

Source: U.S. Environmental Protection Agency "Process Design Manual for Sludge Treatment and Disposal," Municipal Environmental Research Laboratory, EPA 625/1–79–011 (1979).

chemical conditioning as well as costly maintenance. The mechanical methods have the advantage of high sludge capacity per unit of space and are thus often used in large wastewater treatment plants where the sludge volumes and other factors preclude the use of drying beds and sludge lagoons.

Vacuum Filter

The typical vacuum filter, diagrammed in Figure 6.9, consists of a large horizontal drum covered with a metal or cloth belt. In operation the drum rotates partially submerged in a tank of liquid sludge. The drum is sealed at the ends and is divided into sections around its periphery. Each section is sealed from adjacent sections and has its own valving system which permits a vacuum to be applied to the section at an appropriate point in its rotation. As the drum, with its belt, rotates through the liquid sludge a layer of sludge is drawn onto the belt by the applied vacuum. When a given segment of the drum and belt rotate out of the liquid sludge the layer of sludge which adheres to the belt is readily dewatered as the vacuum continues to be applied. As the drum continues to rotate, the belt leaves a given segment and at this point the vacuum to that segment is shut off. The layer of sludge on the belt has lost much water and as the belt moves around smaller shafts the sludge cake on the belt drops into a receiving vessel.

The vacuum filter is able to produce a sludge cake that ranges from 20% to approximately 30% solids. The efficiency of dewatering depends on the nature of the sludge and the type of conditioning used prior to the dewatering step. For example, primary sludges can be dewatered to 28–32% solids, while waste activated sludges may only dewater to 20–25% (USEPA, 1979). Vacuum filters came into wide use in wastewater treatment beginning in the mid-1920s. Since 1970 there has been some decline in their use as other technologies, such as the belt press, have been improved.

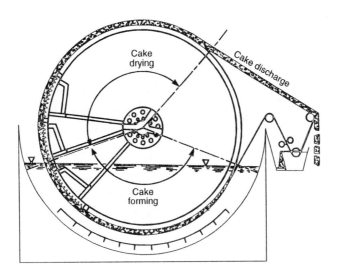

Figure 6.9 A cross sectional view of a fiber cloth-belt rotary vacuum filter (USEPA, 1979).

Belt Filter Press

The belt filter press came into use in the early 1970s and with continuing improvements in its performance has become a major dewatering technology in wastewater treatment plants. Dewatering with belt filter presses consists of three stages: chemical conditioning, removal of water by gravity, and the final pressure removal of water to form a sludge cake (see Figure 6.10).

Chemical conditioning of sludge is important for efficient belt filter press performance, as it is for most mechanical methods of dewatering. The gravity stage is important since it results in removal of a large amount of water to yield a thickened sludge. In some units a vacuum may be applied at this stage to achieve the removal of greater amounts of water. After the gravity stage the sludge moves between porous belts where increasing pressure is applied to force out water.

The dewatering effectiveness of a belt filter press varies greatly. It is especially influenced by sludge characteristics, chemical conditioning, and the design of the machine itself. Sludge cake solids produced from a belt press may range from less than 15% for waste activated sludges to over 40% for some primary sludges.

Centrifugation

The two types of centrifuges, the imperforate basket and the solid bowl (scroll-type) decanter described earlier, can also be operated to achieve dewatering. The advantages and disadvantages of these two dewatering techniques are summarized in Table 6.6.

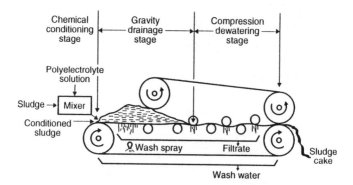

Figure 6.10 A diagrammatic view showing the three stages of the belt press operation during sludge dewatering (USEPA, 1979).

As is true with most dewatering techniques, the characteristics of the sludge are key factors influencing the degree of dewatering achieved with centrifuges. The imperforate basket centrifuge may produce cake solids of 30% with primary sludge, but with waste activated sludge the final solids may be less than 10%. This centrifuge's performance is very good when percent of solids captured is used as a measure (over 95% with primary sludge). It also has a reputation for acceptable performance with so-called difficult sludges. This centrifuge design is not sensitive to grit; therefore, it is often used by small wastewater treatment plants where a grit removal stage is omitted.

The solid bowl centrifuge is capable of effective removal of water, producing a sludge cake with over 35% solids in the dewatering of some primary sludges. The scroll mechanism of this centrifuge design is subject to wear; therefore, grit removal is especially important.

Sludge Drying Beds

Drying beds are the most widely used method of sludge dewatering in the United States. Because of their low cost, simplicity of operation, and high solids content in the final sludge cake, drying beds are especially popular with small waste treatment plants in warmer climates. However, even some larger treatment plants in northern climates have found drying bed technology effective in the dewatering of sludge. The space requirement for drying beds is perhaps the most significant disadvantage, especially for large treatment plants. Drying bed technology has evolved into a variety of approaches, ranging from the conventional sand drying bed that is widely used by small treatment plants serving communities of up to 20,000 population, to more elaborate designs which can handle larger volumes of sludge per unit of land area.

In a typical drying bed, illustrated in Figure 6.11, 20–30 cm of sludge is placed on a large flat sand bed. Depending on climate conditions and odor susceptibility, the bed may be open or covered. The sludge is dewatered by drainage through the sludge solids and the sand, and by evaporation from the large surface area. Since most of the water leaves by drainage, an adequate underdrainage system is required. If drying conditions are favorable, sludge solids may exceed 50% after two weeks. After adequate dewatering, the dried sludge is removed by a scraper or front-end loader.

The area of drying beds needed by a community can often be computed on a per capita basis. In some cases, it may be necessary to estimate the amount of dry sludge solids generated each year in order to obtain a good estimate of the required area. The area needed is also influenced by climate conditions and whether or not the beds are covered. Table 6.7 shows data for the open drying bed area required for two types of sludges.

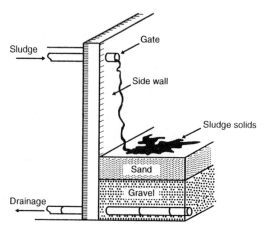

Figure 6.11 An example of sand drying bed design (USEPA, 1979).

The area required for covered beds is approximately 70% of that needed for an open bed.

Various drying bed designs have been developed in order to increase the efficiency of the process especially for communities where space is limited. One approach is the use of stainless steel wedge wire in place of sand. This bed permits more rapid drainage of water; hence, greater dewatering efficiency per unit of bed area is realized. The capital costs of this design are higher.

Table 6.7 Space Required for Open Sludge Drying

Sludge	Sludge Bed Area (m²/person)	Annual Sludge Loading Rate (kg of dry solids/m²)
Primary digested	0.09–0.14	120–150
Primary and waste activated (digested)	0.16–0.23	60–100

Source: U.S. Environmental Protection Agency (USEPA) "Process Design Manual for Sludge Treatment and Disposal," Municipal Environmental Research Laboratory, EPA 625/1–79–011 (Washington, DC: U.S. Printing Office, 1979).

Metcalf & Eddy, Inc. *Wastewater Engineering: Treatment, Disposal, and Reuse* (New York, NY: McGraw-Hill, Inc., 1991).

Vacuums have also been used to accelerate the dewatering process. This can be done by applying a vacuum on the underside of porous plates placed in the drying beds. This design greatly reduces the time necessary for dewatering. The vacuum assisted bed also requires less space but the capital cost is higher.

Lagoons

Lagoons are sometimes used as an alternative to drying beds. Their performance is affected by climate. Low temperatures inhibit dewatering, whereas low humidity in a warm climate accelerates dewatering greatly. The use of lagoons is increasingly more regulated because stringent groundwater regulations may make a lagoon installation unacceptable in many areas. Lagoons are also not considered suitable for dewatering raw sludges or sludges with a high strength supernatant (high BOD) because of the odor and nuisance potential.

Digested sludges are discharged into a lagoon to a depth of 0.75 to 1.25 m (2.5 to 4 ft). Evaporation is a major factor in dewatering. As the sludge solids settle to the bottom of the lagoon the supernatant liquids are removed and recycled back to the treatment plant. Sludge is removed when the solids content reaches about 30%. The cycle time of lagoons varies greatly, ranging from several months to years. For example, sludge might be pumped to a lagoon over a period of 18 months, followed by a rest period of 6 months when no sludge is added. Thus, a treatment plant requires a minimum of two lagoon cells to ensure adequate sludge dewatering capacity on a continuous basis.

DISCUSSION TOPICS AND PROBLEMS

1. (a) What is the average daily volume of municipal wastewater generated in your community?
   (b) What is the volume of sewage sludge generated annually in your community?
   (c) How is this sewage sludge processed?
2. Describe the primary, secondary, and tertiary methods of wastewater treatment in your community. What method(s) are used to remove nitrogen and phosphorus?
3. If your community has an industry which generates a large quantity of wastewater, describe the methods used in treating the wastewater.
4. A sludge has an initial moisture content of 90%. How much water must be removed per tonne if the moisture content is to be reduced to 10%?
5. A waste stream with a concentration of organic material characterized by 1000 mg/L of COD is biologically degraded under aerobic

conditions. The COD content of the waste is reduced by 90%. Approximately how much microbial cell mass is generated per liter of treated waste?

6. Using the equation in Section 6.5 showing the use of calcium to precipitate phosphorus, calculate the amount of ($Ca(OH)_2$) needed to remove phosphate from a wastewater which contains 8 mg of PO_4^{3-}/L.

7. Using information from Table 6.2:
   (a) Calculate the amount of primary sludge generated (dry weight basis) daily when chemical precipitation is used in a community with a wastewater flow of 5×10^6 L/d.
   (b) How much ferrous sulfate and lime will be needed?
   (c) What will be the approximate volume of the sludge generated?
   (d) What is the approximate solids concentration of this primary sludge?

8. Calculate the area and volume of sludge lagoon(s) required by a community if the sludge generated is 50,000 L/d, the sludge depth in the lagoon is 100 cm, and the use cycle of a lagoon is eight months feed followed by four months rest.

9. Five thousand liters of sludge containing 2% solids is dewatered to a sludge containing 6% solids, which is in turn dewatered to a sludge cake with a solids content of 28%. What will be the approximate volume of the sludge containing 6% solids? Of the sludge cake?

10. Why might two communities with the same volume of wastewater generate quite different volumes of primary sludge even when both use chemical treatment?

11. Chlorine has been widely used for many years as a disinfectant in wastewater treatment. Explain why the use of chlorine is likely to decline with ozone or ultraviolet radiation becoming replacement disinfectants.

12. Where in a large modern wastewater treatment plant does gravity play an important role in the treatment processes?

13. (a) What is the primary reason for promoting denitrification reactions in wastewater treatment? (b) Why is it important to have anoxic conditions in a wastewater treatment basin in order to achieve denitrification?

REFERENCES

Grady, C.P.L., Jr., *Biological Wastewater Treatment*, Marcel Dekker, Inc., New York, NY, 1980.

Horan, N.J., *Biological Wastewater Treatment Systems*, John Wiley & Sons, New York, NY, 1990.

Metcalf & Eddy, Inc., *Wastewater Engineering: Treatment, Disposal, and Reuse*, McGraw-Hill, Inc., New York, NY, 1991.

U.S. Environmental Protection Agency (USEPA), Process Design Manual for Sludge Treatment and Disposal, Municipal Environmental Research Laboratory, EPA 625/1-79-011, Washington, DC, U.S. Printing Office, 1979.

Zitomer, D.H. and Speece, R.E., Sequential environments for enhanced biotransformation of aqueous contaminants, *Environmental Science and Technology*, 27(2), 227–244, 1993.

CHAPTER 7

Biological Treatment of Waste Solids

7.1 OVERVIEW

Biological treatment of waste solids encompasses those processes that rely on microbes to change or degrade organic wastes in a controlled manner. Microbes are very sensitive to changes in environmental conditions. Major factors affecting microbial activity are moisture, temperature, pH, oxygen concentration, presence of toxic elements or compounds, and the type and quality of the organic material serving as the microbes' food supply. The challenge for the designer of a biological waste processing system is to be able to exercise some degree of control over these conditions.

The common biological treatment processes are composting, anaerobic decomposition (also called anaerobic digestion), and fermentation. Of these, composting and anaerobic digestion are the most developed. Some examples of organic wastes that are amenable to biological processing include sewage sludge, some industrial wastes, agricultural and food processing residues, and the organic portion of municipal solid waste.

In Chapter 6, it was emphasized that effective municipal wastewater treatment results in large quantities of sludges, or biosolids, that must be dewatered and further stabilized. In particular, a sewage sludge requires further treatment to kill pathogens, to reduce its volume, and to minimize odor in order to achieve an environmentally acceptable material for final disposal or use, such as spreading on land—a topic discussed in Chapter 11. Both composting, usually an aerobic process, and anaerobic treatment can be, and often are, used to treat these biosolids. The high temperatures achieved in aerobic composting are effective in destroying pathogens, while anaerobic digestion generates a useful by-product, methane gas (CH_4).

In addition to its capability for treating sewage sludge, aerobic composting is applicable to other organic sludges, agricultural residues, yard wastes, and the organic fraction of MSW. Composting yard and garden wastes is familiar to many gardeners and farmers, and recently, many communities have established composting sites in an effort to divert the yard and garden wastes from landfills. Yard and garden wastes are estimated to constitute about 18% of MSW (see Table 2.5). Twenty-two states have banned yard waste from landfills (see Appendix E). The compost produced is most often used as a soil amendment or low value fertilizer. The marketability of compost is a factor in determining the feasibility of employing composting as a waste treatment process. It is generally not economically feasible to transport compost over long distances.

The objective of many anaerobic process designs is the efficient production and recovery of methane. One sophisticated approach utilizes engineered reactors in which the biodegradable fraction of MSW is degraded anaerobically (DeBaere, 1987). Methane is recovered and the digested solids used as a compost. A unique concept in anaerobic processing of MSW is to design and operate landfills as anaerobic reactors to optimize anaerobic activity. The methane gas generated by this process can then be captured and utilized.

Another process, fermentation, may also be appropriate for processing some organic waste streams, particularly agricultural residues and food processing wastes. The fermentation process, as it is applied in the production of alcohol from fruits and grains, is a well-developed technology. Although fermentation of the organic portion of MSW has been proposed, problems exist because a major portion of MSW is paper, a highly cellulosic material that is fermentable in theory, but may require potentially costly pre-fermentation processing.

Biological processes are generally viewed more positively by the public than landfilling or incineration. Perhaps the reason is that biological processing is touted as "natural," with the end products being safely returned to the earth. Whether this view is justified or not depends greatly on the feedstocks, the processes, and the chemical analyses of the products. These factors are discussed in more detail in Chapter 11.

7.2 THE ORGANISMS IN BIOLOGICAL TREATMENT

The primary groups of organisms commonly involved in biological waste degradation include bacteria, fungi, protozoa, and algae. The bacteria are generally considered to be the single most important group in the biological degradation of wastes, especially in wastewater treatment.

They are small, single-celled organisms which, as a group, possess a diverse array of enzymes able to degrade many organic and inorganic substances in waste streams. They are also capable of high rates of metabolic activity which makes possible high throughput of waste in well-engineered waste treatment systems.

Most fungi are multicellular, possess a filamentous growth, and are commonly called molds. Fungi require organic substrates for energy and, as a group, also possess enzyme systems capable of attacking a wide variety of organic molecules in nature. In fact, it is among the fungi that we find species that are the most effective in degrading the complex structure of lignocellulose, the major component of wood and of such waste materials as newsprint. The fungi are generally not microbes of significance in wastewater treatment but are important in solid waste treatment processes, such as composting and the degradation of wastes applied to land.

Like bacteria, most protozoa are unicellular organisms. However, their cells are typically much larger and their cell structure and nutritional procurement strategies more diverse. Some protozoa rely primarily on the uptake of dissolved organic compounds from their environment. Others are able to engulf particulate matter, such as free swimming bacterial cells and other suspended matter as encountered in the activated sludge process method of wastewater treatment described in Chapter 6. The activity of this latter group of protozoa is believed to contribute significantly to improved activated sludge process performance. Suspended solids and BOD in the effluent are reduced, while larger numbers of pathogens are removed (Horan, 1990).

The algae comprise a diverse group of photosynthetic organisms ranging from unicellular forms to complex and relatively large multicellular species. The algae are an important part of the complex of organisms in waste treatment systems that rely on waste stabilization ponds (lagoons) or wetlands for wastewater purification. In these environments their photosynthesis provides oxygen for aerobic bacteria and protozoa.

The viruses are noncellular with unique structural features and are defined as obligate intracellular parasites, which means that they can only replicate themselves inside living, cellular organisms. While they do not directly participate in biological degradation they are significant in waste management because many are pathogens for humans or for plants and animals important to humans. Their presence is an important consideration in such waste management decisions as the application of treated wastewater, sewage sludges, and composts to agricultural and other lands.

While the microbes described above are dominant in many waste treatment processes, other organisms, including invertebrates such as nematodes and earthworms, also decompose wastes in soil, and in some composting environments.

7.3 AEROBIC DECOMPOSITION

The secondary stage of municipal wastewater treatment is based upon an aerobic treatment process; consequently, the fundamental concepts of aerobic processing were introduced in Chapter 6. Two important points were made in that discussion. The first point is the importance of supplying adequate oxygen or air for the microbes. The second point is that a large amount of cell mass is produced during the process. In this chapter, the aerobic process will be considered as a means for processing organic solid waste or sludge.

The aerobic decomposition of an organic material $C_aH_bO_cN_d$ is represented by the reaction (Tchobanoglous et al., 1977),

$$C_aH_bO_cN_d + 0.5(ny + 2s + r - c)O_2 \rightarrow$$
$$nC_wH_xO_yN_z + sCO_2 + rH_2O + (d - nx)NH_3$$

where
$$r = 0.5[b - nx - 3(d - nx)],$$
$$s = a - nw,$$
and,
n = number of moles of $C_wH_xO_yN_z$ produced from one mole of $C_aH_bO_cN_d$

The subscripts in the empirical formulas, $C_aH_bO_cN_d$ and $C_wH_xO_yN_z$, represent the relative amounts of carbon, hydrogen, oxygen, and nitrogen in the organic waste and the decomposed product, respectively. The calculation of empirical formulas such as these is demonstrated in Example 7.1.

If there is a complete conversion of the material; that is, the organic product in the above equation is completely oxidized, then the reaction becomes (Tchobanoglous et al., 1977)

$$C_aH_bO_cN_d + 0.25(4a + b - 2c - 3d)O_2 \rightarrow$$
$$aCO_2 + 0.5(b - 3d)H_2O + dNH_3$$

Depending upon the environmental conditions, the ammonia (NH_3) may be oxidized to nitrate (NO_3^-).

Example 7.1: The chemical analysis of an organic waste (on a dry basis) indicates the following composition:

Carbon	45.9%
Hydrogen	6.1%
Oxygen	47.9%
Nitrogen	0.1%
Nonorganic	1.0%

(a) Find the empirical formula for this material. (b) Calculate the amount of oxygen required for the complete conversion of 1 kg of this material.

Solution to Part (a): Find the relative number of moles of each of the elements in the material by assuming a convenient amount of material, such as 1 kg. The nonorganic portion is neglected because it is not involved with the aerobic process. The calculations are shown in the table below. The masses of each element (column 3) in the 1-kg sample are calculated using the percentages above. The number of moles of each element (column 4) is found by dividing the mass of the element in the sample (column 3) by the element's atomic weight (column 2).

	Atomic Weight	Mass of Element (g)	Number of Moles
Carbon	12	459	38.3
Hydrogen	1	61	61
Oxygen	16	479	29.9
Nitrogen	14	1	0.07

Therefore, the empirical formula is $C_{38.3}H_{61}O_{29.9}N_{0.07}$.

Solution to Part (b): Using the coefficients obtained in Part (a); namely, $a = 38.3$, $b = 61$, $c = 29.9$, and $d = 0.07$, the chemical equation for the complete aerobic conversion of this organic material becomes

$$C_{38.3}H_{61}O_{29.9}N_{0.07} + 38.9O_2 \rightarrow 38.3CO_2 + 30.4H_2O + 0.07NH_3$$

One kg of this material contains 990 g of organic material. This mass is converted to moles by dividing the organic mass by the formula mass. The formula mass is calculated as follows:

$$
\begin{aligned}
38.3 \times 12 \text{ (atomic wt. of carbon)} &= 459.6 \\
61 \times 1 \text{ (atomic wt. of hydrogen)} &= 61 \\
29.9 \times 16 \text{ (atomic wt. of oxygen)} &= 478.4 \\
0.07 \times 14 \text{ (atomic wt. of nitrogen)} &= 1.0 \\
\hline
\text{Formula mass} &= 1000
\end{aligned}
$$

Dividing the organic mass by the formula mass yields: 990/1000 = 0.99 moles. From the reaction equation, 38.9 moles of oxygen molecules are consumed for each mole of organic matter that is aerobically converted. For this example, the amount of oxygen consumed is

$$0.99 \text{ mole of organic material} \times \frac{38.9 \text{ moles of } O_2}{1 \text{ mole of organic matter}} = 38.5 \text{ moles of } O_2$$

One mole of any gas occupies a volume of 22.4 L at standard temperature and pressure (STP) of 0°C and 1 atm, respectively. Thus, the volume of oxygen required is

$$38.5 \text{ moles of oxygen} \times \frac{22.4 \text{ L of } O_2}{1 \text{ mole of } O_2} = 863 \text{ L of } O_2.$$

The mass of the oxygen is

$$38.5 \text{ moles of oxygen} \times \frac{32 \text{ g of } O_2}{1 \text{ mole of } O_2} \times \frac{1 \text{ kg of } O_2}{1000 \text{ g of } O_2} = 1.23 \text{ kg of } O_2.$$

Several observations can be made with respect to the aerobic reaction equations above, Example 7.1, and the discussion of aerobic treatment of wastewater in Section 6.4. The first observation is that the primary products are carbon dioxide, water, and, depending upon the extent of the conversion, an organic material. The second observation is that a considerable amount of oxygen is needed for the process. In Example 7.1, the mass of the oxygen required for the complete conversion of the organic matter exceeded the mass of the organic material. The amount of oxygen consumed and the amount of carbon dioxide produced are approximately equal on a molar basis. Finally, aerobic reactions tend to be exothermic reactions. This is an important factor in waste composting because exothermic reactions release energy in the form of heat.

7.4 ANAEROBIC DECOMPOSITION

Anaerobic decomposition is a multistage process occurring in the absence of oxygen. The primary organisms involved in the process are bacteria, although certain protozoa may be a part of the process in some methanogenic ecosystems. Methanogenesis is favored in anaerobic environments where the primary electron acceptor, CO_2, is produced from the fermentation of various organic substrates. It does not occur, or is greatly inhibited, in environments where electron acceptors such as oxygen, nitrate, and sulfate are readily available. The complete anaerobic process

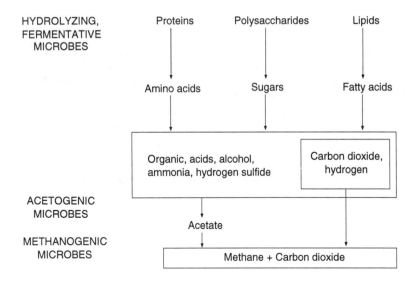

Figure 7.1 Generalized reaction scheme for the anaerobic digestion of complex wastes.

involves a complex mix of microbes, of which only a few actually produce methane. Figure 7.1 outlines the basic biochemical features of the process.

The three major groups of bacteria in anaerobic degradation are: (1) hydrolyzing-fermentative organisms; (2) acetogens; and (3) methanogens. The first group, hydrolyzing-fermentative organisms, produces powerful digestive enzymes which break down biopolymers, such as starch, cellulose, proteins, and others, into their subunit molecules. These, in turn, are fermented to form hydrogen and carbon dioxide; propionic, butyric, and other fatty acids; and ethanol and other short chain carbon compounds. The second bacterial group, the acetogens, are able to take many of the organic compounds resulting from the activity of the first group of organisms and convert them to acetic acid. The third group of organisms, the methanogens, convert acetic acid, hydrogen, and carbon dioxide to methane. Since some nitrogen and sulfur-containing compounds are typically found in many wastes, ammonia (NH_3) and hydrogen sulfide (H_2S) are also produced during anaerobic degradation.

Methanogenesis is unique in that no single organism can completely convert a substrate such as starch to methane. Instead it is the interaction of the bacterial groups described above, comprising a complex microbial ecosystem, that results in the degradation of complex organic compounds into end products such as carbon dioxide, methane, ammonia, and hydrogen sulfide. The interrelationships of the organisms establish thermodynamic conditions that favor methane formation. Some reactions and reaction energies, $\Delta G^{o'}$, are listed in Table 7.1. (Recall from the discussion of the Gibbs function in Chapter 6 that a negative value indicates the reaction is thermodynamically favored.)

Some anaerobic reactions are not thermodynamically favored under standard conditions of temperature, pressure, and chemical concentrations (i.e., 0°C, 1 atmosphere pressure, and 1 molar concentrations) as indicated by a positive $\Delta G^{o'}$, but are favored in in situ conditions as indicated by a negative $\Delta G^{o'}$. For example, the reaction energy for the conversion of butyrate to acetate reaction (shown in Table 7.1)

$$\text{Butyrate} \rightarrow 2 \text{ acetate} + H^+ + 2H_2$$

is +48.1 kJ/mole under standard conditions, but the free energy is –29.2 kJ/mole in the in situ environment of a well-functioning anaerobic reactor. The difference results from the fact that in the anaerobic environment

Table 7.1 Thermodynamics of Selected Reactions in Anaerobic Digestion

	Reaction Energy (kJ/mole)	
Reactions[a]	**Standard Conditions**	**In Situ Conditions**[b]
Hydrolyzing/fermenting organisms		
glucose $\rightarrow$ 2 acetate + $2HCO_3^-$ + $4H^+$ + $4H_2$	–206.3	–363.4
glucose $\rightarrow$ butyrate + $2HCO_3^-$ + $3H^+$ + $2H_2$	–254.8	–310.9
Acetogenic organisms		
butyrate $\rightarrow$ 2 acetate + H^+ + $2H_2$	+48.1	–29.2
propionate $\rightarrow$ acetate + HCO_3^- + H^+ + $3H_2$	+76.1	–8.4
ethanol $\rightarrow$ acetate + H^+ + $2H_2$	+9.6	–49.8
Methanogenic organisms		
$4H_2 + CO_2 \rightarrow CH_4$	–135.6	–16.8
acetate $\rightarrow CH_4 + CO_2$	–31	–22.7

[a]Water is omitted for brevity.

[b]Assumes 0.00001 atm H_2, 0.5 atm CO_2, 0.5 atm CH_4, pH 7.0. The concentrations of propionate, acetate, butyrate, and ethanol are 1 mM each. The concentration of glucose is 10 mM (37°C).

Source: Adapted with permission from Daniels, L., Biological methanogenesis: Physiological and practical aspects, Trends in Biotechnology, 2(4), 91–98, (1984).

the methanogens utilize hydrogen gas, thereby reducing its concentration to very low levels. Low concentrations of hydrogen result in a thermodynamic environment which favors the acetogens' conversion of butyrate or other like compounds to acetate which, in turn, is utilized by methanogens and converted to methane and carbon dioxide.

The overall anaerobic decomposition of an organic material $C_aH_bO_cN_d$ can be represented by this expression (Tchobanoglous et al., 1977)

$$C_aH_bO_cN_d \rightarrow nC_wH_xO_yN_z + mCH_4 + sCO_2 + rH_2O + (d - nx)NH_3,$$

where

$r = c - ny - 2s,$

$s = a - nw - m.$

In this equation $C_aH_bO_cN_d$ and $C_wH_xO_yN_z$ represent the organic substances at the beginning and end of the process, respectively. If it is assumed that the organic matter is completely converted to simple end-products, the reaction becomes (Tchobanoglous et al., 1977)

$$C_aH_bO_cN_d + 0.25(4a - b - 2c + 3d)H_2O \rightarrow$$
$$+ 0.125(4a + b - 2c - 3d)CH_4 + 0.125(4a - b + 2c + 3d)CO_2 + dNH_3.$$

Example 7.2: Calculate the amounts of methane and carbon dioxide produced by the complete anaerobic conversion of 1 kg of the organic matter described by the empirical formula $C_{38.3}H_{61}O_{29.9}N_{0.07}$ (from Example 7.1).

Solution: The expression describing the complete conversion under anaerobic conditions is

$$C_{38.3}H_{61}O_{29.9}N_{0.07} + 8.15H_2O \rightarrow 19.3CH_4 + 19.0CO_2 + 0.07NH_3.$$

In Example 7.1 it was found that 1 kg of the organic matter corresponded to 0.99 g-formula wt. Thus, the moles of CH_4 and CO_2 can be calculated as follows:

$$0.99 \text{ mole of organic matter} \times \frac{19.3 \text{ moles of } CH_4}{1 \text{ mole of organic matter}} = 19.1 \text{ moles of } CH_4$$

$$0.99 \text{ mole of organic matter} \times \frac{19.0 \text{ moles of } CO_2}{1 \text{ mole of organic matter}} = 18.8 \text{ moles of } CO_2$$

Thus, approximately equal amounts of methane and carbon dioxide are produced on a mole basis. (This is generally the case for anaerobic degradation of carbohydrates and proteins.) Using the fact that one mole of a gas occupies a volume of 22.4 L at STP (i.e., 0°C and 1 atm pressure), the volumes of methane and carbon dioxide at STP are calculated to be 428 L and 421 L, respectively. On a mass basis, the 1 kg of organic material yields 0.306 kg of methane and 0.827 kg of carbon dioxide.

Under anaerobic conditions it is possible to degrade organic matter, yet retain about 80% of its energy content in the methane produced. The methane can be subsequently burned to produce heat or generate electricity. The sale or internal use of this energy reduces the overall cost of operating a waste treatment facility.

Because so much of the potential energy in the starting substrate is retained in methane, little free energy is available for cell growth; hence, the cell yield in anaerobic digestion is much less than in aerobic processes. The difference may be as much as a factor of ten. The reduced cell yield in anaerobic waste treatment may represent a considerable advantage over aerobic waste treatment since there is much less waste solids which needs to be disposed.

7.5 ANAEROBIC TREATMENT OF WASTE

Anaerobic degradation can be used in the treatment of sludges, wastewater with high BOD or COD values, and organic solid waste. The challenge is to design a system that provides the proper conditions for maintaining high populations of microbes. In such a design an oxygen-free environment must be maintained and the temperature, pH, and feed rate of the waste must be controlled.

ANAEROBIC REACTOR DESIGNS

The nature of the waste is an important consideration in the choice of an anaerobic reactor design. Beyond this, the molecular composition of the waste may also influence reactor operation. For example, if the waste stream has an easily hydrolyzed and fermentable substrate, such as starch, the hydrolytic/fermenting organisms will be very active, producing large amounts of organic acids, carbon dioxide, and hydrogen. But, if the generation of acid products exceeds the ability of the acetogens and the methanogens to convert them to carbon dioxide and methane, the pH of the reactor contents may drop, leading to an inhibition of the pH-sensitive methanogenic organisms. If this occurs, the reactor may fail because organic acids, such as acetate, propionate, butyrate, and others, accumulate

Table 7.2 Comparing Aerobic and Anaerobic Degradation Processes in Waste Treatment

Characteristic	Aerobic	Anaerobic
Dilute wastes such as municipal wastewater	Very effective in removal of BOD	Generally not as efficient in BOD removal as aerobic
Concentrated waste streams (>3000 mg BOD/L)	Cost of aeration (O_2/air) is significant	Good potential for energy recovery (methane)
Temperature of operation	Effective over a wide range of temperatures, from below 20°C to over 60°C	Best above 30°C, relatively poor efficiency below 20°C
Cell generation	High cell generation per gram BOD removed	Low cell generation per gram BOD removed
Oxygen	Oxygen/air required	No oxygen required

in the reactor. In general, the optimum pH range for anaerobic reactor operation is 6.8 to 7.2.

It is important to note that one of the advantages of the anaerobic digestion process (see Table 7.2), low biological solids generation, may be a handicap in the design and operation of anaerobic reactors. The relatively slow growth rates of some of the microbes in the anaerobic complex mean that a reactor may be subject to failure if the washout rate of microbes exceeds their replacement by new growth.

In Figure 7.2 several reactor designs are illustrated. The operations of the different types of reactors are described below.

STIRRED TANK REACTOR

The stirred tank reactor (Figure 7.2a), also called a high rate digester, is used in many municipal wastewater treatment plants for the digestion of mixed primary and secondary sewage sludges. The sludges are heated to 35° C and kept completely mixed to achieve good contact between the microbes and the waste in the reactor vessel. The average detention time of the sludges in the digester is about 15 days.

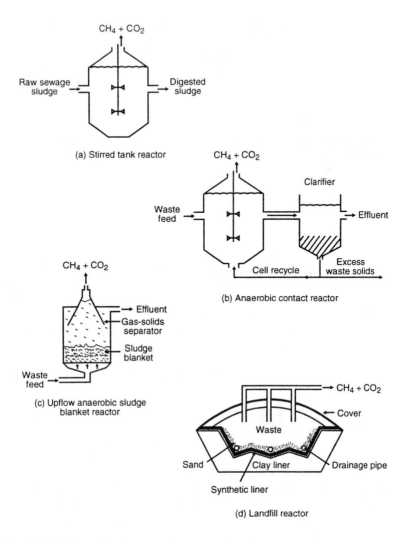

(a) Stirred tank reactor

(b) Anaerobic contact reactor

(c) Upflow anaerobic sludge blanket reactor

(d) Landfill reactor

Figure 7.2 Examples of anaerobic waste treatment reactor designs.

ANAEROBIC CONTACT REACTOR

The anaerobic contact reactor (Figure 7.2b) has been successfully used to treat high BOD industrial wastes; for example, those generated in meat packing operations. In this anaerobic treatment process, raw wastes are mixed with recycled sludge solids and digested in a completely mixed reactor. The detention time may be less than 12 hours and after digestion the mixture is separated in a clarifier or vacuum flotation unit. The supernatant is discharged, typically to an aerobic treatment basin, for further treatment. Much of the settled biological solids are recycled back to the reactor in order to maintain a high rate of biological activity. The synthesis rate of microbes is low in the anaerobic environment; therefore, the amount of waste solids requiring final disposal is small.

UPFLOW ANAEROBIC SLUDGE-BLANKET PROCESS

The upflow anaerobic sludge-blanket (UASB) process (Figure 7.2c) is ideally suited for high strength organic wastes that contain low concentrations of suspended solids. Waste is introduced into the bottom of the reactor where it comes in contact with a concentrated population of anaerobic microbes, the sludge blanket. The organic matter in the waste stream is rapidly degraded to methane and carbon dioxide. As the gases rise in the UASB reactor they carry biological flocs with them. When the gas-solids complexes strike the angled baffles at the top of the reactor the gas separates from the solids, which drop back into the reactor. In this manner a high concentration of biological solids is maintained for efficient anaerobic degradation of the in-coming waste. The detention time of the waste in the UASB reactor may be as little as four hours, depending on the organic content and biodegradability of the waste. The upflow anaerobic sludge-blanket reactor is well suited for the treatment of high strength wastes generated in brewing and certain food industries.

LANDFILL REACTORS

Landfills can also be operated as anaerobic reactors (see Figure 7.2d). In many landfills, the production of landfill gases is considered a nuisance. Gas collection systems are installed to collect the gases in order to prevent them from producing an explosion hazard to off-site structures. The gases are burned or flared to destroy the methane, but at an increasing number of landfills the gases are being collected for beneficial purposes, such as providing heat and generating electricity. At such sites, it is desirable to design the landfill and control the conditions of decomposition to accelerate the decomposition process, thereby increasing the rate of gas generation. There are added benefits: a reduction in leachate strength (i.e., lower BOD) and reduced long-term care for gas and leachate control and site maintenance (Barlaz and Ham, 1990). The production of methane can be enhanced by controlling the types of waste accepted and the placement of the waste, recycling leachate onto the landfill to supply moisture, neutralizing the inhibitory effects of organic acids, and adding old aerobically degraded refuse to increase the population of slow-growing microbes (Barlaz and Ham, 1990).

7.6 COMPOSTING

The term composting is most commonly applied to biological treatment processes in which the waste has a moisture content of 30–50%. Modern definitions of composting further expand on this description to include conditions which promote development of high temperatures (e.g., 60°C) sufficient to destroy pathogens and ending with a compost product sufficiently stable for storage and environmentally safe for application to land. Composting is a potentially effective treatment process for such materials as wastewater sludges, yard wastes, agricultural residues, and the organic fraction of MSW. Because yard and food wastes often constitute 15–30% of MSW and both are amenable to composting, many communities divert yard waste to composting sites to conserve valuable landfill space. Home composting of food, yard, and garden wastes is also encouraged.

During composting, the organic components are converted into a stabilized end product called humus. Compost is primarily used as a soil conditioner and not as a fertilizer because it contains a high organic content but generally low concentrations of nitrogen, phosphorus, and potassium nutrients compared to commercial fertilizers. Compost is comparable to peat moss in its conditioning abilities. Public agencies sometimes use compost in parks or on road embankments. It can be sold to soil dealers, landscapers, and nurseries. Compost may also be used for land and mine reclamation or as landfill cover (Segall and Alpert, 1990).

COMPOSTING PROCESSES

Most composting processes are designed for aerobic operation, but anaerobic composting reactors are also being studied and designed for full-scale implementation. Anaerobic composting is sometimes called solid state fermentation (DeBaere et al., 1987).

The basic biology of composting is relatively well understood. The aerobic composting organisms are the bacteria, fungi, and protozoa naturally occurring in the waste. Materials to be composted should contain at least 25% volatile organics, a moisture content of 50–65%, a carbon-to-nitrogen ratio in the range of 20:1 to 40:1, and a pH between 5 and 8. However, within the context of an economically feasible composting technology, it may be difficult to design a composting environment in which all parts of the waste stream are equally exposed to the high temperatures of the process and converted to a stable compost. Process control is essential in maintaining temperatures at a suitably high level in order to destroy pathogens. This is a particularly important consideration in the composting of sewage sludges.

SEWAGE SLUDGE COMPOSTING

The benefits of sewage sludge composting include: (1) waste stabilization, (2) destruction of pathogens, (3) resource recovery, and (4) moisture removal and volume reduction. Stabilization describes the changes in a waste whereby biological activity converts putrescible components in waste into stable organic and inorganic forms which can be applied to land with minimal concern about its environmental impact. Microbial activity during aerobic composting generates temperatures of 60°C or more, sufficient to destroy or inactivate most pathogens if this high temperature is maintained for several days. This is an important consideration in the management of raw sewage sludges which contain large numbers of potentially pathogenic organisms. Mature compost from sewage sludge contains nitrogen, phosphorus, and potassium, albeit at relatively low concentrations; all essential nutrients for plant growth. In addition, the organic component of compost makes a good soil conditioner.

Sewage sludge composting has attracted considerable attention during the past decade because of its effectiveness in destroying pathogens and the compost's fertilizer value and potential for land application. Table 7.3 shows the status of sewage sludge composting projects using various technologies that will be described below. (See Appendix F for a summary of operating biosolids composting facilities.)

MUNICIPAL WASTE COMPOSTING

Municipal solid waste is a complex highly variable waste stream. To achieve a reasonable quality compost, MSW must be processed prior to

Table 7.3 Summary of 1994 Survey of Biosolids (Sludge) Composting Projects (See Appendix F for more information)

Type	Operating or in Start-Up	Under Construction	Design, Planning, or Permitting Stage
Aerated static pile	92.5[a]	10	4
In-vessel reactor	44.5[a]	6	17
Windrow	43	6	16
Aerated windrow	11	0	0
Static pile	7	1	0

[a] A facility using two different composting technologies is tabulated using 0.5 for each technology.

Source: Goldstein, N. and Steuteville, R., Biosolids composting strengthens its base, *BioCycle* 35(12), 48–57, 1994.

composting. The processing includes shredding or grinding of the paper and other organics to reduce the size to about 2.5 to 5 cm and the removal of metals, glass, plastics, and other noncompostables. Shredding provides an initial aeration to the waste and increases the available surface area for action by microorganisms. Leaves and grass clippings collected by some municipalities may receive little or no processing prior to composting. After composting, the product may require further processing to obtain a uniform size and to remove bits of glass, plastic, or metal missed in the precomposting preparation.

Whereas sewage sludges are high in water and contain significant concentrations of nitrogen and other nutrients, the organic materials in municipal solid waste (MSW) are low in moisture and nitrogen. Therefore, it is possible to combine sewage sludge and the compostable fraction of MSW to achieve a better balance of carbon, nitrogen, and moisture for more efficient composting. The high temperatures that are generated during composting promote evaporation of water from the waste mixture, thus reducing some of the dewatering costs associated with wastewater solids processing.

After the feedstock is prepared, the composting can be accomplished by a variety of processes. Composting systems are often classified as "open nonmechanical systems" or "closed mechanical systems." Some examples are described below.

Open Nonmechanical Composting

Windrows

Open nonmechanical composting refers to a process whereby the waste is piled in heaps or windrows. The piles are typically elongated and dome shaped to shed rain and minimize the accumulation of snow. The piles are often constructed on a paved surface, sometimes with perforated covers on the floor or troughs to drain away excess liquid. This type of composting is simple in design but requires much land area. Once the material is in a heap, decomposition begins quickly. Bacteria and other microbes grow and flourish, utilizing nutrients and generating heat during the rapid breakdown of the waste. Initially, temperatures begin to rise as a result of mesophilic microbial activity. When the temperatures exceed 50°C, thermophilic microbes thrive. Other organisms, including pathogens in the waste, die off as the temperature in the compost pile reaches 60°C or more. As the composting process continues, fungi of the genera *Mucor, Aspergillus, Humicola* and *Actinomyces* become active and feed on dead bacteria and other organic components. This continuing activity leads to a gradual "curing" of the compost (Suess, 1985).

During the composting process the windrows must be aerated by mixing the contents periodically to keep the pile from becoming anaerobic, which may generate serious nuisance odors. Such aeration will also help ensure a more efficient decomposition and produce a more uniform compost product. This can be done with a tractor equipped with a rotary scoop which turns the piles over, or with specialized windrow turning machines. Aeration must be maintained continuously, and during the early stages of composting windrow turning may be required on an every-other-day basis.

In order for decomposition to proceed uninhibited, the moisture content of a composting pile should be kept between 30% and 80%. The optimum moisture content is approximately 50%. The high temperatures generated during composting lead to considerable loss of water; therefore, some composting operations may require the addition of water. On the other hand, if moisture content is high—above 80% for example—it is difficult to maintain an aerobic environment in the composting mass.

Windrow or simple pile composting with occasional turning to promote aeration is viewed by many as a low technology and low cost approach to composting. It may take months for the composting process to be completed, as compared to weeks for closed mechanical composters. However, the simplicity of the windrow/pile process has made it attractive to many communities where it is used in the composting of yard wastes which are now banned from landfills in over 20 states (see Appendix E).

Windrow composting requires extensive land space and may present vermin or other nuisance problems, such as odors and dust. In addition, the compost produced from a windrowing process may be of varying quality due to the inherent difficulty of maintaining a uniform composting environment throughout a windrow.

Static Piles

In a variation of the windrow composting process, aeration is provided to waste in static piles by blowing or drawing air through a system of perforated flexible drainage pipes. Figure 7.3 contains a diagram of the static pile process. The forced aeration system permits good aeration without the problems associated with turning a windrow. The system is simple in principle and design since the blower and distribution system consists of relatively low cost components and usually does not require costly maintenance. With the addition of temperature and oxygen sensors, the system can be made relatively sophisticated in terms of optimizing composting activity and maintaining sufficiently high temperatures to effectively eliminate pathogens.

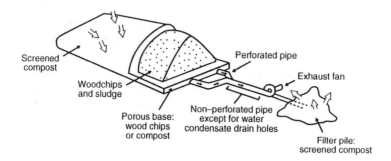

Figure 7.3 Windrow composting with a forced aeration (static pile) system.

Closed Composting

In closed composting, also called in-vessel composting, a closed reactor vessel is used. The biology of the composting process is the same as in the open process; however, this approach permits much greater control of the temperature, moisture, and aeration in the composting process, as well as the final compost quality. The control of these composting variables results in accelerated composting rates. An in-vessel system requires approximately 14 days for composting and 20 days for curing, as compared to at least 21 days and 30 days, respectively, for composting and curing in windrow or static pile systems. The confined composting and greater throughput requires less land area than an open system. Other advantages include better control of odor and dust. The primary disadvantage of in-vessel composting is that it is capital intensive and the operational and maintenance costs also tend to be high.

There are many variations of in-vessel composting designs. These systems achieve the same outcome and offer varying control levels through different design features (Anderson et al., 1986). The types include: (1) agitated bed reactors, (2) silo type systems, (3) tunnel type systems, and (4) the enclosed static pile design. Brief descriptions for these enclosed composting systems are presented below.

Agitated Bed System

One example of the agitated bed system is the circular, covered reactor shown in Figure 7.4. The reactor has a diameter of about 35 m and a composting depth of 2-3 m (Anderson et al., 1986). A typical waste feed stock, containing a mix of 1 part sewage sludge, 1.4 parts recycled compost, and 0.5 part sawdust, is introduced at the perimeter of the reactor and propelled radially through it with the aid of a series of augers. The

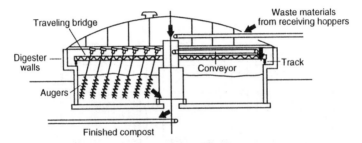

Figure 7.4 In-vessel circular agitated bed composting reactor. Adapted with permission from Anderson, J., Ponte, M., Biuso, S., Brailey, D., Kantorek, J., and Schink, T., Case study of a selection process, *The Biocycle Guide to In-Vessel Composting*, The JG Press, Inc. [Rodale Press], Emmaus, PA, 1986.

augers agitate and aerate the compost while, at the same time, moving it gradually toward the center of the reactor where it is discharged.

The system is designed so that the retention time of the compost is approximately 14 days. Air is forced upward through the reactor via valves which control the amount of air to the composting material. The air valves are controlled by temperature and oxygen probes which permit precise control of the composting environment. Exhaust fans are used to draw off gases generated during the composting process. These gases are passed through an odor control system which may use biological and chemical treatments to eliminate odors. Another version of the agitated bed composting system is based on the use of rectangular concrete basins. The principles utilized in the feeding and general operation are similar to those employed in the circular reactor system.

Silo Composters

Silo composters are usually vertical, circular, or rectangular plug flow systems into which waste is placed at a depth of about 9 m. Waste is fed in at the top and compost is removed at the bottom. Air is forced upward through the composting mixture and exhausted at or near the top of the reactor. Air flow is controlled by instrumentation which monitors carbon dioxide, oxygen, and temperature. A schematic of the process is shown in Figure 7.5. Like the agitated bed reactor, the retention time is also about 14 days. The silo reactor is insulated and contains a roof. Therefore, a building is not needed for the reactor itself.

Tunnel Composting System

The tunnel composting system, displayed in Figure 7.6, is a horizontal, plug flow reactor with a rectangular configuration. An appropriately

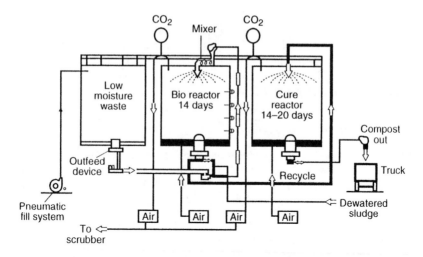

Figure 7.5 In-vessel silo type composting reactor. Adapted with permission from Anderson, J., Ponte, M., Biuso, S., Brailey, D., Kantorek, J., and Schink, T., Case study of a selection process, *The Biocycle Guide to In-Vessel Composting*, The JG Press, Inc. [Rodale Press], Emmaus, PA, 1986.

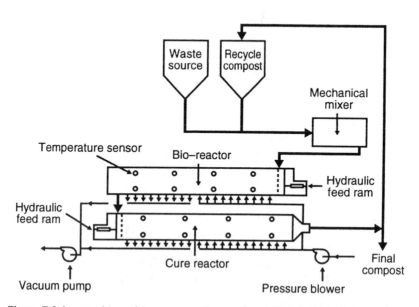

Figure 7.6 In-vessel tunnel type composting reactor. Adapted with permission from Anderson, J., Ponte, M., Biuso, S., Brailey, D., Kantorek, J., and Schink, T., Case study of a selection process, *The Biocycle Guide to In-Vessel Composting*, The JG Press, Inc. [Rodale Press], Emmaus, PA, 1986.

prepared waste mixture is loaded as a batch into one end of the "tunnel reactor" using a hydraulic ram. The ram pushes the new material into the composting mass which, in turn, is moved along the length of the tunnel to the discharge end. The reactor has floor mounted aeration headers that are regulated to deliver air for optimum compost activity throughout the length of the tunnel reactor.

Enclosed Static Pile System

The enclosed static pile system is essentially identical to the open static pile system described earlier except that it is contained within a building. This adds to the cost of static pile composting; however, the effects of precipitation and temperature extremes are minimized so the composting activity can be more easily controlled. The potential odor and dust nuisance problems are also minimized.

The in-vessel systems described above are especially suited to large volume waste flows where economies of scale can offset the higher capital and operating costs associated with these facilities. Recently there has been interest in the development of small in-vessel reactors for use by large institutions and certain commercial, industrial, and agricultural waste generators (Segall, 1994). The emphasis is on simplicity of design and operation. Some units are portable and may not even need a building to house them. For example, one of these small in-vessel composters consists of an overseas shipping container, 12 m × 2.4 m × 2.4 m, fitted with a flail type agitator that runs the length of the container. There is also a low-cost modification to the container body to collect leachate from the composting mass, should this be necessary.

THE UTILIZATION OF COMPOST

Regulations or guidelines regarding compost use are not yet clearly established. Most states apply existing production and use regulations for sewage sludge to MSW compost projects (Goldstein, 1989). The data pertaining to the impact of composted MSW on cultivated and other lands are scanty compared to that for sewage sludge (Gillis, 1992). It is not known with certainty whether MSW compost acts in the same way as sewage sludge when it is applied to land. But, given the high organic content of both sewage sludge and the organic fraction of MSW, it seems reasonable to assume that these materials would have similar impacts when placed on agricultural or other types of soils.

The economics of composting presents a mixed picture. Studies have shown that for some communities, composting yard wastes has resulted in significant savings compared with landfilling. In others, however, the cost is considerably more than landfilling when land costs, labor costs, and the marketability of the compost are taken into account. (Kashmanian and Taylor, 1989).

7.7 FERMENTATION PROCESSES IN WASTE TREATMENT

Microbiologists define microbial fermentations as energy yielding processes in which organic substances are oxidized to various end products in pathways where certain organic molecules serve as electron donors while others serve as electron acceptors. An example is the production of wine alcohol from fruit juices by yeasts, one of the oldest biochemical processes used by humans. Other examples of substances produced in bacterial fermentations include a variety of alcohols, acetic acid, propionic acid, butyric acid, and acetone. Here the discussion will focus on the alcohol fermentation process and how it might be used in waste management to produce ethyl alcohol which can be used as a fuel or blended with gasoline.

The fermentation of glucose or other sugars by yeasts to produce ethyl alcohol is described by the following reaction:

$$\underset{\text{glucose}}{C_6H_{12}O_6} \longrightarrow \underset{\text{ethyl alcohol}}{2\ C_2H_5OH} + 2\ CO_2 + \text{stillage}$$

Each of the end products of this fermentation process has economic value. Stillage is the mixture of yeast cells and the unfermented remains of the organic feed stock. It is high in protein and has good nutritive value for animals. The brewing industry often dewaters stillage to make a high value animal feed called distillers dried grain. Because of its high energy content, the alcohol can be used as a liquid fuel. The carbon dioxide produced during alcohol fermentation also has some value because it can be purified and converted to dry ice or compressed CO_2.

Many waste streams are rich in organic content and have been studied to determine the feasibility of using fermentation processes as a means of recovering products such as alcohol. Examples of potential feedstocks include wood wastes, agricultural residues, and the organic fraction of MSW. As noted earlier, much of the organic fraction in MSW is paper which is largely cellulose, a polymer of glucose. If cellulose is hydrolyzed to glucose, it can be readily fermented to alcohol by yeasts. In theory then, it would appear that wastepapers might be good candidates for alcohol production by fermentation. The process displayed in Figure 7.7 illustrates the key steps that are involved in the fermentation of the organic components of waste newspaper.

The alcohol fermentation process, as presently used in the production of alcohol from grains, faces two obstacles if it is to be applied in an eco-

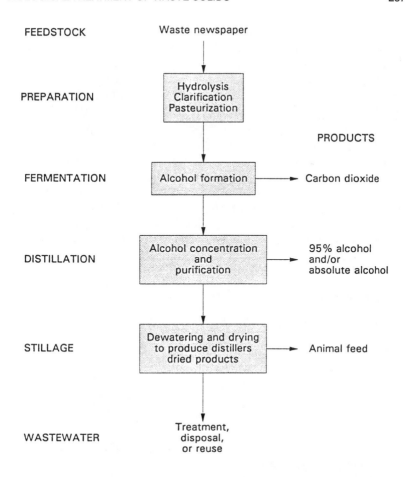

Figure 7.7 A flow diagram for the production of ethanol and other products from waste newspaper.

nomically feasible and environmentally acceptable manner in the conversion of the organic compounds in waste newspaper, or other similar types of wastes, to alcohol. The first obstacle is that the major alcohol producing organisms, the yeasts, cannot ferment cellulose and other complex compounds directly. Hence, extensive pre-fermentation processing is needed to hydrolyze the cellulose and other organic components to fermentable sugars. The second obstacle is that the fermenting organisms function best in dilute environments in which approximately 20% sugars yield a final alcohol concentration of about 10%. This means large volumes of fermentation broth have to be handled during distillation, a major energy-consuming step in the recovery of alcohol. In addition to these obstacles, the costs associated with the processing of the stillage and wastewater treatment for reuse or disposal are high.

The application of the traditional yeast fermentation process in the production of alcohol from the organic components of a complex mixed cellulosic waste such as MSW does not appear promising at this time. Nevertheless, there are some bacteria and fungi capable of degrading cellulosic compounds. The rate of degradation is still considered too slow and variable for large-scale application in the production of alcohol or other compounds. However, the great genetic diversity of bacteria and fungi, coupled with the application of genetic engineering techniques, may lead to future engineered organisms or complexes of enzymes which can efficiently ferment substances, such as cellulose and other waste-derived organic products, to produce alcohol or other products of value (Broda, 1992).

DISCUSSION TOPICS AND PROBLEMS

1. An organic waste has the following composition:

Carbon	45.0%
Hydrogen	6.8%
Oxygen	48.0%
Nitrogen	0.2%

   (a) Determine the empirical formula for this material.
   (b) Write a balanced equation for the aerobic conversion of this waste.
   (c) Calculate the number of moles of oxygen required to react with 1 kg of this material.
   (d) Calculate the volume of air required.

2. An industrial waste with a concentration of 4,000 mg COD/L is to be digested anaerobically in a UASB reactor. The daily waste volume is 50,000 L. The waste detention time in the reactor is 10 hours. Assume that 80% of the COD will be anaerobically degraded to methane and carbon dioxide. In addition, assume that approximately 350 mL of methane are generated for each gram of COD which is degraded.
   (a) What size reactor is needed?
   (b) What will be the approximate volume of methane generated per day?
   (c) What is the energy content of the methane?
   (d) What would be the value of the methane if it had to be purchased from a local gas company?

3. Does your community collect yard wastes for composting? If these materials are collected and composted, describe the composting method. How is the compost used?

4. Backyard composting of yard and garden waste is recommended as a method of MSW source reduction. How should a compost pile be constructed and maintained?

5. Compare the advantages and disadvantages of composting sewage sludge.

6. Sewage sludge and the compostable fraction of MSW are to be mixed to achieve a carbon-to-nitrogen ratio of 35:1. The sewage sludge is 20% solids with a carbon-to-nitrogen ratio of 15:1. The compostable fraction of MSW is 75% solids (25% moisture) with a carbon-to-nitrogen ratio of 100:1.

   (a) What proportions of sewage sludge and compostable MSW should be mixed to obtain the desired carbon-to-nitrogen ratio of 35:1?

   (b) What is the percentage of solids in this mixture?

7. Newspaper is to be converted to ethyl alcohol by a yeast fermentation of the cellulose. If 60% of the paper is cellulose and 70% of the cellulose fraction is converted to alcohol and carbon dioxide, how many liters of ethyl alcohol will be produced from one tonne of paper?

REFERENCES

Anderson, J., Ponte, M., Biuso, S., Brailey, D., Kantorek, J., and Schink, T., Case study of a selection process, *The Biocycle Guide to In-Vessel Composting*, The JG Press, Inc. [Rodale Press], Emmaus, PA, 1986, pp. 39–47.

Barlaz, M.A. and Ham, R.K., Methane production from municipal refuse: A review of enhancement techniques and microbial dynamics, *Critical Reviews in Environmental Control*, 19(6), 557–584, 1990.

Broda, P., Biotechnology in the degradation and utilization of lignocellulose, *Biodegradation*, 3(2/3), 219–238, 1992.

Daniels, L., Biological methanogenesis: Physiological and practical aspects, *Trends in Biotechnology*, 2(4), 91–98, 1984.

DeBaere, L., Van Meenen, P., Debroosere, S., and Verstraete, W., Anaerobic fermentation of refuse, *Resources and Conservation*, 14, 295–308, 1987.

Gillis, A.M., Shrinking the trash heap, *BioScience*, 42(2), 90–93, 1992.

Goldstein, N., States regulations on MSW composting, *Biocycle*, 30(12), 50–53, 1989.

Goldstein, N., and Steuteville, R., Biosolids composting strengthens its base, *Biocycle*, 35(12), 48–57, 1994.

Horan, N. J., *Biological Wastewater Treatment Systems*, John Wiley & Sons, New York, NY, 1990.

Kashmanian, R. M. and Taylor, A. C., Costs of composting yard wastes vs. landfilling, *Biocycle*, 30(10), 60–63, 1989.

Segall, L., In-vessel composting for low volume generators, *Biocycle*, 35(12), 61–66, 1994.

Segall, L. and Alpert, J., Compost market strategy, *Biocycle*, 31(2), 38, 1990.

Suess, M.J., *Solid Waste Management: Selected Topics*, World Health Organization, Copenhagen, Denmark, 1985.

Tchobanoglous, G., Theisen, H., and Eliassen, R., *Solid Wastes: Engineering Principles and Management Issues*, McGraw-Hill Book Company, New York, NY, 1977.

CHAPTER 8

Incineration, Pyrolysis, and Energy Recovery

8.1 OVERVIEW

Incineration is the controlled burning of wastes at high temperatures in a facility designed for efficient and complete combustion. By definition, complete combustion involves the conversion of all carbon to carbon dioxide (CO_2), hydrogen to water (H_2O), and sulfur to sulfur dioxide (SO_2). The by-products of incineration are ash, gases, and heat energy. Wastes are burned for one or more of the following reasons: volume reduction, destruction of certain chemicals or alteration of chemical characteristics, destruction of pathogens, or energy recovery. Specially-designed incinerators burn wastewater and industrial sludges. At the present time, incineration is the only environmentally acceptable method of disposing of some hazardous wastes.

Incineration is more extensively used to manage municipal solid wastes in Europe and Japan than in the United States. In Japan about two-thirds of all municipalities—over 1900—have incinerators (USCOTA, 1989). Table 8.1 shows estimates of the percentage of MSW that is incinerated for selected countries after recyclables, if any, have been removed.

The lack of available land for landfilling and the high cost of energy are two major incentives for including incineration as part of a solid waste management system. In the United States, over 40% of the nation's MSW incinerators are located in New England and the mid-Atlantic states and fewer than 10% are located in the Rocky Mountain states or farther west (USCOTA, 1989). Incineration reduces, but does not eliminate, the need for landfill space. Typically, only about 50% of municipal solid waste is combustible. Incineration reduces the volume of this combustible portion by about 90% and the weight by about 70%. Thus, a residue

Table 8.1   Estimates for Selected Countries of the
            Percentage of MSW (by Weight)
            Incinerated after Recyclables Have Been
            Removed

Country	Percentage of Wastes Incinerated	Year
Denmark	55	1985
France	37	1983
Italy	11	1983
Japan	67	1987
Netherlands	38–42	1985
Sweden	51–55	1985, 1987
Switzerland	75	1985
United Kingdom	9	1983
United States	15	1986
West Germany	22–34	1985, 1986

Source: U.S. Congress, Office of Technology Assessment, Facing America's Trash: What Next for Municipal Solid Waste?, OTA-O-424, U.S. Printing Office, Washington, DC, October, 1989. The data in this report were compiled from other sources.

remains which must be properly handled. It is most often disposed of in a landfill.

Two common types of facilities used to burn unprocessed municipal solid wastes are field-erected mass burn incinerators and prefabricated modular incinerators. They differ in design, construction, processing capacity, air pollution control requirements, service life, and costs. There is extensive operating experience with field-erected mass burn incinerators, both with and without energy recovery, in the U.S., Europe, and Japan. Historically, large cities tend to use mass burn facilities with capacities of 90 to 450 tonnes (100 to 500 tons) per day, whereas smaller communities with quantities of 22 to 90 tonnes (25 to 100 tons) per day choose the modular units. Other incinerator designs exist, including rotary kiln and fluid-bed incinerators, but both require processing of MSW to remove noncombustibles before burning. Municipal waste incinerators are designed to accommodate wastes with widely varying compositions; industrial waste incinerators are designed to burn a more homogeneous fuel.

Most modern incinerators are designed for the recovery of energy. In Europe, the steam and hot water produced by incinerators is commonly used for central district heating of houses and businesses. Approximately 75% of the incinerators in the United States are waste-to-energy plants where the energy is used to produce steam or generate electricity. The revenues received from the sale of steam or electricity partially offset the high capital and operating costs of an incinerator.

An alternative approach to direct burning of wastes for energy recovery is to process the MSW to make refuse-derived fuel (RDF). RDF is a more homogeneous shredded mixture of the combustible organic fraction of municipal solid waste. RDF can be burned in existing industrial or utility boilers as a supplementary fuel or in incinerators designed specifically for RDF. Together, mass burn incinerators and RDF systems account for about 90% of the current incineration capacity in the United States.

Air pollution and ash disposal are major concerns voiced by opponents of incineration. Many communities abandoned incinerators in the 1970s because of the expense of adding air pollution control equipment during a time when air emission standards were becoming more stringent. Modern incinerators must be outfitted with electrostatic precipitators, scrubbers, or other equipment to reduce emissions to levels acceptable to regulatory agencies. The ash from incinerators or facilities burning RDF must be tested to determine whether it is hazardous or not. If the test results indicate that the ash is hazardous, then it is subject to hazardous waste disposal regulations.

Another concern environmental groups and those responsible for implementing recycling programs have about incinerators is the potential impact on recycling efforts. Incineration is viewed as a competitor for recyclable paper and plastic materials. In the hierarchy of choices discussed in Chapter 1, reuse and recycling take precedence over incineration. Recycling is considered a "higher use" for these materials than incineration. An analysis of the waste stream can be used as a basis for obtaining reasonable estimates of achievable recycling rates and the heat value of remaining components of the MSW. This information is important in resolving potential conflicts between recycling efforts and the incineration of waste materials.

8.2 PRINCIPLES OF INCINERATION

Incineration is fundamentally a form of chemical processing, involving the rapid oxidation of materials. In simple terms, the combustion process involves several stages in which the waste is dried (moisture evaporated) as it enters the furnace, the organic compounds are volatilized, and the volatile compounds are ignited in the presence of oxygen.

HEAT VALUE

A major consideration for the selection or design of an incinerator is the fuel value of the material to be burned. The fuel value is described in terms of the gross heat value or higher heat value (HHV) and the net heat value, sometimes called the lower heat value (LHV). The gross heat value is determined by completely burning a weighed sample in oxygen in a bomb calorimeter and calculating the liberated heat by measuring the temperature rise of the surrounding water bath. The American Society for Testing and Materials (ASTM) specifies a procedure for measuring the gross calorific value of refuse-derived fuels (ASTM E711-81). The heat values of various materials in MSW are listed in Table 8.2.

In addition to the heat value, other important fuel properties are the moisture content, the combustible material content, and the ash content. Fuel with a high heat value (greater than 5000 kJ/kg), low moisture content (less than 50%), and low ash content (less than 60%) can be burned without additional fuel, whereas materials with a low heat value, high moisture content, and high ash content require supplementary fuel. An analysis of the waste to determine its heat value, moisture content, and ash content is called a **proximate analysis.** The combustible fraction may be subdivided further as volatile matter and residual or fixed carbon. The determination of these components is accomplished by heating the waste in the absence of air under standardized conditions. Part of the material is volatilized, leaving a charred residue. This residue is called the residual or fixed carbon. The presence of volatile material is closely related to the

Table 8.2 Higher (Gross) Heat Values (HHV) of Selected Materials

Type of Waste	Heat Value (kJ/kg)	Heat Value (BTU/lb)
Mixed MSW	11,600–12,100	5,000–5,200
Refuse–derived fuel (RDF)	13,300–14,400	5,700–6,200
Paper	16,700–18,600	7,200–8,000
Yard	9,800–16,300	4,200–7,000
Food	4,200–18,100	1,800–7,800
Plastics	22,100–41,900	9,500–18,000
Wood	10,900–16,300	4,700–7,000
Rubber waste	27,900–32,600	12,000–14,000
Lignite coal	16,000	6,900
Bituminous coal	20,900–33,700	9,000–14,500
#6 fuel oil	40,500–44,200	17,410–18,990
#2 home heating oil	44,600–45,900	19,170–19,750

Source: Kiser, J.V.L. and Burton, B. K., Energy from municipal waste: Picking up where recycling leaves off, *Waste Age,* 23(11), 39–46, November, 1992.

Table 8.3 Typical Proximate Analyses of Selected Combustible Components of Municipal Solid Waste in the U.S.

Component	Moisture (%)	Volatile Matter (%)	Fixed Carbon (%)	Ash (%)
Paper	10.24	75.94	8.44	5.38
Magazines	4.11	66.39	7.03	22.47
Yard waste (grass)	75.24	18.64	4.50	1.62
Food waste	78.29	17.10	3.55	1.06
Polyethylene	0.20	98.54	0.07	1.19
Wood and bark	20.00	67.89	11.31	0.80
Rubber	1.20	83.98	4.94	9.88
Leather	10.00	68.46	12.49	9.10

Source: Niessen, W.R., Properties of waste materials, in *Handbook of Solid Waste Management*, Wilson, D.G., Ed., Van Nostrand Reinhold Co., New York, NY, 1977. Reprinted with permission.

presence of flames during combustion. The proximate analyses of MSW and selected components found in municipal solid waste are given in Table 8.3.

An elemental or ultimate analysis determines the chemical composition of the combustible portion of the waste in terms of the ash content and the chemical elements, carbon, hydrogen, oxygen, nitrogen, sulfur, and chlorine. The elemental analysis of selected combustible components of MSW is listed in Table 8.4. A more comprehensive table of proximate and elemental analyses and heat values of the components of MSW both as received and dry is included as Appendix H.

Table 8.4 Typical Elemental Analyses of Selected Combustible Components of Municipal Solid Waste (Dry)

Component	C (%)	H (%)	O (%)	N (%)	S (%)	Ash (%)
Paper	43.41	5.82	44.32	0.25	0.20	6.00
Magazines	32.91	4.95	38.55	0.07	0.09	23.43
Yard waste (grass)	46.18	5.96	36.43	4.46	0.42	6.55
Food waste	49.06	6.62	37.55	1.68	0.20	4.89
Polyethylene	87.10	8.45	3.96	0.21	0.20	0.45
Wood and bark	50.46	5.97	42.37	0.15	0.05	1.00
Rubber	77.65	10.35	–	–	2.00	10.00
Leather	60.00	8.00	11.50	10.00	0.40	10.10

Source: Niessen, W.R., Properties of waste materials, in *Handbook of Solid Waste Management*, Wilson, D.G., Ed., Van Nostrand Reinhold Co., New York, NY, 1977. Reprinted with permission.

In the absence of calorimeter data, the heat value of combustible material may be estimated using one of several equations developed for this purpose, based upon the elemental analysis or the waste composition. Most equations are linear combinations of the mass fractions or percents of the elements or waste components with appropriately chosen coefficients.

Hazome et al. (1979) developed an empirical equation specifically for refuse by fitting elemental composition to HHV data:

$$HHV = 0.339(C) + 1.44(H) - 0.139(O) + 0.105(S) \text{ MJ/kg}$$

$$(HHV = 145.7(C) + 619(H) - 59.8(O) + 45.1(S) \text{ BTU/lb})$$

where HHV is the higher heat value, and (C), (H), (O), and (S) are the weight (or mass) percents of carbon, hydrogen, oxygen, and sulfur, respectively. In a more complex approach, D.L. Wilson (1972) approximates the higher heat value (HHV) using an equation developed from thermochemical principles, heating value of carbon and sulfur, types of carbon present, and the formation of water:

$$HHV = 0.3279(C_o) + 1.504(H) - 0.1383(O) - 0.1484(C_i)$$
$$+ 0.09262(S) + 0.02419(N) \text{ MJ/kg}$$

$$(HHV = 141.0(C_o) + 647.8(H) - 59.48(O) - 63.82(C_i) + 39.82(S)$$
$$+ 10.40(N) \text{ BTU/lb})$$

where HHV is the higher heat value, (C_o) is the weight percent of organic carbon, (C_i) is the weight percent of inorganic carbon (typically about 1.2% of the total carbon), and (H), (O), (S), and (N) are the weight percents of hydrogen, oxygen, sulfur, and nitrogen, respectively.

An example of an equation to predict the heat value of municipal solid waste based upon the primary combustible components, paper, plastic, and food (as received), developed by Khan and Abu-Ghararah (1991) is

$$HHV = 0.0535 [F + 3.6 \, CP] + 0.372 \, PLR \text{ MJ/kg}$$

$$(HHV = 23 [F + 3.6 \, CP] + 160 \, PLR \text{ BTU/lb})$$

where HHV is the gross heat value, F is the mass percent of food, CP is the mass percent of cardboard and paper, and PLR is the mass percent of plastic rubber and leather in the dry waste mixture. The estimated heating values of the MSW for major cities in 35 countries based on these components are given in Table 8.5.

Table 8.5 Composition Data and the Estimated Heating Value for MSW by Mass in Different Countries

Country (%)	Paper (%)	Metals (%)	Glass (%)	Food (%)	Plastics	Heat Value (as received) (MJ/kg)
Australia	38	11	18	13	0.1	8.05
Austria	35	10	9	24	6.0	10.26
Bangladesh	2.0	1.0	9.0	40	1.0	2.90
Belgium	30	5.3	8	40	5.0	9.78
Bulgaria	10	1.7	1.6	54	1.7	5.45
Burma	1.0	3.0	6.0	80	4.0	5.96
Colombia	22	1.0	2.0	56	5.0	9.09
Czechoslovakia	13.4	6.2	6.6	41.8	4.2	6.38
Denmark	32.9	4.1	6.1	44	6.8	11.22
England	37.0	8.0	8.0	28	2.0	9.37
Finland	55.0	5.0	6.0	20	6.0	13.90
France	30.0	4.0	4.0	30	1.0	7.76
Gabon	6.0	5.0	9.0	77	3.0	6.39
Germany	20.0	5.0	10.0	21	2.0	5.72
Hong Kong	32	2.0	10	9	11.0	10.74
India	3	1.0	8.0	36	1.0	2.88
Indonesia	10	2.0	1.0	72	6.0	8.01
Iran	17.2	1.8	2.1	69.8	3.8	8.46
Italy	31	7.0	3.0	36.0	7.0	10.50
Japan	21	5.7	3.9	50.0	6.2	9.03
Kenya	12.2	2.7	1.3	42.6	1.0	5.00
Netherlands	22.2	3.2	11.9	50.0	6.2	9.26
New Zealand	28	6.0	7.0	48	0.1	8.00
Nigeria	15.5	4.5	2.5	51.5	2.0	6.49
Norway	38.2	2.0	7.5	30.4	6.5	11.40
Pakistan	2.2	2.2	1.75	52.5	1.2	3.68
Philippines	17.0	2.0	5.0	43.0	4.0	7.06
Saudi Arabia	24.0	9.0	8.0	55.0	2.0	8.31
Singapore	43	3.0	1.0	5.0	6.0	10.78
Spain	18	4.0	3.0	50	4.0	7.63
Sri Lanka	8	1.0	6	80	1.0	6.19
Sweden	50	7.0	8.0	15	8.0	13.41
Taiwan	8	1.0	3.0	25	2.0	3.62
U.S.A.	28.9	9.3	10.4	17.8	3.4	7.78

Source: Khan, M.Z.A. and Abu–Ghararah, Z.H., New approach for estimating energy content of municipal solid waste, *J. Environ. Eng.*, 117(3), 376–380, 1991. Reprinted with permission.

MOISTURE

Moisture significantly lowers the fuel value. As the moisture increases, there is less combustible material per unit mass. In addition, a significant amount of gross heat energy is used to heat and evaporate the water (i.e., moisture) in the waste. (The heat of vaporization of water is 2257 kJ/kg.) Thus, the importance of reclaiming this energy rather than losing it as water vapor in the flue gases is obvious. To account for the effect of the moisture, it is often useful to speak of the net heat value, or lower heat value (LHV). The lower heat value represents the energy that can be realistically captured from the combustion of the waste. It is calculated from the equation (Hougan et al., 1954):

$$LHV = HHV \text{ (in MJ/kg)} - 0.0244 \, (W + 9H) \text{ MJ/kg}$$

$$(LHV = HHV \text{ (in BTU/lb)} - 10.50 \, (W + 9H) \text{ BTU/lb}),$$

where LHV is the lower heat value, HHV is the higher heat value, W is the mass percent of moisture, and H is the percent by weight (or mass) of hydrogen in the dry waste.

COMBUSTION AIR

Another consideration for the design of an incinerator is the quantity of air that must be supplied to achieve the complete combustion of the waste according to the reaction:

$$C_aH_bO_cCl_dF_eN_fS_g + [a + b/4 - (c + d + e - f)/2 + g]O_2 \rightarrow$$

$$aCO_2 + [(b - d - e)/2]H_2O + dHCl + eHF + fNO + gSO_2$$

Frequently, fluorine, chlorine, nitrogen, and sulfur are present in small amounts only and are, therefore, omitted in calculations.

Example 8.1: A cellulosic waste is represented by the empirical formula $C_{38.3}H_{61}O_{29.1}N_{0.07}$. The proximate and elemental analyses of this material are:

Proximate Analysis		Elemental Analysis	
Moisture	5.20%	Carbon	43.73%
Volatile matter	77.47%	Hydrogen	5.70%
Fixed carbon	12.27%	Oxygen	44.93%
Noncombustible	5.06%	Nitrogen	0.09%
		Sulfur	0.21%
		Noncombustible	5.34%

(a) Estimate the gross heat value and net heat value of this waste as received. (b) Calculate the volume of air needed for the complete combustion of 1000 kg of the material.

Solution to Part (a): The higher heat value (HHV) can be calculated using the equation (see page 246),

$$HHV = 0.339(C) + 1.44(H) - 0.139(O) + 0.105(S) \text{ MJ/kg}$$

Substituting the values from the elemental analysis,

$$HHV = 0.339(43.73) + 1.44(5.70) - 0.139(44.93) + 0.105(0.21) \text{ MJ/kg},$$

$$HHV = 16.8 \text{ MJ/kg}$$

The lower heat value (*LHV*) is calculated using the equation

$$LHV = HHV \text{ (in MJ/kg)} - 0.0244 \ (W + 9H) \text{ MJ/kg}$$

$$LHV = 16.8 \text{ MJ/kg} - 0.0244 \ (5.2 + 9(5.7)) \text{ MJ/kg}$$

$$LHV = 15.4 \text{ MJ/kg}$$

Solution to Part (b): When computing the oxygen requirements, the chlorine and sulfur components may be neglected. The combustion equation is (with $a = 38.3$, $b = 61$, $c = 29.1$, $d = 0$, $e = 0$, $f = 0.07$, $g = 0$)

$$C_{38.3}H_{61}O_{29.1}N_{0.07} + 39.0 \ O_2 \rightarrow 38.3 \ CO_2 + 30.5 \ H_2O + 0.07 \ NO$$

The molar mass (formula mass) of the waste is calculated:

Carbon	12×38.3	$= 459.6$
Hydrogen	1×61	$= 61$
Oxygen	16×29.1	$= 465.6$
Nitrogen	14×0.07	$= 1.0$
	Total	$= 987.2$

Therefore the molar mass of the material is 987.2 g or 0.9872 kg.

Of the 1000 kg of the material, 897.4 kg (i.e., 1000 kg less 52.0 kg of moisture and 50.6 kg of inert materials) is combustible. This quantity corresponds to 897.4 kg/0.9872 kg/mole = 909 moles. From the equation, we see that one mole of the material requires 39.0 moles of O_2, so 909 moles of material requires $909 \times 39.0 = 35,500$ moles of O_2. At standard temperature and pressure (i.e., 0°C and 1 atmosphere pressure), one mole of a gas occupies $22.4 \times 10^{-3} \text{ m}^3$. Thus,

$$\text{Volume} = 35,500 \text{ moles of } O_2 \times 22.4 \times 10^{-3} \text{ m}^3/\text{mole of } O_2$$

$$\text{Volume} = 795 \text{ m}^3 \text{ of } O_2$$

Because air contains about 21% oxygen, 3790 m^3 of air are required to supply this volume of oxygen. This is 3.79 m^3 of air per kg of dry combustible material.

Engineers often simplify the calculations by developing formulas such as (Dvirka, 1986):

$$W_a = 0.0431 \ [2.667C + 8H + S - O] \text{ kg of air/kg of waste}$$

where W_a is the mass of dry stoichiometric air (at STP) required to burn 1 kg of combustible waste and C, H, S, and O are the mass (or weight) percents of carbon, hydrogen, sulfur, and oxygen, respectively, of the moisture-free and ash-free material.

The amount of air required for combustion can be computed from stoichiometric equations as demonstrated in the example or by using an equation such as the one above, but both methods require a chemical analysis of the fuel. When the chemical composition is not known, a rule-of-thumb for the volume of combustion air is that 1.05 m^3 of air (STP) (i.e., 1.35 kg of air) is required for each 4200 kJ of heating value. (Buekens and Patrick, 1985)

COMBUSTION CONDITIONS

Efficient combustion of wastes requires sufficiently high combustion temperatures, adequate retention time of the wastes in the combustion chamber, and turbulence to expose unburned surfaces. When wastes are fed into an incinerator, they are heated by contact with hot combustion gases, preheated air, or radiation from the furnace walls. The heat dries the wastes, then thermally decomposes them to produce volatile matter which is generally combustible and ignitable.

Many incinerators provide necessary turbulence by burning the fuel on sloping grates that move in some fashion to agitate the waste. The movement causes the waste to tumble forward through the combustion chamber. The basic moveable grating types commonly used are shown in Figure 8.1. There is no clear advantage of one grate design over another. The grate systems allow ash to fall through the grates as the refuse is transported from the point of induction to a final collection bin. The grates control the speed with which the materials move through the incinerator and also provide some of the turbulence necessary for complete combustion.

Primary air is fed through the wastes from below the grate. This air supports the combustion of the wastes and also cools the grates. The velocity of this air must not be too great or the ash that is created will be forced aloft instead of dropping through to a collection system below.

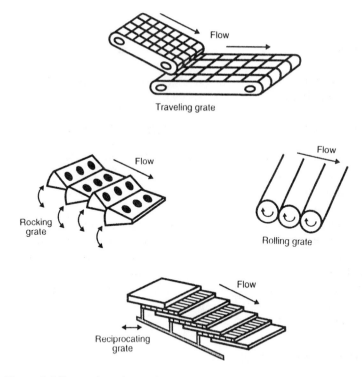

Figure 8.1 Types of moving grates.

Secondary air is blown into the chamber from nozzles above the wastes in order to promote the mixing and burning of the combustion gases. Tertiary air cools flue gases before they reach the air pollution control equipment.

Incinerator designs exist which provide turbulence in other ways. A rotary kiln incinerator burns the waste in a rotating chamber that causes a tumbling action. Another design burns homogeneous or processed waste in suspension. Fluid-bed incinerators use an upward flow of air to keep the waste, along with an inert bed material, in suspension.

The temperature in the incinerator is maintained by controlling the feed rate of the wastes and the primary and secondary air supplies. The amount of air that must be supplied to ensure complete combustion exceeds the stoichiometric amount predicted from the equations describing the burning process. For this reason, the term "excess air" is used. Efficient burning typically requires 100% excess air (i.e., twice as much as the stoichiometric amount) supplied as primary air. This air is introduced

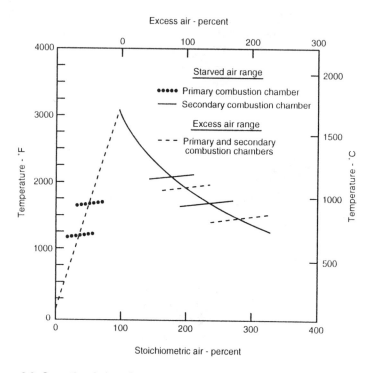

Figure 8.2 Operational air and temperature ranges of controlled air incinerators.

from below the grate. The addition of high pressure secondary air into the combustion chamber serves to control temperature and promotes the mixing of gases. Incinerators are generally operated at a temperature above 750°C to ensure complete combustion, but below 1,000°C to prevent ash from melting and plugging the grates. Figure 8.2 relates the amount of air supplied to the combustion chamber to the temperature in the chamber.

8.3 MASS BURN INCINERATORS

Mass burn incinerators burn unprocessed municipal solid wastes in a single chamber or multiple chambers, with excess air provided to ensure complete combustion. Some mass burn facilities have multiple incineration trains. This design has the advantage of minimizing downtime for repairs and also extending equipment life by idling unneeded trains during slow periods. Wastes are deposited in a pit or tipping floor and fed into the incinerator at a controlled rate. Most large facilities use a pit and overhead bridge crane for waste storage and handling, whereas smaller-scale facilities use a tipping floor where the waste is initially inspected for the purpose of removing objects that are unacceptable for incineration. Then the waste is pushed into a pit or onto conveyors that lead to the feed hoppers. An overhead crane or hydraulic rams feed the waste into the furnace at a regulated rate. The wastes burn on a grate in the combustion chamber with walls covered by a refractory lining or waterwall heat collection tubes covered by a refractory material to protect the tubes from abrasion and corrosion. The general design is shown in Figure 8.3.

A waste heat boiler recovery system uses the flue gas that has left the combustion chamber to heat a series of boiler tubes to convert water to steam. Often more than one series of tubes is used for heat recovery. Efficient generation of electrical power requires steam at higher temperatures and pressures than the steam produced in the steam tubes. Typically a temperature as high as 400°C (750°F), and a gauge pressure as high as 5.2×10^6 Pa (750 psig) are required. To raise the temperature and pressure of the steam to these values, the steam is passed through a gas-to-gas heat exchanger called a superheater, where there is a transfer of heat from the furnace or flue gases to the steam in the superheater. Radiative superheaters that use the radiant heat of the combustion chamber are installed in front of the boiler tubes, whereas convective superheaters that capture heat from the flue gases are installed behind the boiler tubes. The steam is then routed to a collection unit. The purpose of the collection system is to receive steam from the various tubes with their variable

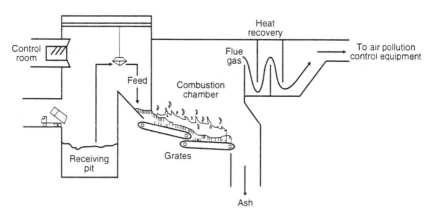

Figure 8.3 Continuous-feed, refractory-lined furnace with traveling-grate stokers (adapted from Buekens, A. and Patrick, P.K., *Solid Waste Management Selected Topics*, World Health Organization, Copenhagen, Denmark, 1985).

temperatures and pressures before it is sent through turbines for the generation of electricity.

Most new mass burn incinerators employ a **waterwall design** for energy recovery and temperature control of the walls. These furnaces have closely spaced tubes in their walls through which superheated water passes. This design is efficient for the recovery of heat because the energy exchange occurs from direct radiation rather than convection. The superheated water is pumped to boiler drums where it produces steam. Waterwall furnaces do not use excess air for temperature regulation. Instead, wastes are burned near stoichiometric conditions. Some excess air is needed to accommodate the fluctuating oxygen demand of the heterogeneous wastes.

Rotary kiln furnaces have a horizontally mounted rotating cylindrical combustion chamber. The kiln is inclined slightly to facilitate the forward movement of the waste. As the refuse burns, ashes fall through screening on the sides of the chamber. Most rotary kiln furnaces are refractory lined, but waterwall designs also exist. The furnaces operate at a high temperature which is conducive to the formation of nitrogen oxides.

Field-erected mass burn incinerators are expensive to build and operate. For this reason, a facility must have a minimum capacity of 300 to 400 tonnes/day to be cost effective. The capital cost of a field-erected incinerator is about \$165,000 to 190,000 per tonne of daily capacity (1992

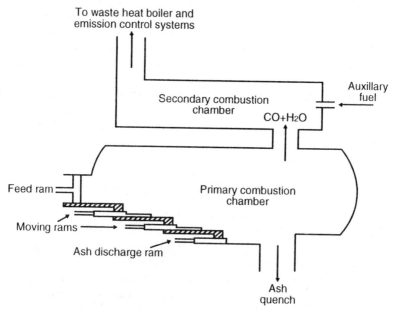

Figure 8.4 Typical modular furnace design.

dollars). Because of the high cost, modern mass burn incinerators are designed for the recovery and sale of energy to partially offset these costs.

8.4 MODULAR INCINERATORS

Modular incinerators are small units, with a capacity of 5 to 100 tonnes per day, that are prefabricated off-site and shipped to the construction site for installation. The on-site construction consists of a building with a tipping floor and utilities. Larger capacity requirements are met by installing additional units.

The typical modular combustion unit has a primary and a secondary combustion chamber as shown in Figure 8.4. Modular incinerators burn the wastes in a refractory-lined primary chamber in a starved-air or substoichiometric environment where there is insufficient oxygen to provide complete combustion of the gases. In a common design the waste is moved through the furnace by a series of transfer rams on the stepped refractory floor, but other designs may use other types of grate systems. The flue gases, primarily a mixture of carbon monoxide and water, enter a secondary combustion chamber (afterburner) where excess air and fuel (if needed) complete the combustion. Normally, additional fuel is not required during the steady state operation because the heat released in the combustion of the gases is sufficient to maintain a temperature of 750°C to 1000°C.

The modular incinerator combustion process is characterized by small gas volumes and low gas velocities. Early modular incinerators attempted to meet air pollution control requirements by controlling combustion air; however, new installations typically must install wet scrubbers, baghouses, or electrostatic precipitators to meet air emission requirements.

Energy is captured from the hot flue gases as they pass waste heat boiler tubes. The steam produced is at low to medium pressure. The steam must be upgraded to high pressure steam if it is to be used for electrical power generation.

8.5 REFUSE-DERIVED FUEL (RDF)

Refuse-derived fuel (RDF) is the combustible fraction of municipal solid waste that has been mechanically processed to produce a more homogeneous product. The processing typically involves shredding the wastes and separating the combustible from the noncombustible fractions in the manner described in Chapter 5. The burnable material that is obtained may be further processed to meet the requirements of a buyer. The material may be densified by compressing it into pellets or briquettes or it may be shredded to a fluff. The RDF may also be dried to achieve a more consistent fuel and to increase the heat value. The RDF may be co-fired with conventional fossil fuels in existing or new boilers, modified or built for feeding the RDF into the combustion chambers. RDF may also be burned in dedicated units, constructed for burning it as the sole fuel. The quality of the RDF depends upon the degree to which noncombustible materials have been removed. The highest thermal values with the least amounts of ash content are found in RDF that represents only 40% or 50% of the amount of MSW initially processed. As noted in Table 8.2, the heat value of unprocessed municipal solid waste in the United States usually averages between 8,100 kJ/kg and 15,100 kJ/kg (3500 and 6500 BTU/lb). When the burnable fraction is processed into RDF, the heat value may range from 14,000 kJ/kg to 18,600 kJ/kg (6550 to 8000 BTU/lb).

RDF was co-fired with coal as a supplemental fuel in a demonstration project at Union Electric Co. in St. Louis, Missouri, in the early 1970s. The initial success of the project prompted other communities to consider co-firing in their municipal electric generation facilities. The largest RDF facility currently in operation is located in Dade County, Florida. It has a design capacity of 2700 tonnes/day (3000 tons/day).

Suspension firing has been found to provide more efficient combustion than stoker (grate) firing and is therefore the preferred method for this fuel. In a suspension boiler, the RDF is pneumatically fed into the combustion chamber from a port above the combustion zone and burns while falling. The ratio of RDF to the other fuel must be limited because its heterogeneous composition, with components having different heat values, can cause fluctuations in heat output and steam pressure. Typically, this ratio is controlled so that the RDF accounts for less than 20% of the total fuel value.

8.6 FLUID-BED INCINERATORS

In contrast to conventional mass burn incinerators that burn wastes on a grate or hearth, fluid-bed combustors burn wastes in a turbulent bed of heated inert materials, such as sand or limestone, kept in suspension by an upward flow of high velocity primary combustion air. Efficient combustion requires a homogeneous fuel. Sludge, wood chips, or other homogeneous combustible waste can be fed into the chamber directly, but municipal solid waste must be processed into refuse-derived fuel (RDF) prior to burning. Heat is transferred to the waste particles by contact and radiation. In the standard design of the fluid-bed reactor, sometimes called the "bubbling fluid-bed" (BFB), the air velocity is 1 to 3 m/s and

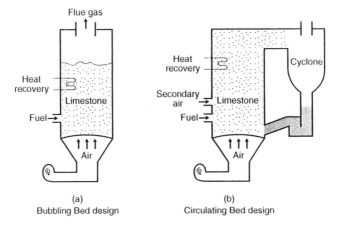

Figure 8.5 Fluid-bed combustor designs.

the bed medium remains in the chamber. A variation of the design, the circulating fluid-bed reactor (CFB), uses air with a velocity of 5 to 10 m/s which continually sweeps bed media and incompletely combusted feedstock from the chamber. This material is separated from the flue gas and reintroduced into the reactor. Secondary combustion air is supplied in the upper part of the chamber. Figure 8.5 depicts the two circulating fluid-bed combustor designs.

Fluid-bed incinerators have the capability of burning a variety of fuels, including municipal solid waste, biomass, and sewage sludge. Fluid-bed combustors have been used to burn municipal solid waste in Japan and Europe for many years, but they have been introduced in the United States only recently. While Japan has over 100 incinerators using the fluid-bed design to burn municipal solid waste, the United States has four that burn refuse-derived fuel (RDF), all built since 1986. These facilities are listed in Appendix G.

Fluid-bed furnaces capable of burning RDF alone, or in combination with other materials, are commercially available with capacities as large as 725 tonnes per day (800 TPD). Experience with the fluid-bed reactors is more limited than with conventional mass burn incinerators; however, the fluid-bed designs have potential advantages over other incinerators. Both the conventional and circulating fluid-bed designs have combustion efficiencies exceeding 99% (i.e., less than 1% unburned carbon), whereas conventional incinerators have efficiencies of 97 or 98%. High combustion efficiency is important for minimizing the emission of dioxins.

Fluid-bed combustors are able to burn a variety of fuels efficiently. These fuels include waste with low fuel value and high ash content.

Fluid-bed combustion may also result in lower emissions of sulfur dioxide (SO_2) and nitrogen oxides. By introducing limestone as a bed material, some of the sulfur dioxide is absorbed in the combustor; however, additional scrubbing is required. The NO_x emissions are lower because the fluid-bed reactor operates at a lower temperature than other incinerators; namely, in a range of 815–925°C (1500–1700°F). The fluid-bed combustors also operate with lower excess air, typically 30% to 90%. The primary disadvantages of the units are the requirement that the waste must be processed to RDF before burning and the limited operational experience, particularly in using MSW for fuel.

8.7 PYROLYSIS

Pyrolysis is the thermal decomposition of materials in an oxygen deficient atmosphere. Pyrolysis has been used commercially for years in the production of charcoal, methanol, turpentine, and coke. By the mid-1970s, several companies, such as Occidental Petroleum, Monsanto, and Union Carbide, experimented with pyrolysis as a process for decomposing MSW. Unfortunately, none of the designs has resulted in a reliable economical processing system for MSW, even though small-scale demonstration plants produced encouraging results. Pyrolysis appears to be best suited for processing organic feedstocks with high heat value.

The pyrolysis process can be represented by the general unbalanced equation (Tillman, 1991):

$$C_aH_bO_c + heat \rightarrow H_2O + CO_2 + H_2 + CO + CH_4 +$$
$$C_2H_6 + CH_2O + tar + char$$

The volatile products tend to remain as gases. The char is a carbon rich solid, and the tar is a thick carbon-rich viscous liquid. The process is endothermic, which means that heat must be supplied.

The composition and yield of the products of pyrolysis can be varied by controlling operating parameters such as pressure, temperature, time, feedstock size, catalysts, and auxiliary fuels. The quality of the char is greatly dependent on the quality of the feedstock. High temperatures, in excess of 760°C (1,400°F), favor the production of gases such as hydrogen, methane, carbon monoxide, and carbon dioxide. Lower temperatures, from 450 to 730°C (850 to 1350°F), produce tar, charcoal, and liquids such as oils, acetic acid, acetone, and methanol. Adding hydrogen to the pyrolysis chamber increases the heat values of the oils and gases formed. The addition of water may produce carbon monoxide and hydrogen that can be converted into natural gas substitutes (Robinson, 1986).

8.8 ENERGY RECOVERY

The burning of solid wastes produces only about 25% as much energy, on a per weight basis, as that resulting from the burning of fossil fuels. Nevertheless, revenues from the sale of the energy can significantly lower the costs of incineration. The primary energy products are hot water, low pressure steam, high pressure steam, and electricity. The specific products generated depend upon the design of the facility, the fuel, and the requirements of the energy buyer. An estimate of the amount of steam that is generated can be found using the rules of thumb that 3 kg of steam are produced from 1 kg of refuse in a water wall incinerator and 2.2 kg of steam are generated from 1 kg of refuse in a modular incinerator (Hecht, 1983).

Steam and hot water are the easiest to generate, but transporting these products requires laying pipelines to the buyer's facility. This construction usually requires a sizable capital cost outlay that may be justified if a long-term arrangement can be instituted with a stable buyer.

Selling the steam requires the identification of buyers who have steam demands that match steam production. Seasonal variations may occur in the demand for steam for heating and cooling that do not match waste generation. In the event that the incinerator is shut down, a backup boiler may be needed to serve the buyer. The primary buyers of steam are industries, institutions, and central heating districts that provide service to multiple buildings. Some district heating systems cover entire downtown areas of cities. Government buildings or universities tend to be the best candidates for energy buyers because they can readily guarantee the purchase of the product over a long time period. Central heating districts are more common in Europe and the republics of the former U.S.S.R. than in the United States. These systems may also be used to cool buildings by having the steam turn turbines to pump chilled water though pipes that are already in place. The best-known project of this type in the United States is in Nashville, Tennessee.

Electricity is generally more marketable than steam. Transportation is not limited to short distances as is the case for steam, and the demand is likely to be less seasonal. Furthermore, in the United States, the Public Utility Regulatory Policies Act (PURPA) requires that electrical utility companies purchase the electricity generated by small, nonutility power generators at rates equal to the estimated cost the utility would incur to generate the electricity itself (i.e., the "avoided cost").

8.9 AIR EMISSIONS

The quantity and composition of air emissions depend upon the composition of the refuse, the design of an incinerator, and the completeness

Table 8.6 Concentrations of Substances in Air Emissions

	Type of Facility		
Substance[a]	Mass Burn	Modular	RDF
Metals ($\mu g/m^3$ at STP)			
Arsenic	0.452–233	6.09–119	19.1–160
Beryllium[b]	0.0005–0.327	0.0961–0.11	20.6[d]
Cadmium	6.22–500	20.9–942	33.7–373
Chromium (total)	21.3–1020	3.57–394	493–6660
Lead[c]	25.1–15400	237–15500	973–9600
Mercury[b]	8.69–2210	130–705	170–441
Nickel	227–476	1.92–553	128–3590
Dioxins/furans (ng/m^3 at STP)			
2,3,7,8–TCDD	0.018–62.5	0.278–1.54	0.522–14.6
TCDD	0.195–1160	1.02–43.7	3.47–258
PCDD	1.13–10700	3.1–1540	53.7–2840
2,3,7,8–TCDF	0.168–448	58.5[d]	2.69[d]
TCDF	0.322–4560	12.2–345	31.7–679
PCDF	0.423–14800	96.6–1810	135–9110
Acid gases (ppm)			
HCl	7.5–477	159–1270	95.9–776
HF	0.620–7.21	1.10–15.6	2.12[d]
SO_3	3.96–44.5	—	—
Criteria pollutants[c] (ppm)			
Particulate matter			
(in mg/m^3 at STP)	5.49–1530	22.9–303	220–533
SO_2	0.040–401	61–124	54.7–188
NO_x	39–376	255–309	263[b]
CO	18.5–1350	3.24–67	217–430

[a] Concentrations normalized to 12% CO_2

[b] National Emission Standard for Hazardous Air Pollutants (NESHAP) promulgated.

[c] National Ambient Air Quality Standard (NAAQS) promulgated.

[d] Data available for only one test.

SOURCE: U.S. Environmental Protection Agency, Municipal Waste Combustion Study, Emission Data Base for Municipal Waste Combustions, EPA/530–SW–87–021b, Washington, DC, June 1987.

of combustion. The major products of incineration of municipal solid wastes are carbon dioxide (CO_2), water (H_2O), sulfur dioxide (SO_2), nitrogen oxides, particulates, and smaller amounts of toxic chemicals such as polychlorinated biphenyls (PCBs), dioxins, and heavy metals (e.g., cadmium, chromium, lead, and mercury). These undesirable materials may also be introduced if sludges from wastewater treatment facilities or industrial wastes are incinerated. Ranges for the concentrations of air emissions from different types of incinerators are shown in Table 8.6 (USEPA, 1987). These data represent the results from a mixture of new facilities and older facilities that lack computerized combustion controls or newer emission control equipment. One would expect RDF facilities to have lower concentrations of heavy metals because most non-combustibles are removed during processing, yet this is not apparent from the USEPA data. Modern well-maintained and operated air pollution control systems are highly efficient, so it is likely that there is little difference in emissions among the various types of incinerators.

PARTICULATES

Three factors may contribute to high particulate emissions during incineration: (1) the entrainment of particles from the burning refuse bed; (2) the cracking in pyrolysis; and (3) the volatilization of metallic salts or oxides. The actual emission rates depend upon the concentration of particulates in the flue gas, the size and mass of the ash particles, and the velocity of the flue gas. Grate systems may yield large amounts of fly ash if the velocity of the underfire air is so great that it sends the ash aloft rather than allowing it to drop into the collection system below. Also, if the chamber temperatures are too high, the presence of metals with low melting points can form aerosols. The highest particulate emission rates, about 60 kg/t (120 lb/ton) of waste burned, are typically found in suspension fired furnaces. Mass burn systems with vigorous stoking, such as in rotary kiln or reciprocating grate furnaces, produce about 15 kg/t (30 lb/ton) of refuse burned. The lowest particulate production, as little as 1.5 kg/t (3 lb/ton) of waste burned, is found with the starved-air modular units.

CARBON MONOXIDE

The incomplete combustion of refuse may also produce an excess of carbon monoxide. An overfire air system is used in mass-burn furnaces to ensure the conversion of this intermediate combustion product to carbon dioxide. Suspension-fired incinerators use designs that encourage intense turbulence so pockets of incompletely combusted gases or materials do not form. Carbon monoxide is measured on a continuous basis as an indicator of combustion efficiency.

SULFUR DIOXIDE

The amount of sulfur in municipal solid waste, 0.1 to 0.2%, is typically much lower than the amount in bituminous coal, 2.5% to 3.5%. About 25% to 50% of the sulfur becomes part of the flue gas and is typically in the form of sulfur dioxide (SO_2) or sulfur trioxide (SO_3). Refuse with a sulfur concentration of 0.1% should yield about 1 kg of SO_2 per tonne of waste. The proportion of SO_2 to SO_3 varies depending upon combustion parameters, but only about 2% to 4% of the sulfur forms SO_3.

NITROGEN OXIDES

High temperatures in a furnace produce nitric oxide from the nitrogen and oxygen in the air as represented by the reaction

$$N_2 + O_2 \leftrightarrow 2NO$$

The amount increases dramatically as the temperatures are raised to the range of 1,000–1,300°C (about 1,800–2,400°F). When the NO reaches the atmosphere, it gradually oxidizes to NO_2. Sunlight may then induce the photochemical reaction that produces ozone (O_3) by liberating one of the oxygen atoms from the NO_2 to be combined with O_2 by the reaction

$$NO_2 + O_2 \overset{\text{sunlight}}{\leftrightarrow} NO + O_3$$

The mixture of nitrogen oxides (NO_x) contributes to the formation of photochemical smog.

HEAVY METALS

Lead is the heavy metal that is most consistently found in MSW and in incinerator air emissions. The lead enters the incinerator from tin cans, electronic components, and plumbing scrap. Mercury, cadmium, and arsenic are also present. The sources of these elements are predominantly batteries, paints, inks, and electronic components.

Most volatilized metals (e.g., mercury, lead, arsenic, cadmium, and zinc) condense onto the flyash particles as the flue gases cool. Less volatile metals (e.g., aluminum, iron, tin, chromium, and nickel) tend to be more concentrated in the ash residue (i.e., bottom ash). The disposal of the ash must be planned so that these metals remain immobilized.

CHLORINE COMPOUNDS

Compared to fossil fuels, the flue gas of incinerated MSW has higher amounts of chlorine compounds. The organic chlorine found in plastics and solvents, such as polyvinyl chloride or methylene chloride, is converted to hydrogen chloride (HCl). Some of the HCl, possibly 20–40%, is absorbed by alkaline particulates within the flue gas. The average concentrations from U.S. incinerators fall between 100 and 150 ppm, which corresponds to a concentration of organic chlorine in the refuse of about 0.2% to 0.3% (Robinson, 1986).

Modern sampling and analytical techniques show the presence of micropollutants in incinerator emissions. Much public attention has been focused on polychlorinated biphenyls (PCBs), polycyclic aromatic hydrocarbons (PAHs), dioxins, and furans, all of which are chlorinated compounds.

Small animal studies indicate that dioxins are extremely toxic to some animals and are carcinogens. The effects on humans have yet to be established. It should be noted that incinerators are not the only source of dioxins. Residential fireplaces, home grills, diesel engines, paper mills, pesticide and wood-preservative manufacturing, and other industrial processes also produce them. The combustion parameter that most affects the rate of dioxin emission appears to be the combustion temperature—low operating temperatures result in increased concentrations of dioxins and furans (USCOTA, 1989). An incinerator testing program in Canada showed that concentrations of dioxins and furans tend to be reduced as the temperature is raised (Finkelstein et al., 1986). At temperatures above 1000°C (1800°F), the concentrations of dioxins and furans are effectively destroyed.

There is evidence that dioxin and furan concentrations in the flue gases are higher in the stack than in the boiler inlet (Lipták, 1991), indicating that these compounds are formed after the gases leave the combustion chamber. It has been postulated that perhaps the formation of these compounds is catalyzed by fly ash particulates (USCOTA, 1989). Laboratory experiments in which dioxins and furans are produced from chlorobenzenes and chlorophenols in the presence of fly ash indicate that the oxygen concentration affects the outcome: fly ash catalyzed the formation when oxygen was in surplus, but it catalyzed decomposition when oxygen was deficient (USCOTA, 1989).

Because dioxins are chlorinated organic compounds, it is to be expected that the quantity of dioxins produced will depend upon the concentration of chlorine in the feed material—particularly PVC plastics and bleached papers; however, there is an apparent lack of correlation between the rate of production of dioxins and the presence of chlorine in the waste (USCOTA, 1989).

8.10 CONTROL OF AIR EMISSIONS

Air pollution control systems in incinerators must conform to current national and state standards. Common air emission control systems use

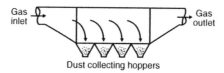

Figure 8.6 Typical settling chamber design.

settling chambers, wet or dry scrubbers, fabric filters, electrostatic precipitators, or some combination of these. Each type of control system is described below.

SETTLING CHAMBERS

A settling chamber is used as a primary collector or first stage collector to remove large particles that may clog filters or foul other equipment. A settling chamber is an enlargement in the flue gas path to slow the speed of the gas to allow gravitational sedimentation to occur. Figure 8.6 shows a diagram of a settling chamber. Only particles with diameters larger than 50×10^{-6} m (i.e., 50 μm) can be removed efficiently. Because a significant fraction of the particles have diameters smaller than this, a settling chamber alone does not provide effective removal.

CYCLONES

Because cyclones have low capital and operating costs, they are commonly used for particulate removal. Flue gases enter a cylindrical chamber tangentially and swirl around the chamber. As the flue gases move in a circular path around the cylinder, inertia carries particulates to the wall, where they move downward in the conical section and are collected in a hopper. Figure 8.7 shows a diagram of a cyclone. The cylindrical chamber is usually small, around 0.25 m in diameter. This small size produces strong centrifugal forces on particles as the gases move in a circular path in the cyclone. Particles of size 10×10^{-6} m (i.e., 10 μm) or larger are effectively removed within this system.

ELECTROSTATIC PRECIPITATORS

Electrostatic precipitators are widely used for the removal of particulates. The flue gases pass through an array of wires and plates such as shown in Figure 8.8. The wires are maintained at a large positive or negative electrical voltage with respect to the plates. The voltage difference produces charged ions and electrons that attach to the surfaces of particu-

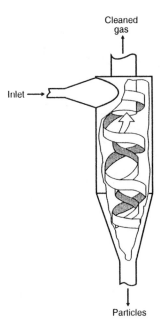

Figure 8.7 Typical cyclone.

lates. These charged particulates are then attracted to the plates or wires and collected. Particulates in the size range 0.1×10^{-6} m to 50×10^{-6} m (0.1 to 50 μm) are effectively removed with electrostatic precipitators (Stern et al., 1984).

FABRIC FILTERS

Fabric filters provide another simple pollution control system. The flue gases are directed through ducts with fabric bags attached to the ends. The bags filter the particulates from the air as it passes through them. The bags are occasionally shaken to release particles to a collection bin below. Filters efficiently remove particles with diameters larger than about 5×10^{-6} μm (i.e., 5 μm). Removal of smaller particles is possible by using a more tightly woven fabric; however, the resistance to the passage of air increases with the smaller weave and may reach unacceptable levels. Because the filters are fabric, the flue gases must be cooled before introduction to the bags or acid condensations may form on the internal surfaces.

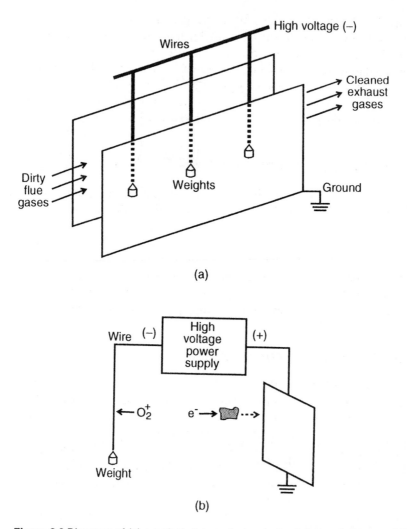

Figure 8.8 Diagrams of (a) a typical plate and wire electrostatic precipitator, and (b) the movement of the ions and charged particulates toward the electrodes.

WET SCRUBBERS

Wet scrubbers are used for removing particulate matter from the gas stream. The common types of wet scrubbers include spray towers, packed towers, cyclone scrubbers, jet scrubbers, venturi scrubbers and mechanical scrubbers. The wet scrubber works through a process in which fine liquid and aerosol particles are intermingled. Large solid particles are captured by impingement and smaller particles by diffusion into the droplets. A high collection efficiency is possible, but this may require a high relative speed between the droplets and the particles and this, in turn, requires large amounts of energy. Effective removal of particles down to a size of 1×10^{-6} m (1 μm) is possible (Stern et al., 1984). Scrubbers produce a wet sludge that may require on-site treatment.

SEMI-WET AND DRY SCRUBBERS

Semi-wet and dry scrubbers control acidic gases (e.g., SO_2 and HCl) by neutralization. The flue gases pass through an alkaline mist of a calcium- or sodium-based slurry. The droplets neutralize the acids as they evaporate. The larger dried particles settle to the floor and the smaller particles exit with the flue gas to be collected in a bag house or an electrostatic precipitator. The unreacted alkaline particles can further neutralize acidic gases as they pass by. Dry scrubbers can neutralize 90% or more of the acids but have little effect on particulate emissions.

SELECTIVE CATALYTIC REDUCTION

Selective Catalytic Reduction (SCR) is a pollution control system capable of removing 70% or more of the NO_x emissions. Ammonia (NH_3) is injected into the flue gas immediately before the gas enters a catalytic chamber. The ammonia reacts with the NO and NO_2 gases to form water in the catalytic chamber. The catalyst enables these reactions to occur at lower temperatures. The temperature of the exhaust gases must be above 233°C (450°F) to operate the catalyst efficiently. Although this condition can be met at waste-to-energy facilities which produce steam, at facilities producing electricity the gases pass through economizers and heat exchangers and exit at temperatures lower than this. In these facilities, the flue gases need to be reheated and then cooled. This requirement adds to the cost of operation.

8.11 DISPOSAL OF ASH

An incinerator produces two distinct types of ash: bottom ash and fly ash. Bottom ash is the residue left from the burned waste and often contains partially burned materials. Bottom ash accounts for about 90% of the volume of ash that is generated. Fly ash consists of the particulates removed from the flue gas. Fly ash is more uniform in composition than bottom ash.

Ash consists mostly of metal oxides. When water leaches through the ash, the soluble metal oxides produce an alkaline solution. In an alkaline solution, the heavy metals are not soluble or mobile. Ash is usually not deposited in a landfill with municipal solid waste (MSW) because the leachate from the MSW tends to be acidic and could release heavy metal ions.

In the United States, the Resource and Recovery Act of 1976 (RCRA) specifically excludes waste-to-energy facilities that burn only municipal waste from regulation as a hazardous waste treatment, storage, or disposal facility. Most states treat the ash as nonhazardous. Recent studies indicate that this is true of bottom ash, but that fly ash may have heavy metal and dioxin concentrations that exceed federal limits (Lipták, 1991). However, on May 2, 1994, the U.S. Supreme Court ruled in the case "City of Chicago v. Environmental Defense Fund" that the exclusion provided to the facilities does not extend to the ash. Consequently, ash must be tested to determine whether or not it qualifies as hazardous. If it is hazardous, it must be treated in accordance with hazardous waste regulations as described in Chapter 12.

Regulatory agencies usually view ash from incinerators that burn municipal waste with more concern than ash resulting from the burning of coal or other homogeneous fuels. If, as a result of this concern, industries are subject to more stringent rules for ash disposal, they will be reluctant to burn waste or RDF as supplemental fuel.

DISCUSSION TOPICS AND PROBLEMS

1. Have any incineration facilities been proposed for your community? What was the outcome?
2. The following table shows the composition of a municipal waste stream and assumed achievable levels of recycling.

Component	Total Stream by Weight (%)	Recovery (%)
Corrugated paper	12.2	70
Newspapers	6.6	70
Magazines	3.0	70
Other paper	15.6	25
Plastics	9.5	35
Rubber, leather	0.0	0
Wood	1.3	0

Textiles	2.9	0
Yard waste	13.5	90
Food waste	13.2	50
Mixed combustibles	5.3	0
Ferrous	6.4	60
Aluminum	1.5	80
Other nonferrous	0.0	0
Glass	6.9	50
Other inerts	2.1	0

(a) If the initial waste stream is 1,000 tonnes per day, what is the quantity of wastes remaining after recyclables are removed?

(b) What is the composition of the waste stream after recyclables are removed?

(c) Using the heat values for the various components (Appendix H), calculate the heat value of the wastes before and after the recyclables are removed. What effect does recycling have on the heat value of the wastes?

3. The receiving area of an incinerator is typically designed to be large enough to accommodate a quantity of waste equal to three days of the nominal daily plant capacity. Calculate the size of the receiving area for a 400 tonne/day facility. Make appropriate assumptions for the density of the waste and the shape of the pile.

4. Estimate the gross heat value of newsprint. Assume the elemental analysis of the newsprint is

	Weight percent
Carbon (C)	49.14
Hydrogen (H)	6.10
Oxygen (O)	43.03
Noncombustible carbon	1.52

5. Incinerators generally burn MSW in the temperature range 750–1,000°C. What are the consequences of (a) burning wastes at temperatures lower than this? (b) burning wastes at temperatures higher than this?

6. Carbon monoxide is routinely measured to monitor combustion efficiency. Why is carbon monoxide the gas chosen for this purpose?

7. Determine the empirical formula for dry mixed paper wastes with the ultimate analysis shown as follows:

	Weight Percent
Carbon	43.41
Hydrogen	5.82
Oxygen	44.32
Nitrogen	0.25
Sulfur	0.20
Noncombustible	6.00

8. What is believed to be the origin(s) of dioxins in the air emissions of an incinerator?

9. What air pollution control systems extract particulates from exhaust gases?

10. Of the pollution abatement systems described in this chapter, which systems extract acids from the flue gas and how do they accomplish this?

11. Prepare an analysis to determine whether thermal processing is appropriate for the wastes in your community. If so, what type of system should be used?

12. Some people believe that incinerator ash is a sterile inert material, while others view the ash as a hazardous material. What do you think? Review recent articles on this issue in a library and prepare a concise position paper supporting your view.

REFERENCES

Buekens, A. and Patrick, P.K., *Solid Waste Management Selected Topics*, World Health Organization, Copenhagen, Denmark, 1985.

Dvirka, M., Resource recovery: Mass burn energy and materials, in *The Solid Waste Handbook: A Practical Guide*, Robinson, W.D., Ed., John Wiley & Sons, New York, NY, 1986.

Finkelstein, A., Hay, D.J., and Klicius, R., The National Incinerator Testing and Evaluation Program (NITEP), National Waste Processing Conference, Denver, CO, ASME, 1986.

Hazome, H., Sato, K., Nakata, K., and Nomura, H., Investigation of elemental analysis of refuse for calorific values, *Nippon Kankyo Eisei Sentra Shoho*, 6, 39–43, 1979 (in Japanese).

Hecht, N.L., *Design Principles in Resource Recovery Engineering*, Butterworth Publishers, Woburn, MA, 1983.

Hougan, O.A., Watson, K.M., and Ragatz, R.A., *Chemical Process Principles Part I: Material and Energy Balances, 2nd Ed.* John Wiley & Sons, New York, NY, 1954.

Khan, M.Z.A. and Abu-Ghararah, Z.H., New approach for estimating energy content of municipal solid waste, *J. Environ. Eng.*, 117(3), 376–380, 1991.

Kiser, J.V.L. and Burton, B. Kent, Energy from municipal waste: Picking up where recycling leaves off, *Waste Age*, 23(11), 39–46, November, 1992.

Lipták, B.A., *Municipal Waste Disposal in the 1990s*, Chilton Book Company, Radnor, PA, 1991.

Niessen, W.R., Properties of waste materials, in *Handbook of Solid Waste Management*, Wilson, D.C., Ed., Van Nostrand Reinhold Co., New York, NY, 1977.

Robinson, W.D., *The Solid Waste Handbook: A Practical Guide*, John Wiley & Sons, New York, NY, 1986.

Stern, A.C., Boubel, R.W., Turner, D.B., and Fox, D.L., *Fundamentals of Air Pollution*, 2nd Ed., Academic Press, Orlando, FL, 1984.

Tillman, D.A., *The Combustion of Solid Fuels and Wastes*, Academic Press, Inc., San Diego, CA, 1991.

U.S. Congress, Office of Technology Assessment (USCOTA), Facing America's Trash: What Next for Municipal Solid Waste?, OTA-O-424, U.S. Printing Office, Washington, DC, October, 1989.

U.S. Environmental Protection Agency, Municipal Waste Combustion Study, Emission Data Base for Municipal Waste Combustions, EPA/530-SW-87-021b, U.S. Printing Office, Washington, DC, June, 1987.

Wilson, D.L., Prediction of heat of combustion of solid wastes from ultimate analysis, *Environ. Sci. Tech.*, 6(13), 1119–1121, 1972.

CHAPTER 9

Landfilling of Wastes

9.1 OVERVIEW

Landfilling wastes is a modern variation of the long-used practice of depositing wastes in a dump site at the outskirts of a community. In the past, this practice served the purpose of isolating the community from the health problems associated with decomposing wastes and nuisance problems, such as vermin, vectors, odors, and litter. Typically the site chosen was land of marginal value for agriculture or urban development.

A modern sanitary landfill is an engineered site, selected, designed, and operated in such a manner as to minimize environment impacts. Municipal wastes are deposited in a confined area, spread in thin layers, compacted to the smallest practical volume and covered at the end of each working day. Certain types of industrial and other nonputrescible wastes may not require daily cover. The design, construction, and operation of the site are subject to state and federal regulations and design or performance standards.

Landfills are a necessary component of any municipal solid waste management system. Waste reduction efforts, recycling, incineration, and composting can reduce the quantity of materials sent to a landfill, but anyone experienced in waste management will acknowledge that there will always be residual materials which require landfilling.

Landfilling is the most commonly used waste disposal method in the United States. Landfills receive about 80% of the discarded municipal waste. Landfilling is also the predominant method of disposal in most other countries; however, as the data in Table 9.1 show, Japan and some European countries utilize landfilling to a lesser extent than the United States.

The number of landfills in the United States is decreasing for three reasons: (1) Many old landfills that do not meet current design and

Table 9.1 Estimates of the Percentage of
Municipal Solid Waste Landfilled, by
Weight in Selected Countries

Country	Percent Landfilled	Year
Denmark	44	1985
France	54	1983
Greece	100	1983
Ireland	100	1985
Italy	85	1983
Japan	33	1987
Netherlands	56–61	1985
Sweden	35–49	1985, 1987
Switzerland	22–25	1985
United Kingdom	90	1983
United States	90[a]	1986
West Germany	66–74	1985, 1986

[a]These figures refer to landfilling after recycling (e.g., of source-separated glass, paper, metals) has occurred. For example, the United States landfills about 80% of all MSW, but about 90% of post-recycled MSW.

Sources: U.S. Congress Office of Technology Assessment, 1989 with data from Franklin Associates. Ltd., Characterization of Municipal Solid Waste in the United States, 1960 to 2000 (Update 1988), final report prepared for U.S. Environmental Protection Agency, Prairie Village, KS, March 1988; Hershkowitz, A., International Experiences in Solid Waste Management, contract prepared for U.S. Congress, Office of Technology Assessment, Municipal Recycling Associates, Elmsford, NY, October, 1988; Institute for Local Self-Reliance (ILSR), Garbage in Europe: Technologies, Economics, and Trends, Washington, DC, 1988; Pollock, C., Mining Urban Wastes: The Potential For Recycling, Worldwatch Paper 76, Worldwatch Institute, Washington, DC, April, 1987; Swedish Association of Public Cleansing and Solid Waste Management, Solid Waste Management in Sweden, Malmo, Sweden, February 1988.

operation standards are being closed; (2) It has become increasingly more difficult to site new landfills; and (3) The costs of site design, construction, operation, leachate and gas monitoring and collection, leachate treatment, administration, and engineering favor the construction of large

Table 9.2 Projected Number of Landfills in the United States

Year	Number of Landfills
1988	5499
1993	3332
1998	2720
2003	1594
2008	1234

Source: U.S. Environmental Protection Agency, Report to Congress: Solid Waste Disposal in the United States, Vol. II, EPA/530-SW-88-011B, U.S. Printing Office, Washington, DC, 1988.

facilities. The projected decline in the number of landfills is indicated by the data in Table 9.2. The number of landfills is a poor indicator of landfill capacity, however, because new landfills tend to be larger than the older landfills.

The difficulty faced by many municipalities in identifying and developing new landfill sites is not necessarily caused by a lack of land at a suitable location with suitable soil and hydrogeological conditions, but rather by public opposition. People object to landfills because of the nuisance issues (e.g., dust, noise, traffic, odor), aesthetics, and environmental concerns (e.g., groundwater pollution from landfill leachate, migration of landfill gases to adjacent properties, and use of agricultural land). Although public opposition and more restrictive governmental regulations make the solid waste manager's job more challenging, these factors have the beneficial effect of forcing communities and industries to investigate waste reduction, recycling, and processing alternatives more seriously.

In September 1991, the U.S. EPA promulgated regulations for municipal solid waste landfills in the United States, including those that receive sewage sludge or municipal incinerator ash (*Federal Register*, October 9, 1991). Because these regulations were developed in response to directives contained in Part D of the Resource and Conservation Act of 1976, they are commonly referred to as Subtitle D regulations. They set forth the minimum criteria for location, design, operation, groundwater monitoring and corrective action, closure, and post-closure care of a landfill.

9.2 LANDFILL SITE SELECTION

Prior to initiating a search for a landfill site, it is necessary to determine the size of the site that will be required by a service region for a designated time period, typically ten to twenty years. Smaller sites rarely justify the investment in development and operating costs. The site size is determined by the quantity of refuse to be deposited during the lifetime of the site. The quantity of refuse depends on the population and the industrial, institutional, and commercial characteristics of the service area, the waste generation rates, preliminary design, and the potential for future expansion.

A reasonable "rule-of-thumb" estimate of the annual volume requirement of a landfill for municipal solid waste is 1.7 ha·m per 10,000 persons (14 acre-ft/10,000 persons). The origin of this number and the assumptions used to compute it are explained in Example 9.1. (Several other "rules-of-thumb" are given in Problem 3 at the end of this chapter.)

Example 9.1: Determine the volume of landfill space needed to accept the MSW from a community of 10,000 persons for one year. Assume an average daily per capita waste generation rate of 1.8 kg (4 lb).

Solution: The quantity of waste generated by one person in a year, q, is

$$q = 1.8 \text{ kg/person/day} \times 365 \text{ days/year} = 657 \text{ kg/year}.$$

The quantity of waste generated by 10,000 people in one year, Q, is calculated as follows:

$$Q = 10,000 \text{ persons} \times 657 \text{ kg/person} = 6.57 \times 10^6 \text{ kg}$$

This quantity can be converted to metric tonnes (1 tonne = 1000 kg) with the following calculation:

$$Q = 6.57 \times 10^6 \text{ kg} \times 1 \text{ tonne}/1000 \text{ kg} = 6.57 \times 10^3 \text{ tonnes}.$$

This quantity can be converted to an equivalent volume in a landfill by using the density of the compacted wastes. A reasonable conservative estimate is 0.60 tonne/m^3 (0.5 ton/yd^3). Using this estimate, the volume, V, is

$$V = 6.57 \times 10^3 \text{ tonne} \times 1 \text{ m}^3/0.60 \text{ tonne},$$

or

$$V = 11,000 \text{ m}^3.$$

The area depleted each year depends upon the depth to which the wastes are to be deposited. A convenient unit of volume is hectare·meter (ha·m), where 1 hectare·meter = 10,000 m^3. Refuse occupying a volume of 1 ha·m corresponds to an area of 1 ha with the refuse 1 m deep, an area of 2 ha with refuse 0.5 m deep or any other combination of area in hectares and depth in meters having a product equal to 1 ha·m. (The unit commonly used in the English system is acre·ft; the relationship between the two units is 1 ha·m = 8.107 acre·ft.)

Converting the volume of wastes to hectare·meters, we have:

$$V = 11,000 \text{ m}^3 \times 1 \text{ ha·m}/10,000 \text{ m}^3$$

or

$$V = 1.10 \text{ } ha·m.$$

Additional landfill space is required to account for the soil used as cover material (intermediate and final), usually in a ratio of 1 part soil to 4 or 5 parts refuse. This would increase the volume by 20% to 25%. Thus, the adjusted volume

$$V_a = 1.10 \text{ ha·m} \times 1.25$$

or

$$V_a = 1.38 \text{ ha·m (about 11.2 acre·ft)}$$

This volume number does not take into account the unused land needed for buffer areas and berms or access roads and auxiliary facilities such as the scalehouse, equipment storage, and leachate collection or treatment facilities. Robinson (1986) suggests a multiplier of 1.25 to 2 be used to obtain the total land area requirement, including this "waste land." Multiplying 1.38 by 1.25 yields 1.7, the "rule-of-thumb" estimate stated earlier.

The search for a suitable landfill site often begins by ruling out environmentally unacceptable sites. In general there are six types of land that are environmentally unsuitable for a landfill. These are floodplains, wetlands, land near airports, geological fault zones, seismic impact areas, or other unstable areas. Local regulatory agencies may specify additional restrictions, such as maintaining specified distances from highways, parks, ponds, lakes, or water supply areas; or prohibiting the selection of sites which pose a threat of groundwater contamination or to critical habitat

Table 9.3 Suggested Distances from a Proposed Landfill or Expansions of Landfills to the Critical Areas or Constructions

Area or Construction	Distance
Navigable lake or pond	300 m (1000 ft)
River	90 m (300 ft)
Highway	300 m (1000 ft)
Public park	300 m (1000 ft)
Water well supply	360 m (1200 ft)
Airports[a]	3030 m (10000 ft)
Wetlands	prohibited within borders
Critical habitat areas	prohibited within borders
100 year floodplain	prohibited within borders

[a]The 1991 U.S. regulations require notification to the Federal Aviation Administration and the affected airport if the distance is less than 5 miles (8050 m).

Source: Bagchi, A., *Design, Construction, and Monitoring of Sanitary Landfill*, John Wiley & Sons, New York, NY, 1990.

areas of endangered plant or animal species. When a site does not meet all of the siting criteria, it may be possible to seek permission from a regulatory agency to modify or waive some restrictions. In the absence of other regulatory requirements, Bagchi (1994) suggests that the distances from a proposed landfill or expansions of landfills to critical areas or construction sites be those given in Table 9.3.

Subsurface conditions, such as the types of soil, underlying rock strata, and groundwater conditions are important factors for determining whether an environmentally safe landfill can be economically designed for a specific site. The primary concern is the protection of groundwater from contamination. In addressing this concern, information must be obtained regarding the distance from the bottom of the proposed fill to the groundwater, the type of soils and other unconsolidated materials as well as bedrock beneath the site, the volume and direction of flow of the groundwater, and the existence of any impervious bedrock or clay layers between the fill and the groundwater. Impervious bedrock or clay layers are important because of their potential for isolating the leachate produced by the landfill from important groundwater aquifers. Using this information, the landfill designer can determine the impact of leachate leaving the fill area and, if remedial action is required, how best to address it. Some of the preliminary information can be gleaned from topographic maps, soil maps, and geologic maps and other U.S. Geological Survey or State Geological Survey reports, but much of the detailed information about the soil stratigraphy and groundwater hydrology can only be obtained from borings and test wells. The challenge is to characterize the subsurface conditions under a large land area utilizing data obtained from a limited number of discrete sampling points.

Regulatory guidelines may specify or suggest the minimum number of borings required. In the absence of such guidelines, Bagchi (1994) suggests five borings be completed for the first two hectares (5 acres) or less, and two additional borings in each additional hectare. The borings should be well-distributed to cover an area at least 25% larger than the planned waste disposal site and extend to at least 7.5 m (25 ft) below the proposed base of the landfill. The borings are used to determine soil stratification, hydraulic conductivity of the soil layers (i.e., ease with which moisture moves through the material), and depth to bedrock.

Water monitoring wells must be drilled to determine the depth to the water table and the volume and direction of groundwater flow. Water samples can also be extracted and analyzed to determine the background quality of the water. A more detailed discussion of hydrogeology is presented in the next section.

Economic factors of importance are hauling distances from the waste sources, the availability of suitable soils for liners, daily cover and final cover materials, and special design, construction, and operating considerations. The hauling costs are borne by the users of the facility, whereas the other costs are part of the construction and operating costs of the site.

A carefully planned site selection process must not only address the engineering feasibility of developing the site and the environmental constraints, but also the social, economic, aesthetic, and political concerns. A typical siting process involves collecting data about each of these concerns and implementing a screening process to eliminate sites that do not meet minimum criteria.

Social criteria include factors, such as current land use; location of residences, schools, and businesses; archeological, historical, and aesthetic considerations; final use of the site; and any others pertaining to the degree to which a landfill would disrupt the human environment.

The process of selecting a landfill site is complex and often politically controversial. Many urban communities lacking suitable land areas look to rural locations for a landfill site. This may be done as part of a county or regional approach to waste management. Those living in rural areas are reluctant to accept the wastes of their urban neighbors. It is likely that legal issues, zoning restrictions, and compensation to affected individuals and municipalities will need to be addressed. The process requires public education regarding the dangers and benefits of a landfill, open communication about the process of identifying a suitable site, and a willingness to address the concerns raised by those affected by a potential facility.

A structured site selection process needs to account for the fact that most communities will be hesitant to accept a landfill. The process should reject unacceptable sites, and it may be appropriate to include a ranking system for evaluating acceptable sites. The areas that are unsuitable can be identified by using map overlays with the unacceptable areas shaded. Examples of unacceptable areas are wetlands, floodplains, and regions with shallow bedrock. This procedure can sometimes be performed by utilizing geographic information systems (GIS) or other types of information stored in computer databases. Suitable sites of the required size may be identified from the portion of the map that is not eliminated by one of the overlays.

If a ranking or scoring system is developed for a group of potential sites, it should be simple, understandable, logical, and accurate. It should be structured so that it can be easily explained at public meetings and to the news media. Even if the system is understandable, it will not be accepted unless it is also logical and consistent. Furthermore, data must be accurate so that if the results are questioned, others may test the system themselves.

9.3 CONCEPTS OF GROUNDWATER HYDROLOGY

In order to predict the fate of the leachate that leaves a landfill site and its potential for contaminating groundwater, an understanding of the flow of water through subsurface materials is necessary. Much of the information about the hydrogeological conditions is obtained from soil borings and wells installed at the site. The purpose of this section is to discuss how the information obtained from wells can be used to calculate subsurface flow when the underlying conditions are simple. In practice, much more complex conditions may exist; these conditions require analytical methods and mathematical modeling techniques beyond the scope of this book.

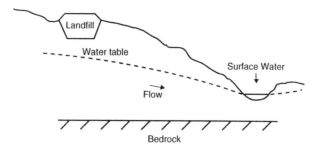

Figure 9.1 Groundwater flow beneath a landfill.

Groundwater is the subsurface water that resides in the zone of saturation. In the zone of saturation, the voids (pores) between the soil particles are filled with water. The **water table** is the upper boundary of this zone of saturation. The water table is also defined as the surface at which the fluid pressure in the pores is equal to the atmospheric pressure. Above the water table is the zone of aeration, or unsaturated zone.

Qualitatively, it can be said that the water table surface in a humid climate tends to follow the topography of the land surface, provided that the aquifer is unconfined (i.e., there are no impermeable horizontal clay, rock, or other layers to hinder vertical movement of the water). A sloping water table implies that groundwater is flowing from a point of higher elevation to one of lower elevation. The water table intersects the land surface where there is surface water. Figure 9.1 depicts a typical condition for unconfined groundwater flow beneath a landfill site.

When the aquifer is confined by an overlaying impermeable layer, the water will be under pressure and will rise in a well that is cased to the aquifer. Such a well is sometimes called a **piezometer**. The elevation of the water in the pipe is a measure of the pressure at the base of the well pipe. The three-dimensional surface depicting the height to which water will rise in a piezometer is called a **potentiometric surface**. In general, a potentiometric surface is not influenced by surface topography and surface water features. The groundwater will flow from locations of higher pressure toward those at lower pressure (i.e., downhill on the potentiometric surface). Groundwater flow in both unconfined and confined aquifers will be discussed in more detail later in this section.

To deal with groundwater flow quantitatively it is necessary to discuss three important principles or laws. They are the Bernoulli equation, which is a statement of conservation of energy of fluids; the continuity principle, which is a statement of conservation of mass for an incompressible fluid, such as water; and Darcy's law, which describes the flow of water through a porous medium such as soil.

The flow of an incompressible fluid with negligible viscosity along a smooth line, commonly referred to as laminar or nonturbulent flow, is described by the Bernoulli equation which states that the energy per unit volume, E, of the fluid is constant. The equation is:

$$E = 1/2\ \rho v^2 + \rho gz + P = constant$$

where
   ρ = mass density of the fluid
   v = velocity of the fluid
   g = acceleration of gravity (= 9.8 m/s^2)
   z = vertical position
   P = pressure of fluid

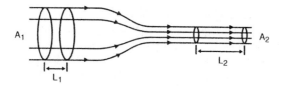

Figure 9.2 Flow of an incompressible fluid in a tube with a changing cross-sectional area.

The first term on the right side of the equation, $1/2\ \rho v^2$, represents the kinetic energy per unit volume; the second term, ρgz, represents the gravitational potential energy per unit volume; and the last term is related to the work done per unit volume. The derivation of this equation is found in most introductory physics texts.

Let us apply the Bernoulli equation to the simple condition of an incompressible nonviscous fluid with steady flow in a tube of flow with variable cross section as shown in Figure 9.2. The reader may visualize the tube of flow more concretely by thinking of the tube of flow as a pipe, but should realize that the streamlines used to describe laminar flow never cross and by themselves serve as an imaginary boundary for the liquid. Let A_1 and v_1 be the cross-sectional area and velocity, respectively, at the left end of the tube, and A_2 and v_2 the values at the right end of the tube. By the conservation of mass, the mass of the fluid entering the tube at the left, $\rho A_1 v_1$, and the mass leaving the tube at the right, $\rho A_2 v_2$, are equal. Thus

$$\rho A_1 v_1 = \rho A_2 v_2$$

This expression is the equation of continuity. The product $\rho A_1 v_1$ is the flow rate in units of mass per second. For an incompressible fluid, ρ is a constant. By dividing both sides of the equation by the density, ρ, the equation becomes

$$A_1 v_1 = A_2 v_2 = constant$$

The product $A_1 v_1$ is the flow rate expressed as the volume per second. If the cross-sectional area of the flow becomes smaller, there must be a corresponding increase in the velocity of flow, and vice versa.

Often the Bernoulli equation is written in a form where each term is divided by ρg, yielding the following:

$$E/\rho g = 1/2\ v^2/g + z + P/\rho g$$

Each term of this equation expresses a form of energy per unit weight. Dimensionally, each term is a length, and is commonly called a "head". Thus this equation can be written as

$$h_t = h_v + h_z + h_p$$

where h_t $(= E/\rho g)$ is the total or hydraulic head, h_v $(= 1/2\ v^2/g)$, is the velocity head, h_z $(=z)$ is the elevation head, and h_p $(= P/\rho g)$ is the pressure head.

In the laboratory, pressure can be measured with a manometer, an instrument which essentially consists of a column of liquid in a vertical tube. The basic principle of operation is that the absolute pressure at the bottom of a column of liquid open to the atmosphere equals the weight per unit area of the liquid, that is, $P = \rho g h$ for a liquid column of density ρ and height h plus the pressure of the atmosphere. Thus,

$$P_{absolute} = \rho g h + P_{atmosphere}$$

In the field, gauge pressure, the pressure in excess of atmospheric pressure, is commonly used, that is,

$$P_{gauge} = P_{absolute} - P_{atmosphere} = \rho g h.$$

Thus the gauge pressure at the base of a column of liquid can readily be determined from the height of the column.

In the case of groundwater flow, the velocities are extremely small and the velocity head term can almost always be neglected. Let us assume that this is the case here and that we have a hydrostatic situation where water is contained in a tube with manometers installed to measure pressure. These are shown in Figure 9.3. The Bernoulli equation, expressed in terms of heads, with the condition that $h_v = 0$, becomes

$$h_t = h_z + h_p = h_{z1} + h_{y1} = h_{z2} + h_{y2}.$$

The hydraulic head is the same for both vertical tubes, but the left end has a smaller elevation head and larger pressure head. There is no flow when the hydraulic head, h_t, is the same at both ends. If we were to increase the pressure at the left, by turning on a pump for example, the pressure head (and hydraulic head) would increase at the left and the water would flow to the right.

The French hydrologist Henry Darcy investigated the flow of water through beds of sand using an apparatus similar to that shown in Figure 9.4. Darcy measured the volumetric flow, Q, of water passing through a bed of sand and related it to the dimensions of the bed and the pressure

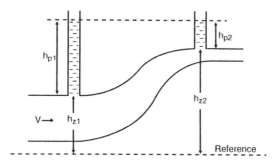

Figure 9.3 The elevation and pressure heads at different points in slow flowing liquid confined in a tube.

difference, as measured by the difference in the levels of water in a pair of manometers (i.e., differences in hydraulic head across the bed of sand). Darcy discovered that the velocity, v, of flow of water through the sand, is proportional to the hydraulic gradient. The hydraulic gradient expresses the rate at which the hydraulic head changes with distance. The hydraulic gradient for this apparatus is $(h_1 - h_2)/L$, where L is the distance between the points where the manometers measure the pressure. The volume of water passing through the sand is given by Darcy's law and is expressed by the equation,

$$Q = kA(h_1 - h_2)/L$$

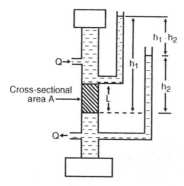

Figure 9.4 Schematic of Darcy's apparatus for measuring the flow volume of water through a soil sample (dark shading) as a function of hydraulic gradient.

Table 9.4 Typical Values of Hydraulic Conductivity

Material	Hydraulic Conductivity (m/s)
Gravel	$10^{-4} - 10^{-2}$
Well–sorted sand	$10^{-5} - 10^{-3}$
Fine sands, silty sands	$10^{-7} - 10^{-5}$
Silt, sandy silts, till	$10^{-8} - 10^{-6}$
Clay	$10^{-11} - 10^{-8}$

Source: Adapted from Fetter, C.W., Jr., *Applied Hydrology*, Third Ed., Macmillan, New York, 1994.

The constant k is called the **hydraulic conductivity** or **coefficient of permeability**. It depends on the particle size of the soil, the viscosity and density of the liquid, and the degree of saturation of the soil. The values of k are determined experimentally for each type of soil and typically vary from about 10^{-2} m/s in gravel to 10^{-11} m/s in clay. Table 9.4 provides a detailed listing of the typical values of k.

A more general form of Darcy's law, using concepts from calculus, is

$$Q = -kA(dh/dl)$$

where dh represents a small change in the hydraulic head between two nearby points a distance dl apart. The minus sign is included because the flow is in the direction of decreasing head. Thus the dh/dl term is negative. An implicit assumption is that the fluid flow is laminar; that is, the water flow is characterized by smooth streamlines. This condition is usually satisfied when the discharge velocity is less than 0.0023 m/s (Fetter, 1994).

When water flows through an open channel or pipe, the volume of water per unit time—the discharge, Q—is equal to the product of cross-sectional area of flow, A, and the velocity, v; that is,

$$Q = Av$$

A rearrangement of this equation yields an expression for the discharge velocity:

$$v = Q/A$$

Using the same reasoning, Darcy's law can be solved for Q/A to provide an expression for the velocity for the flow through a porous medium:

$$v = Q/A = k(h_1 - h_2)/L = -k(h_2 - h_1)/L$$

The velocity of flow calculated in this manner is called the specific discharge. Synonymous terms are specific discharge velocity, Darcian velocity, or Darcy flux. This velocity is an apparent velocity, calculated as though the flow were in an open conduit of cross-sectional area A. But in porous material, water can move only through the connected pore spaces that act as channels of flow, thereby reducing the "effective" cross-sectional area to some fraction, η, of A. Thus the actual velocity of flow—the average linear velocity, v_s, (also called the seepage velocity)—is calculated by dividing the specific discharge by the effective porosity η as expressed by the equation

$$v_s = Q/(\eta\,A) = v/\eta = k(h_1 - h_2)/(\eta L)$$

Let us apply the concepts developed above to groundwater hydrology problems. In the laboratory, as noted above, the tube to measure pressure is called a manometer, but in the field a water observation well or a piezometer serves this purpose. The relationship between the elevation head, pressure head, and total head for a water observation well installed in a groundwater flow field is shown in Figure 9.5. The elevation head represents the work required to increase the elevation of a unit weight of water from a reference level (datum), typically sea level, to the base of the well. The pressure head, determined by the length of the water column in the well, indicates the work the water is capable of doing because of its sustained pressure. When the total hydraulic head varies, water flows from a point where the total hydraulic head is higher to where it is lower. It is instructive to consider two general cases of groundwater flow, steady flow in a confined aquifer and steady flow in an unconfined aquifer.

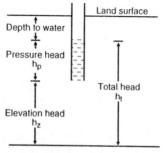

Figure 9.5 Relationship of total hydraulic head, h_t, pressure head, h_p, and elevation head, h_z, at a piezometer.

STEADY FLOW IN A UNIFORM CONFINED AQUIFER

A confined aquifer is a porous aquifer layer bounded above and below by impervious strata. Assume there is a steady flow of groundwater, Q, in the aquifer of uniform thickness t and width w. Furthermore, the aquifer is both homogeneous and isotropic between the impervious confining layers. From Darcy's law, and the fact that Q, k, and A are constants, we can conclude that the gradient must also be constant. Hence,

$$Q/kA = constant = -dh/dx$$

where x indicates the horizontal distance. Since the derivative, dh/dx, is constant, h is a linear function of x. Therefore the hydraulic head decreases linearly in the direction of flow as shown in Figure 9.6.

The values of the hydraulic heads at different locations define the potentiometric surface. Contours can be drawn on a two-dimensional map to designate points with the same value of the total hydraulic head. The water flow is in the direction of the negative of the gradient, flowing from higher to lower hydraulic head levels. The groundwater flow, Q, is a vector quantity having both a direction and magnitude. Graphically these flow lines are drawn so that they are perpendicular to the potentiometric contour lines.

STEADY FLOW IN AN UNCONFINED AQUIFER

Unconfined flow occurs when the zone of saturation is bounded below by an impervious layer and above by the water table. The cross-sectional area through which the water flows is

$$A = hw$$

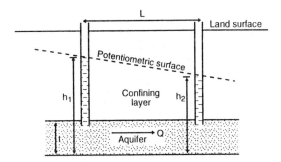

Figure 9.6 Change of hydraulic head for steady groundwater flow in a confined aquifer of uniform thickness.

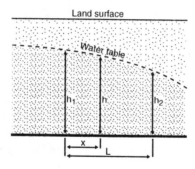

Figure 9.7 Change of hydraulic head for steady groundwater flow in a unconfined aquifer underlain with an impermeable layer.

where h is the hydraulic head (i.e., height of the water table above the impervious layer) and w is the width of the flow. The flow in an unconfined aquifer is shown in Figure 9.7. Observe that the vertical thickness of the water layer is not a constant, as was the case for the confined aquifer. Darcy's law, with the substitution of hw for A becomes:

$$Q = -kA \ dh/dx = -kwh \ dh/dx$$

Here the hydraulic gradient is not constant but is given by the varying slope of the water table. With the assumptions that Q is constant, which means that no gain of water through infiltration nor loss through evaporation occurs, and that the flow is horizontal, the volume of flow can be related to the hydraulic heads measured at two wells a distance L apart. Upon separating the variables and performing a simple integration with the boundary conditions h_1 at $x = 0$ and h_2 at $x = L$, the result is:

$$Q/kw = -h \ dh/dx$$

$$\int_0^L (Q/kw) \ dx = \int_{h_1}^{h_2} -h \ dh$$

$$(Q/kw) x \Big|_0^L = -h^2/2 \Big|_{h_1}^{h_2}$$

$$(Q/kw) \ L = -(h_2^2 - h_1^2)/2$$

$$Q = -(kw/2L) \ (h_2^2 - h_1^2)$$

The problem of finding the hydraulic head at a point between two wells is left as an exercise for the reader (Problem 8).

An example where information at two wells is used to determine the volume of groundwater flow follows.

Example 9.2: An unconfined sand aquifer has a hydraulic conductivity of 3×10^{-4} m/s and porosity of 0.23. The aquifer is underlain by an horizontal impervious layer 25 m below the land surface. Two wells are installed 100 m apart. The water level in Well 1 is 12 m below the land surface and the water level in Well 2 is 18 m below the land surface. (a) What volume of water is flowing in the aquifer? (b) What is the seepage velocity at Well 1?

Solution: (a) First calculate the hydraulic heads at the wells.

At Well 1 the hydraulic head = 25 m – 12 m = 13 m.
At Well 2 the hydraulic head = 25 m – 18 m = 7 m.

The width of the aquifer is not given, so for simplicity, we calculate the flow per unit width of aquifer. If the width is subsequently specified, the total volume can be easily calculated by multiplying by the width.

Substitute the values

$$h_1 = 13 \text{ m}$$
$$h_2 = 7 \text{ m}$$
$$k = 3 \times 10^{-4} \text{ m/s}$$
$$L = 100 \text{ m}$$

into the equation for unconfined flow,

$$Q = -(kw/2L)(h_2{}^2 - h_1{}^2)$$

to get

$$Q/w = -[(3 \times 10^{-4} \text{ m/s})/(2 \times 100 \text{ m})][(7 \text{ m})^2 - (13 \text{ m})^2]$$

or

$$Q/w = 1.8 \times 10^{-4} \text{ m}^3/\text{s per unit of width of the aquifer}$$

(b) The average linear velocity is calculated from the equation

$$v_s = Q/(\eta A) = Q/(\eta\, h_1 w) = (Q/w)/(\eta\, h_1)$$

Evaluating this expression with $Q/w = 1.8 \times 10^{-4}$ m^3/s, porosity $\eta = 0.23$, and $h_1 = 13$ m,

$$v_s = (1.8 \times 10^{-4}\ \text{m}^3/\text{s})/\ (0.23 \times 13\ \text{m} \times 1\ \text{m})$$

$$v_s = 6.0 \times 10^{-5}\ \text{m/s}$$

In this section, we have presented the physical principles describing groundwater flow using the simplifying assumptions that the aquifers are homogeneous and isotropic, water is not being added by infiltration nor lost by evaporation, and the aquifers are continuous and infinite to a real extent. Real applications are more complex, but the same physical principles apply. However, additional mathematical tools from the field of numerical analysis are needed. These methods require that a grid of polygonal elements be imposed on the map of the region of interest, and equations developed for each finite element or node, depending upon the technique being used. The equations are then solved with the assistance of a computer. For an overview of these techniques the reader is referred to an article by Wang and Anderson (1982).

9.4 LANDFILL DESIGN

Landfills must be designed to protect groundwater quality. The majority of dumps that were prevalent in the United States in the 1960s and 1970s provided little groundwater protection. Several landfill design concepts have evolved to manage leachate and afford some degree of groundwater protection.

Many landfills designed to receive nonhazardous MSW have been constructed as natural attenuation type landfills. A **natural attenuation landfill** is constructed without a liner, thereby allowing the leachate to percolate through the landfill base with the expectation that it will be rejuvenated as it passes through the unsaturated soil beneath the site. The ideal geophysical setting for a natural attenuation landfill can be conceptualized as one where there is a thick soil with low permeability (e.g., clay), an impermeable underlying bedrock structure, no outflowing aquifer above the bedrock and a large unsaturated zone above the watertable. Such a setting provides a suitable environment for those chemical, physical, and biological processes to occur which reduce the concentrations of undesirable components in the leachate. However, studies by Bagchi (1983) and others indicated that no matter how thick the soil is, it cannot attenuate all of the contaminants. Consequently, the use of natural attenuation landfills has become restricted in some countries and in some states in the United States. Germany and the state of Wisconsin are examples. A natural attenuation design may be acceptable in a dry climate where minimal amounts of leachate are produced.

A second type of landfill—a containment landfill—is designed and constructed with liners and a leachate collection system. At one time, lined landfills were used only when hazardous waste was to be landfilled or site conditions did not permit the construction of a natural attenuation landfill, perhaps because of insufficiently thick underlying soils, or fractured bedrock. An obvious disadvantage of the containment landfill is the necessity to collect and treat the leachate both during the active site life and after the site is closed.

Some operating containment landfills have been constructed with the base of the landfill located below the water table. The primary advantage of this "zone-of-saturation design" is that a hydraulic gradient is created toward the landfill. Thus, in the event of a liner leak, the tendency is for uncontaminated groundwater to leak in (to be collected and treated) rather than for leachate to leak out. With the advent of the Subtitle D design criteria, a true zone-of-saturation landfill design is no longer allowed in the United States.

The USEPA Subtitle D regulations issued in 1991 establish "dry tomb" landfilling as a national standard and include landfill design criteria to minimize the formation of leachate and to prevent leachate from leaving the landfill site. The Subtitle D design standard specifies a composite liner and a leachate collection system. The composite liner must consist of two components: an upper flexible membrane liner of at least 30-mil thickness (60-mil thickness if high density polyethylene is used), and a lower liner consisting of at least 0.6 m (2 ft) of compacted clay soil with a hydraulic conductivity of no more than 1×10^{-9} m/s as shown in Figure 9.8. The leachate collection system must be capable of maintaining leachate at a depth of less than 0.3 m (12 in.) over the liner. A typical

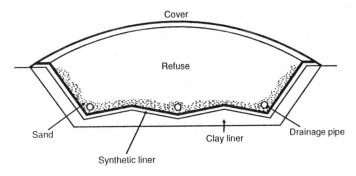

Figure 9.8 Composite liner and leachate collection system.

design might have the liner sloping at 2–4% toward perforated PVC collection pipes and the collection pipes sloping at a minimum of 0.5% toward a collection sump, from which the leachate can be pumped to a tank truck or sent via sewer lines to a treatment facility. A coarse soil drainage blanket is spread over the liner to provide a pathway for leachate flow across the base to the collection system. Sand is a typical material used for this purpose.

A final cover must be installed over the completed landfill to minimize infiltration of water. A suitable configuration described by Bagchi (1994) consists of a base or grading layer 15–60 cm (6–24 in.) thick over which a barrier layer of clay, synthetic clay liner, or synthetic membrane is installed. Above the barrier layer is a drainage layer covered by a protective soil layer between 30 and 105 cm (1 and 3.5 ft) thick and topsoil.

States having a regulatory agency, such as a Department of Natural Resources, may seek permission for that agency to assume the power to implement and enforce the provisions of the Subtitle D regulations. States which are granted this power are designated as approved states. These states may have different design criteria than states that do not seek or receive EPA approval. The standards in approved states are required to be at least as stringent as the provisions in Subtitle D. In those states not seeking approval, known as unapproved states, the landfill owners and operators are responsible for documenting a facility's compliance with the Subtitle D design standards. New landfills and expansions of existing landfills in unapproved states must conform to the national design standard.

New landfills and expansions of existing landfills in approved states must conform to a set of performance-based design standards. The liner and leachate collection system design must ensure that the concentrations of 24 organic and inorganic constituents in the uppermost aquifer beneath the area of compliance—which, depending upon the local hydrogeological conditions, can range from the waste management unit boundary to as far as 150 meters from this boundary—do not exceed maximum contaminant levels.

Gas ventilation must also be addressed in the landfill design. The gas of greatest concern is the methane (CH_4) produced from the decomposition of organic wastes. Methane gas is explosive in concentrations between 5% and 15%. There is insufficient oxygen for an explosion to occur if the concentration is above 15%, but such concentrations produce a condition of extreme danger. When an impermeable cover material is placed over the wastes to prevent water infiltration, or during the winter season when the ground may be frozen in cold climates, the methane, which is lighter than air, cannot escape upward. Therefore it migrates laterally. The nature of the movement depends upon whether the gases are impeded by a landfill liner or impervious soils. Methane may move several hundred meters through a porous sandy soil. If buildings are located near the landfill, they should be monitored for methane gas. Gas collection systems, clay cutoff walls, and venting trenches are common remedies for sites where methane is a problem. If its volume is low, methane is commonly burned using specially designed burners. If its volume is high, energy recovery may be economically feasible.

Not everyone agrees that the Subtitle D "dry tomb" approach is environmentally or economically the best alternative. Lee and Jones (1990) and others contend that constructing landfills in this manner prolongs the decomposition process so that the site will require perpetual care, a time period which extends beyond the expected life of the liners and the designated 30 year long-term care period. Rather than the dry approach, they propose stabilizing the MSW by shredding it, then adding moisture in a controlled manner to optimize methane production. Treated leachate provides the moisture for the fermentation and leaching of the wastes. After the process is complete—probably 5 to 10 years—the residue can be moved to a lined landfill for permanent burial.

Readers interested in the engineering design principles and construction guidelines are referred to Bagchi's book *Design, Construction, and Monitoring of Sanitary Landfills* (1994).

9.5 LINERS

Liners are impervious or semi-impervious barriers designed to contain both biological and chemical contaminants within a landfill. Materials used to construct a liner include soils (particularly clays), polymeric membranes, sprayed-on liners, soil sealants, admixed liners, and chemical absorptive liners. A liner is designed to prevent environmental contamination in one of two ways. It may be intended to prevent or retard liquid from flowing through the base of the landfill or it may be intended to filter out biological or chemical contaminants from the water it allows to pass through. The filtering of chemicals is not expected to be completely successful, but the amounts allowed through are projected to keep concentrations below harmful levels. Although clay liners do not effectively filter many chemical contaminants, they do trap some metals and also prevent harmful microorganisms, such as bacteria and fungi, from escaping.

Clay liners are by far the most common and the least expensive type of liner employed in landfill design. Clay liners are composed of the minerals kaolinite, illinite, or montmorillonite (Robinson, 1986). The clays are sometimes mixed with other fine-grained soil materials such as silt. Once the layer of clay is in place, it is packed by heavy equipment to make it as impermeable as possible. Well packed clays have permeabilities that range as low as 10^{-11} m/s.

Clay liners are most effective if they are installed in several thin layers rather than a single thick layer because the thin layers compact more thoroughly. In addition, the moisture content in the clay affects its compactibility. If it is either too wet or too dry, the clay will not compact easily. Different procedures for compaction are used, depending on the characteristics of the clay and the equipment that is available at the facility. Vibrating baseplate compactors, power tampers, crawler tractors, or heavy equipment with smooth roller wheels may be used.

Admixed liners are formed in place by mixing the liner material with the soil at the base of the landfill. These materials may include asphalt, concrete, soil cement, soil concrete, and bentonite clay. Admixed liners are not commonly used.

Polymer membranes are manufactured with thicknesses ranging from 0.5 to 3.0 mm (0.020 to 0.120 in.) and lengths of hundreds of meters. The membrane sheets are placed over a bedding material such as sand and the sheets are then sealed together with adhesives or heat. Usually the liner is extended up the sides of the landfill and anchored to the ground.

All liners must be able to withstand weathering, physical and chemical attacks, and temperature extremes that may be associated with the site. These stresses include tensions from the weight of the refuse, differential settling, hydrostatic pressure, abrasion, creep and freeze-thaw cracking, and chemicals deposited or created in the landfill.

9.6 LANDFILL OPERATIONS

A sanitary landfill is operated to minimize nuisance conditions arising from litter, dust, odors, and fire, and the attraction of rats, flies, mosquitoes, and birds. The standard method of operation consists of: (1) spreading waste in thin layers in a confined area; (2) compacting the waste to the smallest practical volume; and (3) covering the waste daily with soil or another type of material.

Waste compaction is accomplished by repeated trips over a waste layer with a heavy compactor vehicle. Figure 9.9 depicts the increase in density resulting from multiple trips over deposited wastes with a compactor vehicle.

In some cases, soil cover is required less frequently than at the end of each day's operation. When the refuse layers have reached a thickness of 2–3 m (6–9 ft), a layer of soil 0.15 to 0.3 m (6–12 in.) thick, or a layer of other approved cover, is spread over the refuse.

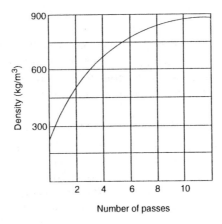

Figure 9.9 Relationship of the density of deposited municipal wastes to the number of passes of the compactor vehicle (adapted from Robinson, W.D., *The Solid Waste Handbook: A Practical Guide*, John Wiley & Sons, New York, 1986).

The compacted and covered material constitutes the basic building unit of the landfill, the cell. A group of adjacent cells of the same height constitute a lift. Cover material is hauled from adjacent land or borrow areas to complete cells and lifts as the landfilling progresses. Figure 9.10 illustrates the structure of a landfill.

Only a fraction of a landfill is developed at a time. These developed portions or phases can accept waste for a year or two after which the

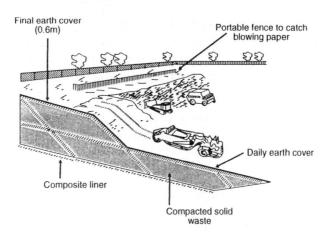

Figure 9.10 Construction of a landfill.

completed phase will be closed and another opened. The landfill designer identifies the sequence in which liners are to be laid, wastes deposited, cover materials excavated, and access roads constructed. These areas are designed and constructed so that liners are contiguous and the liner slopes are maintained. The designer also tries to minimize the excavation tasks and effort involved in moving cover materials.

9.7 LANDFILLING PROCESSED WASTE

Many communities and industries are reducing the volume of wastes sent to landfills by promoting waste reduction and recycling. In some communities, compostable materials, such as yard wastes, are being diverted to separate facilities that are not required to meet the stringent regulations of a sanitary landfill. Also, separate facilities may be provided for construction and demolition wastes. In addition to these types of programs, some communities further process the waste to be landfilled.

Experiments in France and in Madison, Wisconsin (Reinhardt and Ham, 1974) showed that shredded MSW placed in a landfill can be compacted to a greater density than unshredded wastes, and the need for daily cover material may be reduced. The shredded refuse can be compacted more quickly than uncompacted waste. Also, shredding promotes a more even settling of the fill material. The concept of fermenting and leaching of shredded MSW proposed by Lee and Jones (1990) and discussed in Section 9.4 is another example of processing prior to final landfilling.

Baling is another process which may facilitate landfill operations. Because of its greater density, baled refuse can extend a landfill lifetime by as much as 50%. Baling may reduce transportation costs, particularly when long haul distances are involved. It also minimizes litter, odor, vermin, and vector problems, as well as uneven settling of wastes.

9.8 CODISPOSAL OF WASTES

Codisposal refers to the disposal of two or more different forms of waste in the same landfill facility. These mixtures might consist of municipal solid waste and nonhazardous industrial process residues, ashes, or sludges. The specific nature of the industrial wastes, ashes, or sludges, and the design and operation of a landfill, may limit the combinations and the proportions of the components in the waste mixture. If sludges are accepted, the solids content should be at least 40% and the volume should be restricted to 25–30% of municipal waste (Bagchi, 1994). In some landfills, the wastes and sludges are mixed, while in others the sludge disposal area is segregated.

9.9 MONOFILLS

A monofill is a facility designed to receive a relatively homogeneous refuse such as fly ash or sludge. The chemical nature of the material may require the use of a monofill. For example, many ashes are alkaline, and as such, render the metal ions contained in them insoluble. On the other hand, municipal solid waste tends to produce leachate that is acidic. A mixture of the ash and MSW may result in more neutral conditions and free otherwise bound metal ions.

While it is necessary to cover a sanitary landfill daily, a monofill may require cover less frequently, provided nuisance conditions such as odor and blowing of material can be controlled adequately. An ash monofill does not require a daily cover because ash does not attract vectors nor produce odors. Intermediate cover at intervals may be required to reduce leachate production.

9.10 SITE CLOSURE AND POST-CLOSURE CARE

When a landfill has reached full capacity and no longer receives waste, a final cover is put in place to minimize infiltration of moisture and erosion. In the United States, the current EPA Subtitle D regulations (Subpart F) require an erosion layer consisting of a minimum of 0.15 m (6 in) of earthen material capable of sustaining natural plant growth. The erosion layer must be underlain with an infiltration barrier layer having a minimum thickness of 0.45 m (18 in.). The infiltration layer must be composed of earthen material that has a permeability less than or equal to the permeability of any bottom liner system or natural soils present, or a permeability of 1×10^{-7} m/s, whichever is smaller. An alternative design may be used if it provides equivalent infiltration and erosion protection.

A closed site has limited potential for construction of buildings because of potential settling and the dangers posed by methane gas. Consequently, the final uses of completed sites are likely to be as recreational or conservancy areas. The protection of the monitoring wells and equipment is a concern that may warrant limiting public access to a site.

To protect the integrity of the final cover and prevent infiltration of moisture, erosion should be minimized by establishing a vegetative cover. Other potential problems include damage to the surface by burrowing animals and penetration by roots of trees and other plants. Studies by Flower et al. (1978) and Gilman et al. (1981) address the problem of establishing trees and vegetation on and around a landfill. The main threat to plants at a completed landfill site is the presence of gases, primarily methane and carbon dioxide, in the root zone. These gases are not toxic to plants per se, but they displace the oxygen required by the roots. Gas barriers made of clay and plastic sheeting have been successfully used to counter this problem. Studies have also indicated that slow-growing trees are more suited to landfill sites than fast-growing trees because they are less likely to suffer water stress from lack of moisture. Examples of slow-growing trees are white pine, ginko, Norway spruce, and American basswood. Trees and shrubs with shallow and extensive root systems, such as the Japanese black pine and Norway spruce, are preferred to trees with deep root systems.

9.11　COSTS OF CLOSURE AND POST-CLOSURE CARE

Landfills continue to produce leachate and gases for many years after the site is closed. Consequently, it is important that there be a post-closure care plan to maintain the site, monitor groundwater and landfill gases, collect and treat leachate, and address other problems that might arise at the site. Subtitle D regulations require the site owner to be financially responsible for these obligations for 30 years after closure of the site.

The costs of site closure and post-closure (long-term) care represent a significant financial obligation. In the past, some landfill owners have abandoned sites or declared bankruptcy to avoid these costs. Because of these unethical practices, many states, and now the USEPA, require that owners provide proof of financial responsibility or assurance that money will be available to pay expenses as they arise. One approach to addressing this obligation is to establish segregated funds or escrow accounts into which deposits are paid while the site is operational and generating revenue. Another is to buy a payment guarantee from an insurance company or bank.

It is beyond the scope of this book to explore the complete range of financial mechanisms available to landfill operators. We will consider only the case of an interest-bearing segregated or escrow account for closure and post-closure care costs, and restrict our discussion to examples where equal annual payments are deposited in or withdrawn from interest-bearing accounts. For a more extensive treatment of the subject, including derivations of relevant formulas, the reader is referred to Bagchi (1994). Computer spreadsheets are invaluable for modeling complex schedules for cash deposits and payments, or in situations where an appropriate formula is not available.

FUNDS REQUIRED FOR CLOSURE

Assume that the cost of closing a landfill is estimated to be C, measured in dollars in the year when the operation of the facility begins. Assume also that the annual inflation rate is constant over the lifetime of the facility and has a decimal value of f. Then the money required to close the landfill, measured in dollars at the end of year 1 is $C(1 + f)$. The cost measured in dollars at the end of year 2 is $C(1 + f)^2$, and so on. The cost, A_c, measured in dollars at the end of SL years when the facility has reached the end of its operational lifetime is

$$A_c = C(1 + f)^{SL}$$

If this sum is to be accumulated in SL years by depositing equal annual payments in an interest-bearing account, the annual payment can be calculated using a standard annuity formula:

$$P = A_c i / [(1+i)^{SL} - 1]$$

where P is the annual payment to be made at the end of each year and i is the annual interest rate (expressed as a decimal).

Example 9.3: Assume a new landfill is opened with a projected lifetime of 20 years. It is estimated that closing the site will cost $100,000 in current dollars. (a) If the annual inflation rate is 3%, how much money will be needed to close the site in 20 years? (b) If equal annual deposits are made to an account paying 4% interest, what amount must be deposited? (c) If the facility is to accept 100,000 tonnes of waste annually, what user tonnage charge must be imposed for closure?

Solution: (a) The amount A required for closure in 20 years ($N = 20$) with the inflation rate of 3% ($f = 0.03$) can be calculated using the formula $A = C(1 + f)^{SL}$. The result is

$$A_c = \$100,000 (1 + 0.03)^{20} = \$180,611$$

(b) The annual payment can be calculated using the annuity formula with $i = 0.04$.

$$P = A_c i / [(1 + i)^{SL} - 1]$$
$$P = (\$180,611 \times 0.04) / [(1 + 0.04)^{20} - 1]$$
$$P = \$6065.23$$

(c) The charge which needs to be levied to generate the amount in (b) is $\$6065.23/100,000$ t = $\$0.061$ /t.

FUNDS REQUIRED FOR POST-CLOSURE CARE

In contrast to a site closure where the expenditures arise within a short time period, post-closure care often requires a series of annual expenditures for tasks such as site inspections and maintenance, leachate and gas monitoring, leachate treatment, and remedial actions over many years. Subtitle D requires site care for a period of 40 years after a facility has closed operation. Sufficient funds must be accumulated in an account during the active site lifetime to cover these annual expenditures.

The calculation of the amount required to fund the post-closure care must account for inflation as well as the interest that will be earned by the account. The calculation requires that some assumptions be made about interest and inflation rates and also the amounts of the annual care payments. For simplicity we will assume that the same tasks are to be accomplished each year and the annual payments for these tasks will be constant at L_c when measured in dollars at the beginning of the first post-closure year. Thus, the expenditure at the end of the first year of care will have increased to $L_c(1 + f)$, where f is the annual rate of inflation expressed as a decimal. If an amount of money, a_1, is earmarked at the beginning of the care period to pay for the first year of care, by the time it is to be spent it will have earned interest at an annual rate i (expressed as a decimal) and increased to the value $a_1(1 + i)$. We can equate the cost and the value of the account to obtain

$$a_1(1 + i) = L_c(1 + f)$$

and then solve for a_1 to get

$$a_1 = L_c(1 + f)/(1 + i).$$

Thus, the factor $(1 + f)/(1 + i)$ accounts for the effect of inflation on L_c and the interest earned on the amount a_1.

Using the same reasoning, the amount of money, a_j, that must be on deposit to cover the cost of care in year j is $Lc[(1 + f)/(1 + i)]^j$. We conclude then that the amount of money that should be in an interest-bearing account at the beginning of the post-closure care period, A_{LTC}, to pay for LTC years of care is given by the sum

$$A_{LTC} = a_1 + a_2 + \ldots + a_{LTC}$$

or

$$A_{LTC} = L_c[(1 + f)/(1 + i)] + L_c[(1 + f)/(1 + i)]^2 + \ldots + L_c[(1 + f)/(1 + i)]^{LTC}$$

By factoring L_c from each term and letting $x = [(1 + f)/(1 + i)]$, the equation becomes

$$A_{LTC} = L_c x \, (1 + x + x^2 + \ldots + x^{LTC-1})$$

By using the series identity

$$1 + x + x^2 + \ldots + x^{N-1} = (x^N - 1)/(x - 1)$$

which is valid if the value of x is not 1 (i.e., the inflation rate is not equal to the interest rate), the equation can be written

$$A_{LTC} = L_c x \, [(x^{LTC} - 1)/(x - 1)]$$

If L_c is the annual cost of care as measured in dollars in the year at the beginning of the long-term care period and L is the comparable figure measured in dollars in the year when the facility operation begins, then

$$L_c = L \, (1 + f)^{SL}$$

By substituting this expression for L_c and $(1 + f)/(1 + i)$ for x in the equation for A_{LTC}, we get the formula for the amount that should be on deposit at the beginning of the long-term care period:

$$A_{LTC} = L(1 + f)^{SL}[(1 + f)/(1 + i)]$$
$$\{[(1 + f)/(1 + i)^{LTC} - 1]/[(1 + f)/(1 + i) - 1]\}$$

This formula cannot be used if the inflation and interest rates are equal.

The amount of the annual payments during the active site lifetime that produces a sum of A_{LTC} in SL years is calculated using the same annuity formula that we used for determining annual deposits during the active site life for site closure, which in this case becomes

$$P_{LTC} = A_{LTC} \, i/ \, [(1 + i)^{SL} - 1]$$

DISCUSSION TOPICS AND PROBLEMS

1. Why should landfills not be located near airports?
2. Using the assumptions and results from Example 9.1, estimate the landfill volume required to serve a region with 240,000 persons?
3. In the *Handbook of Solid Waste Management* (Robinson 1986), several "rules of thumb" are described for calculating area requirements

for a landfill. These include: (1) 2 yd^3/person/yr; (2) 2 acres/10,000 persons/yr (8 ft lift); and (3) 1.25 acre·ft/1000 persons/yr. Determine how consistent these rules of thumb are with each other and with the result found in Example 9.1.

4. A potential landfill site has an area of 60 ha, of which 48 ha is to be used for waste disposal. Using the guideline of 5 borings for the first 2 hectares (5 acres) or less and 2 additional borings for each additional hectare, what is the minimum number of borings required for this site?

5. The degree of compaction of refuse can be described in terms of compaction factors. The compaction factor is defined as the ratio of the compacted volume V_f to the initial uncompacted volume V_i. Typical compaction factors for wastes discarded in a landfill (Tchobanoglous, 1977) are listed in the table below.

Component	Compaction Factors (Normal Compaction)
Paper	0.2
Cardboard	0.25
Yard waste	0.25
Food	0.33
Glass	0.6
Ferrous metals	0.35
Tin cans	0.18
Nonferrous metals	0.18
Plastics	0.15
Wood	0.3
Rubber	0.3
Leather	0.3
Textiles	0.18
Dirt, ash, etc.	0.85

Calculate the density of mixed municipal wastes in the United States using the composition and density data from Tables 2.5 and 2.13 and the compaction factors given above. How does your result compare to the value assumed in Example 9.1 in this chapter; namely, 0.60 t/m^3 (0.5 ton/yd^3)?

6. Calculate the area requirements for a landfill to serve 10,000 people if all yard and food wastes are composted at a different location and 60% of the paper, 80% of the metals, 40% of the plastic, and 40% of the glass are recycled. Use the composition and density data from Tables 2.5 and 2.13 and the compaction factors from Problem 5 above.

7. Show that the dimensional units of each term in the equation

$$E/\rho g = 1/2 \; v^2/g + z + P/\rho g,$$

reduce to units of length.

8. Two water observation wells are installed in an unconfined aquifer. Well 1 has a hydraulic head of h_1 and Well 2, located a distance L from the first, has a hydraulic head of h_2. Develop an equation to calculate the hydraulic head at a location between the wells at a distance x from Well 1. Assume there is no infiltration or evaporation.

9. Three piezometers are installed in the same confined aquifer. The piezometers are 500 m apart, with piezometer A located due west of piezometer B and piezometer C located due south of piezometer B. The surface elevations of piezometers A, B, and C (above sea level) are 260m, 275m, and 300m, respectively, and the depths to water from the surface are 5m, 20m, and 25 m, respectively. (a) What is the direction of groundwater flow through the triangle ABC? (b) What is the value of the hydraulic gradient? (c) If the aquifer is 6 m thick and is composed of sand ($k = 2.5 \times 10^{-6}$ m/s), what is the volumetric flow rate of the groundwater water? (d) If the effective porosity $\eta = 0.28$, what is the seepage velocity of the water?

10. Design a scoring system to compare and rank the suitability of sites for development as a landfill. Identify appropriate criteria to evaluate a site. Are your criteria to be weighted, and if so, how?

11. What limitations and concerns pertain to the construction of a building on a completed landfill?

12. A newly constructed landfill is to receive 500,000 tonnes of waste annually for a period of 20 years. During the active life of the site, money is to be deposited into an account where it will earn 5% interest. Assume an inflation rate of 3% during both the active life of the site and the long-term care period. (a) Using the assumed inflation rate, calculate the closure cost in 20 years. (b) What annual deposits must be made at the end of each year in order to accumulate adequate funds to close the site? (c) What annual deposits must be made at the end of each year in order to provide adequate funds for post-closure care? (d) What tonnage charge must be levied for site closure and post-closure care?

13. Sixteen wells are arranged in a four well by four well square grid as shown in the diagram below. The well locations are identified by x-y coordinates, indicated in parentheses and referenced to the origin $(0,0)$ in the lower left corner. The hydraulic head is determined for each well and shown next to the well. Draw the contours of equal hydraulic head and the groundwater flow lines. If you have a computer program available for constructing contours from discrete data (e.g., the SURFER program from Golden Software), you may wish to use it for this problem

230m	216m	192m	184m
(0,300)	(100,300)	(200,300)	(300,300)
x	x	x	x
220m	212m	194m	180m
(0,200)	(100,200)	(200,200)	(300,200)
x	x	x	x
215m	206m	186m	175m
(0,100)	(100,100)	(200,100)	(300,100)
x	x	x	x
210m	200m	180m	170m
(0,0)	(100,0)	(200,0)	(300,0)
x	x	x	x

14. Prepare a computer spreadsheet showing the yearly growth of the account described in Example 9.3. This account receives annual deposits of $6,065.23 for a period of 20 years and pays interest at 4%.
15. Review and discuss the regulations pertaining to the siting and design of landfills in your state.

REFERENCES

Bagchi, A., *Design, Construction, and Monitoring of Sanitary Landfills*, John Wiley & Sons, New York, NY, 1994.

Bagchi, A., Design of natural attenuation landfills, *J. Environ. Eng. Div. (Am. Soc. Civ. Eng.)*, 109, No. EE4 800–811, 1983.

Brunner, D.R., and Keller, D.J., Sanitary Landfill Design and Operation, Environmental Protection Agency Publication SW-65ts, U.S. Government Printing Office, Washington, D.C., 1972.

Federal Register, Solid Waste Disposal Criteria; Final Rule, 40 CFR Parts 257 and 258, pp. 50978–51119, Oct. 9, 1991.

Fetter, C.W., Jr., *Applied Hydrology*, Third Edition, Macmillan College Publishing Co., New York, NY, 1994.

Flower, F.B., Leone, I.A., Gilman, E.F., and Arthur, J., A Study of Vegetation Problems Associated With Refuse Landfills, EPA 600/2-78-094, Washington, DC, U.S. Printing Office, 1978.

Gilman, E.F., Leone, I.A., and Flower, F.B., The adaptability of 19 woody species in vegetating a former sanitary landfill, *Forest Science*, 27(1), 13–18, March, 1981.

Golden Software, Inc., 807 14th Street, P.O. Box 281, Golden, CO 80402.

Lee, G.F. and Jones, R.A., Managed fermentation and leaching: An alternative to MSW landfills, *BioCycle*, 31(5), 78–80, 83, 1990.

Lin, H., New regs will spell doom for 600 sites in Wisconsin, *Waste Age*, 20(9), 52–54, 262, 1989.

National Center for Resource Recovery, *Sanitary Landfill*, Lexington Books [D.C. Heath and Co.], Lexington, MA, 1974.

Reinhardt, J.J. and Ham, R.K., Solid Waste Disposal on Land Without Cover, U.S. Environmental Protection Agency, NTIS Publication PB 234930, 1974.

Robinson, W.D., *The Solid Waste Handbook: A Practical Guide*, John Wiley & Sons, New York, NY, 1986.

Salimando, J., A future in older landfills, *Waste Age*, 20(8), 46–52, 1989.

Tchobanoglous, G., Theisen, H., and Eliassen, R., *Solid Wastes: Engineering Principles and Management Issues*, McGraw-Hill Book Company, New York, NY, 1977.

U.S. Congress, Office of Technology Assessment (USCOTA), Facing America's Trash: What Next for Municipal Solid Wastes?, OTA-O–424, U.S. Printing Office, Washington, DC, 1989.

U.S. Environmental Protection Agency, Report to Congress: Solid Waste Disposal in the United States, Vol. II, EPA/530-SW-88-011B, U.S. Printing Office, Washington, DC, 1988.

Wang, H.F. and Anderson, M.P., *Introduction to Groundwater Modelling—Finite Difference and Finite Element Methods*, W.H. Freeman, San Francisco, CA, 1982.

CHAPTER 10

Landfill Leachate and Gas Management

10.1 OVERVIEW

Leachate is the contaminated liquid that results from the percolation of water through a landfill. The origins of the liquid are rainfall, moisture, and other liquids contained in the wastes deposited in the landfill, and by-products of decomposition. The chemical and biological characteristics of the leachate are dependent upon the type of refuse deposited in the land-fill and the degree of decomposition. The quantity of leachate is determined by rainfall, temperature, humidity, surface runoff, underground water migration, and the age of the landfill.

Much of the leachate in a new landfill is retained by the refuse until field capacity is achieved. Field capacity is the maximum amount of water that the waste can retain against gravitational forces. In the early stages of the landfill the leachate produced is primarily liquid that has channeled its way through the wastes and surface water runoff from the disposal area. Later, when portions of the wastes are saturated and decomposing, the concentrations of chemical species and biological materials will increase markedly. Controlling leachate production and containing and collecting leachate are major considerations in designing a landfill. The collected leachate requires treatment, either on-site or off-site. There have been experiments in which the leachate has been recycled back onto the wastes in an attempt to accelerate the decomposition of the wastes.

If leachate leaves a landfill area it poses a threat to groundwater. A knowledge of the hydrogeology of a particular site is required in order to fully understand the nature of this threat. Hydrogeologic information must be available during the siting phase so that a specific location can be rejected if the potential hazard is too great. Once a site has been chosen, the landfill operation must be designed to minimize the possibility of groundwater contamination from leachate. Groundwater protection is accomplished by maintaining a sufficiently thick soil layer between the waste and the groundwater and installing a liner at the bottom of the landfill site with a full leachate collection system. Modern landfills are designed with monitoring wells and ongoing monitoring programs to detect contamination of groundwater and leachate movement resulting from failure in a liner.

Another consideration in designing a landfill is the need to control landfill gases produced by the anaerobic decomposition of the organic wastes. As described in Chapter 7, organic wastes are biodegradable and can decompose to produce carbon dioxide (CO_2), methane (CH_4), ammonia (NH_3), organic acids, water (H_2O) and other chemicals. In contrast, the inorganic materials are not subject to biological activity and do not decompose. The methane produced by anaerobic decomposition of organic materials is the most dangerous of the gases because of its explosive nature. Cases of methane migrating through the soil into building basements have been documented. Methane is also the gas with the greatest economic value. At some facilities, methane is collected and used as an energy source for heating or producing steam or electricity.

Recently, methane emissions have received attention because of their potential effect on global warming. Current estimates of methane emissions to the atmosphere from municipal landfills around the world vary considerably, but one source (Bingemer and Crutzen, 1987) places it in the range of 27–63 million tonnes (30–70 millions tons) annually. Because methane traps more infrared energy than carbon dioxide, it may be more desirable to flare the methane to convert it to carbon dioxide rather than vent it into the atmosphere (USCOTA, 1989).

There has been concern about other landfill gas constituents, primarily volatile organic compounds, many of which are potential carcinogens. The U.S. Environmental Protection Agency (USCOTA, 1989) estimates that 200,000 tonnes of volatile organic compounds (VOCs), excluding methane, are released annually. Table 10.1 lists the concentrations of volatile organic compounds measured at municipal solid waste landfills.

10.2 THE DECOMPOSITION OF LANDFILLED ORGANIC WASTES

The organic wastes in a landfill decompose by aerobic and anaerobic processes. These processes were described in Chapter 7 in terms of engi-

Table 10.1 Concentrations of Volatile Organic Constituents of Gases from Municipal Solid Waste Landfills

Constituent	Range of Concentration (ppm)	Median
Benzene	0–32	0.3
Carbon dioxide	342,000–470,000	350,000
Carbon monoxide	0.011	
1,2 Dichloroethane	19–59	
Ethylbenzene	0–91	1.5
Heptane	0–11	0.45
Hexane	0–31	0.8
Isopentane	0.05–4.5	2.0
Methane	440,000–587,000	500,000
Methyl–cyclohexane	0.017–19	3.6
Methyl–cyclopentane	0–12	2.8
Methylene chloride	0–118	0.83
Nonane	0–24	0.54
Perchloroethylene	0–186	0.03
Toluene	0–357	6.8
1,1,1-Trichloroethane	0–3.6	0.03
Trichloroethylene	0–44	0.12
Trichloromethane	0.61	
Vinyl chloride	0–10	2.2
Xylene	0–111	0. 1
m–Xylene	1.7–76	4. 1
O–Xylene	0–19	1.8

Sources: U.S. Environmental Protection Agency, Report for Congress: Solid Waste Disposal in the United States, Vol. II, EPA/530–SW–88–011B, Washington, DC, October 1988; Wood, J., and Porter, M., Hazardous pollutants in Class II landfills, *J. Air Pollution Control* 37(5), 609–615, May, 1987.

neered methods for biologically treating waste. In a landfill, the processes are largely uncontrolled.

The decomposition of landfilled wastes is characterized by three decomposition stages (Ham, 1979). In the initial stage (stage 1), the organic wastes decompose aerobically, with the production of carbon dioxide (CO_2), water, nitrates, and other products. The temperature of the waste rises. As the oxygen becomes depleted, facultative and anaerobic microorganisms predominate (stage 2), producing organic acids and carbon dioxide which lower the pH and, as a result, inorganic salts are solubilized. The rise in ionic conductivity is evidence that solubilization has taken place. Methane production is suppressed because the low pH conditions are toxic to the methane-producing organisms. After a time, the methane-producing microorganisms predominate (stage 3), degrading the organic acids. The pH rises and the specific conductance of the leachate

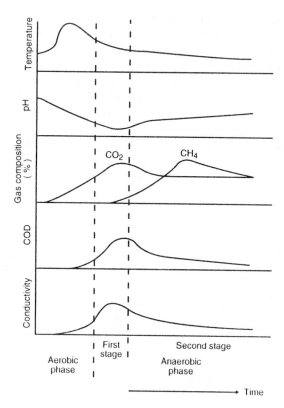

Figure 10.1 Variation of chemical parameters during the decomposition of landfilled municipal wastes (Ham, 1979).

decreases as a result of lower concentrations of organic acids and fewer
inorganic salts being solubilized. Figure 10.1 depicts the variation of the
chemical parameters for a hypothetical decomposition in a landfill.

The overall decomposition occurs over many years. In fact, recent
evidence indicates that the decomposition process proceeds even more
slowly than previously thought (Rathje et al., 1988). In dry conditions,
materials may remain unaltered for decades.

10.3 LEACHATE COMPOSITION AND CHARACTERISTICS

The characteristics of leachate are determined by the types of wastes
deposited in the landfill. If the wastes are primarily organic, as municipal
wastes typically are, then the decomposition proceeds in the manner described in the previous section. However, the leachate from an ash or
sludge monofill may be very different. Ash, for example, consists of a number of metal and nonmetal oxides. Many of the metal oxides, such as those

of sodium (Na), potassium (K), and calcium (Ca), are soluble in water. When dissolved in water, hydroxides are formed and the pH is raised.

The leachate in municipal waste landfills acquires significant BOD and COD concentrations as a result of the biological decomposition processes. There are also physical and chemical mechanisms at work, such as the leaching of matter by water passing through the wastes and dissolving of materials by the leachate. The overall result of all of the processes is a liquid with organic and inorganic constituents that may pose a threat to groundwater quality if not collected and treated. Table 10.2 lists median concentrations of substances found in municipal landfill leachate and existing exposure standards.

10.4 ESTIMATING LEACHATE QUANTITY

The quantity of leachate depends most directly upon the amount of precipitation falling on the site. If the site relies on natural attenuation for the leachate, as many older landfills do, an estimate is needed to assist the hydrologist in determining whether or not the groundwater system will be affected. If the site has a liner, then an estimate is important for determining the spacing and size of the leachate collection pipes at the base of the landfill and the capacity of the holding tanks.

A simple estimate of leachate quantity generated after a steady state condition is attained can be developed using a water balance approach. Basically, the amount of water that percolates into the wastes (i.e., the infiltration) is the difference between the amount of water received by the site from precipitation, surface water run-on, and water entering through the sides and bottom of the fill, and the water leaving the site from surface run-off, evaporation, and transpiration by plants. (Present design requirements should make the quantities of surface water run-on and water entering through the sides and bottom of the fill negligible.) The water loss by evaporation from the soil and the loss of water resulting from plant uptake are often described using the single term, evapotranspiration.

This water balance is summarized by the equation

$$L = P + R_{on} + U - E - R_{off}$$

where,

L = leachate (infiltration)

P = precipitation

R_{on} = run-on surface water

U = underflow of groundwater into the wastes

E = evapotranspiration

R_{off} = run-off surface water

Table 10.2 Median Concentrations of Constituents of Substances Found in Municipal Landfill Leachate and Existing Exposure Standards

Substance[a]	Mean Concentration (ppm)	Exposure Standards Type[b]	Value (ppm)
Inorganics			
Antimony (11)	4.52	T	0.01
Arsenic (72)	0.042	N	0.05
Barium (60)	0.853	N	1.0
Beryllium (6)	0.006	T	0.2
Cadmium (46)	0.022	N	0.01
Chromium (total)(97)	0.175	N	0.05
Copper (68)	0.168	T	0.012
		W	0.018
Cyanide (21)	0.063	T	0.7
Iron (120)	221	W	1000
Lead (73)	0.162	N	0.05
Manganese (103)	9.59	W	0.05
Mercury (19)	0.002	N	0.002
Nickel (98)	0.326	T	0.07
Nitrate (38)	1.88	W	10
Selenium (18)	0.012	N	0.01
Silver (19)	0.021	N	0.05
Thallium (11)	0.175	W	0.04
Zinc (114)	8.32	W	0.110
Organics			
Acrolein (1)	270	W	21
Benzene (35)	221	T	5
Bromomethane (1)	170	T	10
Carbon tetrachloride (2)	202	T	5
Chlorobenzene (12)	128	T	1000
Chloroform (8)	195	C	5.7
bis (Chloromethyl) ether (1)	250	C	0.0037
p–Cresol (10)	2394	T	2000
2,4–D (7)	129	T	100
4,4–DDT (16)	0.103	C	0.1
Di–n–butyl phthalate (5)	70.2	T	4000
1,2–Dichlorobenzene (8)	11.8	W	763
2,4–Dichlorobenzene (12)	13.2	T	75
Dichlorodifluoromethane (6)	237	T	7000
1,1–Dichloroethane (34)	1715	N	7
		C	7
1,2–Dichloroethane (6)	1841	T	5
1,2–Dichloropropane (12)	66.7	W	5700
1,3–Dichloropropane (2)	24	C	0.19
Diethyl phthalate (27)	118	T	30,000

Table 10.2 (Continued)

Substance[a]	Mean Concentration (ppm)	Exposure Standards Type[b]	Value (ppm)
Organics (continued)			
2,4–Dimethyl phenol (2)	19	W	2120
Dimethyl phthalate (2)	42.5	W	313,000
Endrin (3)	16.8	T	0.2
Ethyl benzene (41)	274	W	1400
bis (2–Ethylhexl)			
phthalate (10)	184	T	70
Isophorone (19)	1168	W	5200
Lindane (2)	0.020	T	4
Methylene chloride (68)	5352	C	4.7
Methyl ethyl ketone (24)	4151	W	2000
Naphthalene (23)	32.4	W	620
Nitrobenzene (3)	54.7	T	20
4–Nitrophenol (1)	17	W	150
Pentachlorophenol (3)	173	T	1000
Phenol (45)	2456	T	1000
1,1,2,2–Tetrachloroethane (10)	210	C	1.75
Tetrachloroethylene (18)	132	C	6.9
Toluene (69)	1016	T	10,000
Toxaphene (1)	1	N	5
1,1,1–Trichloroethane (20)	887	N	200
1,1,2–Trichloroethane (4)	378	C	6.1
Trichloroethylene (28)	187	N	5
		T	3.2
Trichlorofluoromethane (10)	56.1	T	10,000
1,2,3–Trichloropropane (1)	230	T	20
Vinyl chloride (10)	36.1	N	2

[a]Number of samples in parentheses.

[b]Types of exposure standards.
 C = U.S. EPA Human Health Criteria, based on carcinogenicity
 N = National Interim Primary or Secondary Drinking Water Standard
 T = U.S. EPA Human Health Criteria, based on systemic toxicity
 W = Water Quality Criteria

Sources: U.S. Congress Office of Technology Assessment (USCOTA), Facing America's Trash: What Next for Municipal Solid Waste?, OTA–O–424 U.S. Printing Office, Washington, DC, 1989. Data are from U.S. Environmental Protection Agency, Office of Solid Waste, Summary of Data on Municipal Solid Waste Landfill Leachate Characteristics: Criteria for Municipal Solid Waste Landfills (40 CFR Part 258), EPA/530–SW–88–038 U.S. Printing Office, Washington, DC, July, 1988,

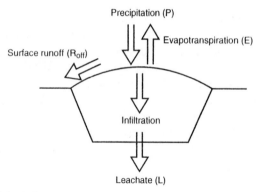

Figure 10.2 Water balance.

In a well-designed and operated landfill, surface water will be diverted from the wastes and R_{on} will be zero ($R_{on} = 0$). Also, for a landfill located above the water table with lined sides, there will be no underflow ($U = 0$). This equation can then be simplified to

$$L = P - E - R_{off}$$

These flows are represented in Figure 10.2. When applying this equation, the challenge is to obtain good estimates for the values of the runoff and evapotranspiration.

The amount of runoff depends upon the soil permeability, the slope of the surface, the type of vegetation, duration and frequency of precipitation, and whether the precipitation is in the form of rain or snow. The

Table 10.3 Runoff Coefficients for Various Slopes and Soil Permeabilities

Surface	Slope	Runoff Coefficient
Grass–sandy soil	Flat – 2%	0.05–0.10
	2 – 7%	0.10–0.15
	Over 7%	0.15–0.20
Grass–heavy soil	Flat – 2%	0.13–0.17
	2 – 7%	0.17–0.25
	Over 7%	0.25–0.35

Source: Fenn, D.G., Hanley, K.J., and Degeare, T.V., Use of the Water Balance Method for Predicting Leachate Generation from Solid Waste Disposal Sites, EPA publication EPA–530/SW–168, U.S. Printing Office, Washington, DC, 1975.

fraction of precipitation that becomes runoff is expressed by the runoff coefficient. Table 10.3 shows that the fraction of precipitation that becomes runoff is in the range 0.05 to 0.35.

Bagchi (1994) discusses several empirical and theoretical methods for determining the value of the evapotranspiration term. In applying the water-balance method for predicting leachate at landfills, Fenn et al. (1975), use the Thornwaite equation and tables (Thornwaite and Mather, 1957) for estimating evapotranspiration.

Example 10.1 provides a sample calculation of leachate quantity with some modest assumptions.

Example 10.1: Estimate the annual quantity of leachate per hectare for a site located in a region with a temperate climate where the average rainfall is 0.75 m/yr (30 inches/yr) and evapotranspiration is about 50%. Assume that the wastes are covered so that the run-off from the site is 25% and that there is no run-on surface water nor underflow of groundwater into the wastes (i.e., $R_{on} = 0$ and $U = 0$).

Solution: The problem gives precipitation in terms of a depth, but the depth can be interpreted in terms of a volume, because this is the depth of water received over an area. The precipitation is then 0.75 hectare·meter. The quantity of leachate can be calculated using the equation,

$$L = P - E - R_{off}$$

Substituting in the equation we obtain

$$L = 0.75 \text{ ha·m} - (0.5) \cdot (0.75 \text{ ha·m}) - (0.25) \cdot (0.75 \text{ ha·m})$$

or

$$L = 0.19 \text{ ha·m } (1.5 \text{ acre·ft})$$

The calculation of the equivalent number of liters (and gallons) is left as an exercise for the reader.

The explicit assumption behind the water balance equation described above is that the moisture content of the waste is not changing. If the moisture content is changing, then a term must be included to account for the fact that the wastes will absorb water for some period of time without releasing leachate. It is estimated (Fenn et al., 1975) that the moisture content of MSW at saturation is 20–30% by volume. There is also an implicit assumption that the waste is uniformly saturated. In reality, however, water often channels through the waste, leaving some portions dry while others are saturated.

Computer modeling is often used to calculate water balances. One such model is the Hydrologic Evaluation of Landfill Performance (HELP) Model, developed by the USEPA and the Army Corps of Engineers Waterways Experiment Station (USEPA, 1984). The model calculates runoff, evapotranspiration, and infiltration using temperature and precipitation records and simulates the movement of water into and through various cover layers and liner designs.

10.5 LEACHATE MOVEMENT

Leachate that escapes from a landfill and reaches groundwater, either by design or by accident, is not diluted by the entire body of groundwater, but forms a plume as depicted in Figure 10.3. The plume travels in the direction of the groundwater flow. Some of the constituents of the plume move at a rate which is greater than the average groundwater velocity, with some diffusion and dispersion occurring at the boundaries of the plume. Because the leachate moves primarily with the groundwater, any contamination of groundwater by leachate should be detected in down-gradient wells but not in the up-gradient wells. This is the basis for regulatory agencies requiring more wells in down-gradient than up-gradient locations.

In addition to dilution of the leachate with the groundwater, several other mechanisms reduce the concentrations of the constituents in the plume. The soil provides filtration of suspended solids and pathogens. The effectiveness of the filtration depends upon the type of soil and its thickness. Microbial activity by microorganisms in the soil may utilize some of the organic and inorganic contaminants. Some ions, such as heavy metal ions, may precipitate as the pH conditions change in the plume. Both anions and cations may be adsorbed on soil particles. Also,

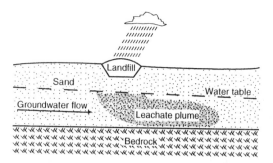

Figure 10.3 The flow of leachate for simple subsurface conditions.

ionic exchange reactions may occur in which the ions in the leachate replace another type of ion in a mineral structure. Exchange reactions are most significant in clay soils. Chloride, nitrate, bicarbonate, and sulfate ions are highly soluble and are not attenuated by any of the methods just described. The attenuation of these ions is by dilution with the groundwater. Table 10.4 summarizes the mobilities and attenuation mechanisms in clay of some of the pollutants found in leachate.

10.6 LEACHATE MONITORING

An effective groundwater monitoring program requires a sufficient number of sampling wells, installed at appropriate locations and depths, to provide samples from the aquifers underlying the landfill site; a schedule for periodically retrieving samples from the wells; and sample analyses by a competent staff at an analytical laboratory. The wells should be distributed to include up-gradient sites for establishing background water quality and down-gradient sites for sampling groundwater that has passed beneath the landfill. Well clusters, called nests, are also installed. A nest is a set of wells at a single location in which the individual wells are drilled to varying depths.

Figure 10.4 illustrates the construction of a sampling well. The well is constructed with a casing. After the casing is installed, the space around the casing is sealed with a low permeability material to prevent surface water from entering the well. When a well is sampled, the stagnant water in the well is removed by pumping or bailing. After the well recharges, a sample is retrieved for analysis.

The water testing schedule should include several rounds of sampling, typically monthly or quarterly, prior to the opening of a landfill to provide baseline values for the chemical parameters at the site. The water testing is continued during the active lifetime of the site and the post-closure care period.

The chemical parameters that are typically monitored in municipal solid waste landfills include the metals and volatile organic compounds (VOCs) designated by the USEPA in the Subtitle D regulations. Additional parameters may be specified by state regulatory agencies. The three major categories contained in Subtitle D are Indicator, Public Welfare, and Public Health Parameters. A complete listing of the parameters is presented in Table 10.5.

The Indicator Parameters include chemical species whose presence do not pose a health hazard and are easily detected with various analytical methods. A major change in concentration of an indicator chemical over time may indicate a problem in controlling leachate at a landfill. The presence of a Public Welfare Parameter in moderate concentrations is ob-

Table 10.4 The Mobilities and Attenuation Mechanisms of Selected Components of Leachate

Leachate Constituent	Attenuation Mechanism	Mobility in Clay
Aluminum	Precipitation	Low
Ammonium	Exchange, biological uptake	Moderate
Arsenic	Precipitation, adsorption	Moderate
Barium	Adsorption, exchange, precipitation	Low
Beryllium	Precipitation, exchange	Low
Boron	Adsorption, precipitation	High
Cadmium	Precipitation, adsorption,	Moderate
Calcium	Precipitation, exchange	High
Chemical oxygen demand (COD)	Biological uptake, filtration	Moderate
Chloride	Dilution	High
Chromium	Precipitation, exchange, adsorption	Low (Cr^{+3}) High (Cr^{+6})
Copper	Adsorption, exchange, precipitation	Low
Cyanide	Adsorption	High
Fluoride	Precipitation, exchange, adsorption	High
Iron	Precipitation, exchange	Moderate
Lead	Precipitation, exchange	Low
Magnesium	Exchange, precipitation	Moderate
Manganese	Precipitation, exchange	High
Mercury	Adsorption, precipitation	High
Nickel	Adsorption, precipitation	Moderate
Nitrate	Biological uptake, dilution	High
Potassium	Adsorption, exchange	Moderate
Selenium	Adsorption, exchange	Moderate
Silica	Precipitation	Moderate
Sodium	Exchange	Low to high
Sulfate	Exchange, dilution	High
Virus	Unknown	Low
Volatile organic compounds	Biological uptake, dilution	Moderate
Zinc	Exchange, adsorption, precipitation	Low

Source: Bagchi, A., Natural attenuation mechanisms of landfill leachate and effects of various factors on the mechanisms, *Waste Management and Research*, Denmark, 1987. Reprinted with permission.

jectionable because of the associated color and odors, but it generally does not pose a serious problem to human health. The presence of a Public Health Parameter poses a danger to human health. These species include heavy metals and carcinogenic organic compounds. The acceptable

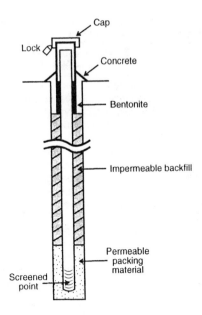

Figure 10.4 Water sampling (observation) well.

methods for handling, storing, and analyzing the samples are prescribed in the report, "Test Methods for Evaluating Solid Waste" (USEPA, 1986). A description of each of the parameters and methods of analysis is given in Appendix I.

If the results of the chemical analyses show that there is a statistically significant difference between the up-gradient and down-gradient wells for one or more of the chemical species or that there are significant changes in concentrations of a species at a particular well, then an assessment of the potential impact on the groundwater is made. If the results indicate that the landfill is leaking and poses a threat to groundwater quality, a remedial action plan may be implemented.

Lysimeters are devices used to detect leakage from landfill base liners. Two basic types of lysimeters are available: basin lysimeters and suction lysimeters. A basin lysimeter is installed below a landfill liner to collect leachate. Bagchi (1994) suggests that if one basin lysimeter is installed per phase in a landfill, the lysimeter should be installed below a crest of the base liner. If two are installed, the second one should be placed at a location midway between the crest of the liner and the collection trench. Figure 10.5 contains a diagram showing the construction of a basin lysimeter.

Suction lysimeters are collection vessels which are also installed below a landfill liner. The construction of a typical suction lysimeter is

Table 10.5 Chemical Parameters Typically Monitored at Municipal Solid Waste Landfills and Methods of Analysis

Parameter	Suggested EPA Methods of Analysis[a]
Indicator Parameters	
Alkalinity, filtered	
COD, filtered	
Field conductivity (@25°C)	
Total hardness, filtered	
Field pH	
Field temperature	
Groundwater elevation	
Public Welfare Parameters	
Chloride	9250, 9251, 9252
Copper	6010, 7210, 7211
Iron	6010, 7380, 7381
Manganese	6010, 7460
Sulfate	9035, 9036, 9038
Zinc	6010, 7950, 7951
Public Health Parameters	
Arsenic	6010, 7061
Antimony	6010, 7040, 7041
Barium	6010, 7080, 7081
Beryllium	6010, 7090, 7091
Cadmium	6010, 7130, 7131
Cobalt	6010, 7200, 7201
Chromium	6010, 7190, 7191
Fluoride	
Lead	6010, 7420, 7421
Mercury	7470
Nickel	6010, 7520
Nitrate + nitrite (as N)	9200
Selenium	6010, 7740, 7741
Silver	6010, 7760,
Thallium	6010, 7840, 7841
Vanadium	6010, 7910, 7911
Volatile Organic Compounds (VOCs)	8021 or 8260

[a]See Appendix I for description of methods.

Source: U.S. Environmental Protection Agency, Test Methods for Evaluating Solid Waste, Report SW–846, U.S. Printing Office, Washington, DC, 1986.

shown in Figure 10.6. The basic principle of operation is that a vacuum is applied to the collection chamber to draw liquid through the porous walls of the chamber. The liquid in the lysimeter is retrieved by removing the vacuum and applying air pressure (above atmospheric pressure), thereby

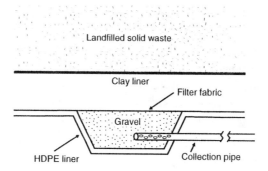

Figure 10.5 Basin lysimeter.

forcing the liquid up through the other tube (shown on the right in Figure 10.6) to a collection container.

In addition to water sampling instruments, there are instruments which serve as indirect leakage monitors. These devices are used in situ to detect changes in the moisture content or in the chemical composition of the moisture in the unsaturated zone above the water table (vadose zone). One such device is a conductivity meter that measures electrical conductivity between electrodes installed in the soil around the site perimeter. Increases in the conductivity of the soil occur when there is leachate leakage because of an increase in the concentration of dissolved ions.

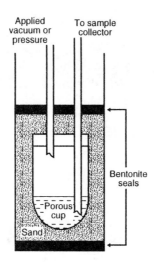

Figure 10.6 Suction lysimeter.

10.7 LEACHATE TREATMENT

The most common method of treating leachate collected from municipal landfills is to use conventional wastewater treatment processes. Typically the leachate is hauled to a local municipal wastewater treatment plant where it is diluted by the large quantity of municipal wastewater and treated biologically and chemically as described in Chapter 6. Large municipal landfills and landfills designed for hazardous wastes may have on-site treatment or pretreatment facilities. The on-site treatment facilities may employ biological (aerobic and anaerobic stabilization), chemical (neutralization, precipitation, coagulation, etc.), physical (sedimentation or reverse osmosis) processes, or various combinations of these technologies.

An alternative approach to leachate management is to recirculate leachate back into the landfill. This procedure is attractive for several reasons: (1) recirculation generally costs less than hauling and treating the leachate off-site; (2) the volume of leachate may be reduced by evaporation or evapotranspiration; and (3) the added moisture accelerates the decomposition of the waste, thereby reducing the long-term care costs. Lee and Jones-Lee (1994a) point out that the landfill leachate recycle process has potential limitations and problems. A major problem is the retention of hazardous and otherwise deleterious organic and nonorganic substances in the landfill. At some future time, when the waste is considered "stabilized" and leachate is no longer recycled, water will eventually enter the landfill, even a Subtitle D-type "dry-tomb" landfill. This leachate will eventually breach a liner system and possibly pollute the groundwater. Lee and Jones-Lee (1994b) believe that the recirculation of untreated leachate should be allowed only at landfills where potential groundwater pollution by leachate is considered to be of little consequence or at landfills constructed with a double-composite liner in which the lower liner is used as a lysimeter under the entire area of the landfill. The lower liner serves as a leak detection system for the upper liner.

10.8 LANDFILL GAS PRODUCTION

Landfill gas production depends upon the organic content of the refuse, moisture, pH, and the age of the landfill. Methane production typically begins anytime from several months to a year or two after deposition. At standard temperature and pressure (STP), the theoretical volume of methane generated from organic materials in a MSW landfill is about 0.25 m^3/kg (4 ft^3/lb) of organic waste. The actual amount generated is about one-quarter of this. Example 10.2 illustrates the method for calculating gas quantities from the anaerobic decomposition of organic wastes.

Example 10.2: Calculate the volume of methane produced by the decomposition of MSW having the composition of the discarded MSW in the United States in 1990 (from Table 2.5) and a typical moisture content indicated below. The rightmost column shows the dry weight of solids in 1 kg of MSW.

	Composition % by wt.	Moisture (%)	Dry Solids (g)
Inert Materials			
Glass	6.7	2	66
Ferrous metals	6.3	3	61
Nonferrous metals	2.0	2	20
Plastics	8.3	2	81
Rubber and leather	2.4	2	24
Misc	3.0	8	28
Decomposable Materials (DM)			
Paper	37.5	6	353
Yard waste	17.9	60	72
Food waste	6.7	70	20
Wood	6.3	20	50
Textiles	2.9	10	26

Solution: The total amount or dry decomposable material (DM) in 1 kg of MSW is determined by adding the values in the "Dry Solids" column for paper, yard waste, food waste, wood, and textiles. The result is 521 g of dry DM.

An empirical formula for the dry DM can be calculated from elemental analysis data for representative types of the paper, yard waste, etc. found in Appendix H and the relative amounts of each in the DM. A representative formula is $C_{59.4} H_{94.4} O_{42.5} N$.

Assume that complete decomposition occurs by anaerobic processes and proceeds according to the following equation (from page 223):

$$C_a H_b O_c N_d + [(4a - b - 2c + 3d)/4] H_2O \rightarrow [(4a + b - 2c - 3d)/8]CH_4 + [(4a - b + 2c + 3d)/8]CO_2 + dNH_3$$

Substituting a = 59.5, b = 94.4, c = 42.5, and d = 1, the equation becomes

$$C_{59.5} H_{94.4} O_{42.5} N + 15.4 H_2O \rightarrow 30.6 CH_4 + 29.0 CO_2 + NH_3$$

Starting with 521 g of dry decomposable material (DM), the volume of the methane produced is calculated by first calculating the number of moles of methane produced, then converting from moles of the gas to volume at STP using the fact that a gas occupies a volume of 22.4 L at STP.

$$\text{Volume of CH}_4 = 521\,\text{g of DM} \times \frac{1\,\text{mole DM}}{1502\,\text{g DM}} \times \frac{30.6\,\text{mole CH}_4}{1\,\text{mole DM}} \times \frac{22.4\,\text{L CH}_4}{1\,\text{mole CH}_4}$$

$$\text{Volume of CH}_4 = 238\,\text{L} = 0.238\,\text{m}^3\ (3.8\ \text{ft}^3)$$

This calculation optimistically assumes that all the DM does actually decompose. But in real landfills, not all materials are provided with sufficient moisture for the microbial activity. Furthermore, the degradation of the various components of the DM progresses at very different rates. Consequently the net result is that the actual quantity of methane produced is less than the value calculated here.

10.9 GAS MOVEMENT AND CONTROL

As discussed in the previous section, the quantity and composition of landfill gases, predominantly methane and carbon dioxide, depend upon the conditions that affect the decomposition process; primarily, temperature, moisture, and pH conditions within the waste. In general, the movement of the gases depends upon diffusion, pressure gradients, and gas density. Carbon dioxide is more dense than air and tends to sink in a landfill, while methane is less dense than air and tends to rise. When an impermeable cover material is placed over the wastes to prevent water infiltration, or if during the winter season in cold climates the ground is frozen, the methane cannot escape upward. Therefore it is forced to migrate laterally.

Changes in the barometric pressure have been observed to affect the rate of gas movement. When the barometric pressure drops, the pressure difference between the gas in the landfill and the atmosphere increases, leading to an increased movement of the gas from the landfill into the surrounding soil and to gas vents. When the barometric pressure rises, gas flow is suppressed.

Passive or active venting systems can be used to intercept and control landfill gas. In passive systems, gas vents, or gravel-filled cut-off trenches provide a route for gases to escape into the atmosphere. A gas vent is typically a vertical pipe, the base of which is located in the fill, surrounded by gravel. The vents may be isolated or several may be connected together to a main header pipe. At some landfills the collected gases are burned. Bagchi (1994) suggests that one vent per 7,500 m³ (approx. 10,000 yd³) of wastes is probably sufficient. Because of increasingly stringent air emission regulations, passive gas venting systems are generally not acceptable for large MSW landfills. Large landfills require active gas extraction systems, coupled with gas treatment.

In active systems, blowers or compressors are used to reduce the gas pressure in a gas collection system to a level below atmospheric pressure in effect sucking the gas into a collection system. The gas collection system may consist of a group of gas vents, much like those used in a passive system, connected to a main header or perforated PVC pipes installed in a cut-off trench. The collected gases may be flared or burned for energy recovery. Of the more than 1,500 municipal solid waste landfills that vent or collect methane, only 123 actually collect methane to recover energy (USCOTA 1989).

10.10 GAS MONITORING

Gas monitoring traditionally implies monitoring of methane. The presence of volatile organic compounds described in the overview of this chapter and listed in Table 10.1 is a recent concern that cannot be ignored. These gases also need to be monitored. To adequately monitor gases it is important to install monitoring wells both in the landfill site and around the perimeter of a landfill. The well pipes are slotted or screened at the bottom, surrounded by gravel and sealed and capped as shown in Figure 10.7.

Periodically, the gases in the wells are monitored by drawing a sample into a measuring instrument or sorbant tube. In a liner containment type MSW landfill designed for venting or collecting methane, the amount of methane or pollutant gases detected in perimeter wells should be negligible. On the other hand, the methane concentrations measured in gas wells inside the fill site are expected to be high. These wells will also have measurable quantities of the other volatile gases listed in Table 10.1.

10.11 GAS UTILIZATION

A landfill may produce sufficient methane to warrant recovery. In the past, the rule of thumb was that only a landfill having over 900,000 tonnes (one million tons) of wastes in place could achieve efficient gas recovery; however, recent experience indicates that gas utilization may be economical for smaller landfills. The interest in gas recovery is evidenced by the growing number of landfills with gas recovery capabilities. As of 1989, there were 87 existing and 68 planned landfill gas facilities, whereas the total number of existing and planned facilities was 16 in 1982, 75 in 1984, and 138 in 1986 (Berenyi, 1989).

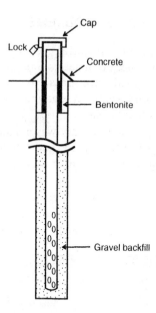

Figure 10.7 Gas monitoring well.

The energy content of landfill gas is about 1.86×10^6 J/m^3 (500 BTU/ft^3), which is about half the energy content of pipeline-quality natural gas. If the landfill gas can be burned directly in a boiler, then minimal treatment is required; usually it is sufficient to remove the moisture. The removal of moisture can be accomplished by condensing the water vapor by passing the gas through a cold compartment (condenser or chiller). Another method of removing the water vapor is to pass the gas through a dehydrating chemical compound. If the landfill gas is to be upgraded to pipeline-quality natural gas or is to be used to generate electricity by using the gas as fuel in an internal combustion engine connected to a generator, then in addition to moisture, some of the carbon dioxide must also be removed. Carbon dioxide can be removed with the aid of a molecular sieve. A molecular sieve contains material with pores of a sufficiently small size that allow the methane molecules to pass through, but not the larger carbon dioxide molecules.

If a methane source generates 1000 m^3 (35.3×10^3 ft^3) of gas per day, the fuel content is sufficient for a turbine-generator to produce an equivalent of 45–55 kilowatts of electrical power. The economic feasibility of energy recovery from landfill gas depends upon matching the quantity and quality of the gas to the needs of an accessible user.

DISCUSSION TOPICS AND PROBLEMS

1. During a storm, a rainfall measuring 5 cm (2 in.) falls on a 40 hectare (98.9 acres) landfill site. (a) What volume of water falls on the site? (b) If the site is capped with grass-covered clay having slope of 2–7% and the evapotranspiration is assumed to be 50%, what volume of water percolates into the site?

2. The chemical composition of a dry waste material is 47.50% carbon, 6.28% hydrogen, 45.29% oxygen, and 0.93% nitrogen. Derive the empirical chemical formula for this material that is stated and used in Example 10.2.

3. Describe the methods for treating leachate.

4. How does climate affect the quantity of leachate produced?

5. Probes to measure the electrical conductivity of soil are often installed around the perimeter of a landfill. What is the purpose of soil conductivity measurements?

6. Show that the equation describing anaerobic decomposition

$$C_aH_bO_cN_d + [(4a - b - 2c + 3d)/4] H_2O \rightarrow$$
$$[(4a + b - 2c - 3d)/8]CH_4 + [(4a - b + 2c + 3d)/8]CO_2 + dNH_3$$

is a balanced chemical equation.

7. The actual volume of methane generated in a landfill is about 25% of the amount predicted if all organic material, including paper, wood, food, and yard wastes, biodegrades. One of the observations from Project Garbage (see Section 1.3) is that only the food and yard wastes biodegrade at an appreciable rate. Does this fact account for the decreased methane generation? Justify your answer with appropriate calculations.

8. (a) What can be done to accelerate the decomposition process in a landfill? (b) What are the advantages of accelerating the decomposition process in a landfill? (Note: In answering this question you should refer to Chapter 7.)

9. What environmental factors and waste characteristics determine leachate composition?

10. Review the regulations pertaining to the monitoring requirements for landfill leachate and gases in your state. Do the regulations specify how many monitoring wells must be installed at a site? If so, how many monitoring wells are required for a 50 hectare (124 acre) site? How often must the wells be sampled? Must the samples be analyzed at certified laboratories?

REFERENCES

Bagchi, A., Natural attenuation mechanisms of landfill leachate and effects of various factors on the mechanisms, *Waste Management and Research*, Denmark, 1987.

Bagchi, A., *Design, Construction, and Monitoring of Sanitary Landfill*, John Wiley & Sons, New York, NY, 1994.

Berenyi, E., Landfill gas has a future, *Waste Age*, 20(8), 173–176, August, 1989.

Bingemer, H., and Crutzen, P., The production of methane from solid wastes, *J. Geophysical Research*, 92(D2), 2181–2187, 1987.

Fenn, D.G., Hanley, K.J., Degeare, T.V., Use of the Water Balance Method for Predicting Leachate Generation from Solid Waste Disposal Sites, EPA publication EPA-530/SW-168 U.S. Printing Office, Washington, DC, 1975.

Ham, R.K., Recovery, Processing, and Utilization of Gas from Sanitary Landfills, EPA publication EPA-600/2-79-001 U.S. Printing Office, Washington, DC, 1979.

Kennedy. M.D. and York, R.J. Leachate treatment: No shortage of options, *Waste Age*, 18(5), 66–70, 1987.

Lee, G.F. and Jones-Lee, A., Leachate recycle process offers pros and cons, *World Wastes*, 37(8), 16,19, 1994a.

Lee, G.F. and Jones-Lee, A., MSW landfill leachate recycle and groundwater quality protection, preprint submitted to *Waste Management and Research*, June, 1994b.

O'Leary, P. and Tansel, B., Landfill gas movement, control and uses, *Waste Age*, 17(4), 104–110, 1986.

Rathje, W.L., Hughes, W.W., Archer, G., and Wilson, D.C., Source Reduction and Landfill Myths, paper presented at ASTSWMO National Solid Waste Forum on Integrated Municipal Waste Management, Lake Buena Vista, FL, July 17–20, 1988.

Robinson. H.D. and Maris, P.J., The treatment of leachate from domestic wastes in landfill sites, *Journal of Water Pollution Control Federation*, 57(1), 30–37, 1985.

Stone, R., Reclamation of landfill methane and control of off-site migration hazards, *Solid Waste Management*, 21(7), 52–69, 1978.

Tchobanoglous G., Theisen, H., and Eliassen, R., *Solid Wastes: Engineering Principles and Management Issues*, McGraw Hill Book Company, New York, NY, 1977.

Thornwaite, C.W. and Mather, J.R., Instructions and Tables for Computing Potential Evapotranspiration and the Water Balance, Publ. Climatol. Lab., Drexel Institute of Technology, Centerton, NJ, 1957.

U.S. Congress Office of Technology Assessment (USCOTA), Facing America's Trash: What Next for Municipal Solid Waste?, OTA-O-424, U.S. Printing Office, Washington, DC, 1989.

U.S. Environmental Protection Agency (USEPA), Decomposition of Residential and Light Commercial Solid Waste in Test Lysimeters, U.S. Printing Office, Washington, DC, 1980.

U.S. Environmental Protection Agency (USEPA), The Hydrologic Evaluation of Landfill Performance (HELP) Model: User's Guide for Version I, Publication EPA/530/SW-84-009, U.S. Printing Office, Washington, DC, 1984.

U.S. Environmental Protection Agency (USEPA), Test Methods for Evaluating Solid Waste, Report SW-846, U.S. Printing Office, Washington, DC, 1986.

Waste Age, Can small modular plants expand landfill gas horizons?, 18(3), 87–88, 1987.

Wegner. R.W. and Lekstutis, P.E., LFG system design for the long term, *Waste Age*, 20(4), 188–195, 1986.

CHAPTER 11

Application of Wastewater and Biosolids to Land

11.1 OVERVIEW

Municipally-owned wastewater treatment plants in the United States treat over 29 trillion liters of wastewater annually, producing approximately 7.7 million tonnes of biosolids (*U.S. Federal Register*, 1989). The disposal and/or use of biosolids is described in Table 11.1. As the population increases and advanced treatment equipment is installed, the amount of solids will increase, perhaps even double by the year 2000.

Landfilling and incineration have been acceptable sludge disposal methods. Ocean dumping has been used since the 1920s by many coastal cities, but this practice was banned in the United States by 1992 (Lipták, 1991). Landfilling and incineration are widely used, but the costs of these disposal methods have risen significantly in recent years. Furthermore, landfills are often limited in the quantity of sludge that can be accepted because of its high moisture content. The limits may be imposed not only by regulatory agencies, but also by practical difficulties in handling a mixture of other wastes with sludge in a landfill. Given some of the drawbacks of incineration and landfilling as waste disposal methods, there is an interest in processing sewage sludge using methods such as anaerobic treatment and composting (discussed in Chapter 7) so the stabilized product can be benefically used on land as a soil amendment.

There is also interest in applying wastewater to land as an alternative to some treatment steps. For example, while tertiary treatment—typically an expensive process—is needed to remove nitrogen and phosphorous from treated water before it can be discharged to rivers or other surface waters, these same elements are valuable nutrients for crops. Thus, savings may be realized by not removing the nitrogen and phosphorous from

Table 11.1 Amount of Municipal Sewage Sludge Generated by Size of Treatment Plant and Disposed by Various Uses or Disposal Practices[a]

Use/Disposal Practice	Quantity of Dry Sludge (in thousands of dry tonnes/year)					
	Size of Plant (ML/d)					
	0<0.76	0.76<3.8	3.8<38	38<228	>228	Total
Land application	23.1	103.0	438.3	321.5	316.2	1,202.2
Distribution and marketing	0.1	4.0	36.3	97.5	567.5	705.5
Municipal landfills	56.5	278.6	1043.1	899.5	884.7	3,162.3
Monofills	0.1	2.8	25.5	44.5	28.5	101.4
Surface disposal	27.6	33.5	40.9	79.9	15.5	197.5
Incineration	0	0.6	94.1	383.6	1,173.1	1,651.4
Ocean disposal	0	0.3	1.2	35.4	387.4	424.4
Other	37.8	45.8	56.0	109.3	21.1	270.0
Total	145.3	468.6	1,735.5	1,971.2	3,394.0	7,713.6

[a]Assumptions: the amount of sewage sludge generated is identical to the amount of sewage sludge disposed of, and that a facility uses a single practice to dispose of its sludge.

Source: *U.S. Federal Register*, 54(23), 5745, U.S. Printing Office, Washington, DC, Feb. 6, 1989.

wastewater, while simultaneously reducing the amounts of commercial fertilizer which need to be applied.

The desirability of applying wastewater and sludge to land was emphasized in the Clean Water Act of 1977, where under Title II (Grants for the Construction of Treatment Works) the Act stated:

> Administrator shall encourage waste treatment management which results in the construction of revenue producing facilities providing for (1) the recycling of potential sewage pollutants through the production of agriculture, silviculture, or aquaculture products, or any combination thereof . . .

There are a number of factors that need to be addressed in order to apply wastewater or biosolids to land in an acceptable manner. These include: the amount of heavy metals present, the amount of wastewater or sludge the land is capable of accepting, when or how often applications can be made, and the type of equipment used to spread the wastewater or sludge.

With the curtailment of alternative forms of waste disposal, it appears that land application of wastewater and wastewater sludges is likely to increase in the future. The recently issued USEPA 503 regulations (*U.S. Federal Register*, Feb. 19, 1993) encourage wastewater and sludge treatment technologies that promote land application as a final end use of sewage sludge and provide guidance to communities interested in implementing such programs.

11.2 WASTEWATER TO LAND

Under the right circumstances, land application of wastewater can be a more effective and less energy-intensive treatment alternative than conventional systems such as activated sludge processes, trickling filters, and aerated lagoons. However, transportation distances to application sites, land area of application sites, soil types, and public attitudes may be limiting factors in choosing land application as a method of wastewater treatment. Land application is generally not suitable in communities with large urban populations. In these settings, suitable sites for land application of wastewater within a reasonable distance from the wastewater source are usually limited. An appropriate soil environment is essential for removal of wastewater nutrients and other constituents. Climate also plays a key role. The preferred land sites are usually located in arid or semi-arid areas where evaporation and evapotranspiration rates exceed the natural precipitation rates. However, with an appropriately chosen site and a suitable engineering design, it is possible to achieve effective wastewater treatment and a high quality effluent in almost any region regardless of climatic conditions.

In addition to soil and climate considerations, other factors affecting the suitability of land application include the nature of the wastewater itself, types of crops grown in the application area, and preapplication treatment. Preapplication treatment is especially important and must accomplish the following:

1. Pathogenic microbes must be destroyed or inactivated to an acceptable level in order to minimize health risks to workers and grazing animals and deleterious impacts to the human food chain.
2. Operational problems resulting from blockages of spray nozzles and distribution pipes must be minimized.
3. Clogging of soil pores must be minimized.
4. Nuisance conditions must be avoided.

Pretreatment usually includes screening and other handling equivalent to conventional primary treatment as described in Chapter 6. Depending on the site and use of the land (for example, raising crops from which food products are consumed directly by humans), secondary or even tertiary treatment may be necessary. Treatment may also include disinfection and, if nitrate contamination is a problem, ammonia stripping, or some form of nitrate removal through vegetative uptake or denitrification may be required.

Because of potential cost savings, especially for small communities or generators of certain industrial wastewaters, a variety of land treatment methods have been studied and put into full-scale operation in the past two decades. These may be conveniently categorized as follows: (1) Slow Rate (SR), sometimes referred to as irrigation, (2) Rapid Infiltration (RI), (3) Overland Flow (OF), and (4) Constructed Wetlands.

The choice of a land treatment process depends on treatment objectives and soil conditions. If it has a good design, a land treatment system can yield a high-quality effluent, provided BOD loadings are much lower than the BOD removal capacity of the system itself. The SR, RI, and OF processes are distinctly different in their design and operation. Performance data and other descriptive information in Tables 11.2 and 11.3 indicate they are capable of effective BOD, nitrogen, and phosphorus removal. The SR and RI processes are capable of reducing fecal coliform bacteria to quite low levels.

It is also possible to combine these processes. For example, if the rapid infiltration process precedes the slow rate (irrigation) process, the recovered water will likely meet the restrictive requirements for use on food crops. Alternatively the RI process could be used after overland flow (OF) to further increase the quality of renovated water by reducing BOD, suspended solids and phosphorus to even lower levels. This approach would also allow for higher rates of wastewater application in the OF stage of treatment.

11.3 SLOW RATE (IRRIGATION) PROCESS

Slow rate (SR) systems are typically designed for the controlled application of wastewater to vegetated land at a rate of a few centimeters per week. When determining the rate, emphasis is placed on the moisture needs of the plants. In effect, most SR systems are managed like irrigation systems (see Figure 11.1) by providing optimal amounts of water for crop needs. Treatment takes place on the surface and in the passage of the wastewater through the plant/soil matrix. Some water may reach underground aquifers but most is taken up by plants. Sites for SR systems are typically designed to minimize surface runoff. If site conditions require it,

Table 11.2 Design Features for Land Treatment Processes

Factor	Land Treatment Process		
	Slow Rate	Rapid Infiltration	Overland Flow
Application techniques	Sprinkler or surface[a]	Usually surface	Sprinkler or surface
Annual loading rate, m	0.5–6	6–125	3–20
Field area, ha[b]	23–280	3–23	6.5–44
Typical weekly loading rate, cm	1.3–10	10–240	60–40[c]
Minimum preapplication treatment	Primary sedimentation[d]	Primary sedimentation[e]	Grit removal and comminution[e]
Disposition of applied wastewater	Evapotranspiration and percolation	Mainly percolation	Surface runoff and evapo–transpiration, some percolation
Need for vegetation	Required	Optional	Required
BOD_5 loading[f], kg/ha/y	370–1,830	8,000–46,000	2000–7500

[a]Includes ridge–and–furrow and border strip.

[b]Field area in hectares not including buffer area, roads, or ditches for 3785 m^3/day flow.

[c]Range includes raw wastewater to secondary effluent, higher rates for higher level of preapplication treatment.

[d]With restricted public access; crops not for direct human consumption.

[e]With restricted public access.

[f]Range for municipal wastewater.

Source: U.S. Environmental Protection Agency, Process Design Manual for Land Treatment of Municipal Wastewater, U.S. EPA–625/1–81–013, U.S. Printing Office, Washington, DC, 1981.

percolated water can be collected through underdrains or from recovery wells. The primary objectives of the SR process are treatment of the applied wastewater, and conservation of water through irrigation to support crop production.

While the SR process may achieve an excellent level of treatment, it does have some limitations. Considerable land area is required and the

Table 11.3 Expected Quality of Treated Water from Land Treatment Processes[a] (units in mg/L unless otherwise noted)

Constituent	Slow Rate[b] Average	Slow Rate[b] Upper Range	Rapid Infiltration[c] Average	Rapid Infiltration[c] Upper Range	Overland Flow[d] Average	Overland Flow[d] Upper Range
BOD_5	<2	<5	5	<10	10	<15
Suspended solids	<1	<5	2	<5	10	<20
Ammonia nitrogen as N	<0.5	<2	0.5	<2	<4	<8
Total nitrogen as N	3[e]	<8[e]	10	<20	5[f]	<10[f]
Total phosphorus as P	<0.1	<0.3	1	<5	4	<6
Fecal coliforms number/100 mL	0	<10	10	<200	<200	<2,000

[a]Quality expected with loading rates at the mid to low end of the range shown in Table 11.1.

[b]Percolation of primary or secondary effluent through 1.5 m (5 ft.) of unsaturated soil.

[c]Percolation of primary or secondary effluent through 4.5 m (15 ft) of unsaturated soil; phosphorus and fecal coliform removals increase with distance of wastewater flow.

[d]Treating comminuted, screened wastewater using a slope length of 30–36 m (100–120 ft).

[e]Concentration depends on loading rate and crop.

[f]Higher values expected when operating through a moderately cold winter or when using secondary effluent at high rates.

Source: U.S. Environmental Protection Agency, Process Design Manual for Land Treatment of Municipal Wastewater, EPA–625/1–81–013, U.S. Printing Office, Washington, DC, 1981.

operating cost may be relatively high. Wastewaters with high concentrations of salt and/or boron may inhibit productivity of some crops, especially in arid regions. If the wastewater is applied by spraying, disinfection of the wastewater may be required since aerosols from wastewater which has not been disinfected could pose a health hazard.

The wastewater application rates range from about 1.5 to 10 cm/week. Loading rates are dependent on climate, soil permeability, vegetation, and desired level of treatment quality. Since crop production is a pri-

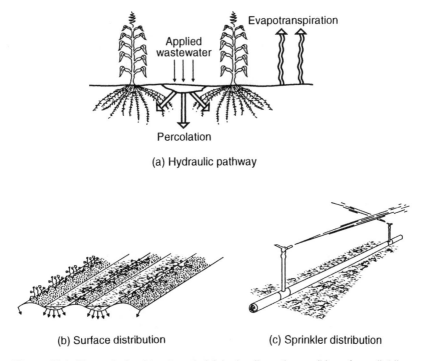

Figure 11.1 Slow rate land treatment: (a) hydraulic pathway, (b) surface distribution, (c) and sprinkler distribution (adapted from USEPA, 1981).

mary consideration in the SR process, nitrogen and moisture requirements for the crop become significant factors in the application rate. Data on nutrient requirements for selected crops are listed in Table 11.4.

Managing wastewater nitrogen loading on lands designed primarily for crop production is influenced by a number of factors: ammonia volatilization, ammonia uptake by plants, ammonia conversion to nitrate (nitrification), nitrate uptake by plants, and nitrate conversion to N_2 (denitrification). These individual factors vary according to soil and other types of conditions. As an example, denitrification is favored by finely-textured soils containing large amounts of organic matter. Frequent wetting of soils (anaerobic conditions), vegetative cover and warm temperatures also contribute to denitrification. Depending on the manner in which a particular site is managed, denitrification losses can range as high as 50% of nitrate. Additional details on the biological processes of nitrification and denitrification are provided in Chapter 6.

Nitrogen loading rates are especially critical because of the role nitrogen plays in crop production, and the potential for nitrate

Table 11.4 Nutrient Uptake Rates for Selected Crops

Crop	Uptake Rate (kg/ha/y)		
	Nitrogen	Phosphorus	Potassium
Forage crops			
Alfalfa[a]	225–540	22–35	175–225
Bromegrass	130–225	40–55	245
Reed canarygrass	335–450	40–45	315
Ryegrass	200–280	60–85	270–325
Field crops			
Corn	175–200	20–30	110
Potatoes	230	20	245–325
Soybeans[a]	250	10–20	30–55
Wheat	160	15	20–45

[a]Legumes are also able to utilize atmospheric nitrogen through the process of nitrogen fixation.

Source: Adapted from U.S. Environmental Protection Agency, Process Design Manual for Land Treatment of Municipal Wastewater, U.S. EPA–625/1–81–013, U.S. Printing Office, Washington, DC, 1981.

contamination of groundwater. Nitrates are often present in the wastewater itself but they may also be generated in the soil as a result of bacterial nitrification activity which converts ammonia into nitrate. Nitrate not removed by plant growth or denitrification processes will readily migrate with the wastewater and percolate into groundwater aquifers. In such aquifers nitrate may accumulate to the extent that the water is considered unsafe for human and animal consumption. For drinking water to be safe, it is generally accepted that a concentration of 10 mg/L is the upper limit. Nitrate concentrations exceeding 10 mg/L may place infants of less than six months of age at risk because nitrate interferes with the oxygen carrying capacity of the hemoglobin. This leads to a condition known as blue baby disease.

While nitrogen and other mineral nutrients derived from treated wastewater may be quite beneficial to plants, it is the water itself that may have the most value in regions where shortages exist for crop production. A wastewater reclamation study which took place in Monterey, California (Sheikh et al., 1990) provides an example of an application of the SR process. This ten-year field-scale project was designed to evaluate the safety and feasibility of irrigating food crops with reclaimed wastewater.

The initial years of the study were devoted to the selection of a site and an environmental assessment. In the second stage, beginning in late 1980 and continuing through 1985, a five-year field study to assess the

**Table 11.5 Estimated Costs of Reclaimed Water
Using Tertiary Treatment**

Treatment Process	Estimated Cost[a]
Filtered effluent	$0.05/m^3
Filtered effluent with flocculation	$0.06/m^3
Title 22 with 50 mg/L alum	$0.09/m^3
Title 22 with 200 mg/L alum	$0.13/m^3

[a]In 1985 dollars.

Source: Adapted with permission from Sheikh, B., Cort, R.P., Kirkpatrick, W.R., Jaques, R.S., and Asano, T., Monterey wastewater reclamation study for agriculture, *Res. J. Water Pollut. Con. Fed.* 62, 216–226, 1990.

impacts on crops was conducted. Crops grown during this period included artichokes, celery, broccoli, lettuce, and cauliflower. An existing wastewater treatment plant was upgraded to produce an effluent considered suitable for application to food crops. Effluent from secondary treatment (an activated sludge process) was pumped to two parallel tertiary treatment facilities: a Title 22 (T-22) process (State of California, 1978) and a direct filtration process. Both tertiary treatment processes produced reclaimed water of high quality, but the T-22 process which included coagulation, clarification, filtration, and disinfection was more costly (see Table 11.5). The reclaimed water was applied in conjunction with standard agricultural practices, including sprinkler and furrow irrigation, fertilization, and spraying with herbicides and pesticides.

The results obtained from this five-year study showed that the use of reclaimed water was safe for food crop application. Excellent yields of high-quality produce were obtained and, in the case of broccoli and cauliflower, yields were significantly improved when irrigated with the reclaimed water. No problem was observed with an accumulation of heavy metals; in fact, the conventional fertilizers were found to add far greater quantities of heavy metals (Sheikh et al., 1990). It was also observed that the chlorine residuals in the reclaimed wastewater had no observable impact on crops. Hence, the cost of dechlorination of a chlorine-disinfected treated wastewater could be avoided.

11.4 RAPID INFILTRATION PROCESS

The rapid infiltration process has proved to be a successful, cost-effective method of wastewater treatment. It can be designed specifically for wastewater treatment or as a component in a system to recover treated water for reuse or storage in an aquifer (USEPA, 1984). Approximately 320 municipal RI systems were in existence in the United States in 1981, 30% of which began operation after 1971.

The RI process involves the high-rate application of wastewater in earthen basins containing highly permeable soils (see Figure 11.2). Treatment is accomplished through biological, chemical, and physical interactions that occur in the soil matrix, especially in the portion near the soil surface. Wastewater moves through the soil matrix by a combination of infiltration and lateral flows. The application of wastewater is timed so that a drying period follows a flooding period. In this way the surface returns to an aerobic state, enabling the percolate to drain more efficiently.

The hydrogeology of an RI site is critical. Knowledge of the soil and groundwater environment is essential for efficient and environmentally-safe system performance. Coarse-textured soils with moderate to rapid permeabilities are best for RI processes. In successful RI designs, wastewater application rates have ranged from 15 m per year to 120 m per year. In contrast, the loading rate in an SR system typically falls in a range one-tenth the size of this one.

Figure 11.2 shows the schematic details of an RI design. By its very nature much more wastewater percolates to the groundwater from an RI system than from an SR process. The percolated water can be collected

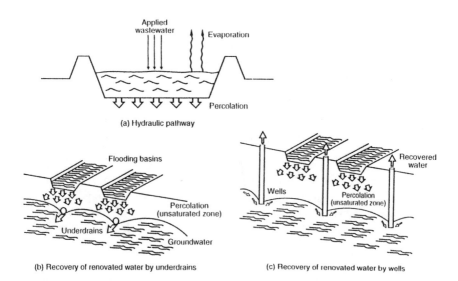

(a) Hydraulic pathway

(b) Recovery of renovated water by underdrains

(c) Recovery of renovated water by wells

Figure 11.2 Rapid infiltration: (a) hydraulic pathway, (b) renovated water recovery by underdrains, and (c) renovated water recovery by wells. (Adapted from U.S. Environmental Protection Agency, Process Design Manual for Land Treatment of Municipal Wastewater, EPA-625/1–81–013, 1981).

for reuse with underdrains or recovery wells. If geologic conditions allow, the water will move through underground aquifers into nearby surface waters.

Rapid infiltration systems are designed for high rates of infiltration with the primary objective being efficient wastewater treatment. Vegetation is not usually a significant part of the system, therefore most of the wastewater is subjected to a combination of physical, chemical, and biological treatment as it moves through the soil matrix. In effect the RI system is similar to a trickling filter (see Chapter 6). The renovated water can be used on the surface for irrigation of nearby lands or as recharge for surface waters. Alternatively, it could be used to change hydraulic gradients in groundwater to stop, or even reverse, the intrusion of saline waters.

Initial planning for an RI system requires that sites with suitable soil and groundwater conditions be identified. The land area needed is computed according to the following equation (USEPA, 1984):

$$A = \frac{0.0365 \, Q}{L}$$

where
  A = area of treatment field, (ha)
  Q = design daily flow, (m^3/d)
  L = annual design percolation rate, (m/y)

In this equation a year-round operation is assumed, the usual case for RI systems.

The hydraulic loading used in an RI system is determined by field or laboratory tests. By definition, hydraulic conductivity at a selected RI site is the amount of clean water that can move through a cross-section of soil having unit area. For example, a soil shown to have an effective vertical conductivity of 5 cm/h would transmit 438 m^3/y (= 0.05 m/h × 24 h/d × 365 d/y) through a cross-section of one square meter. A small portion, usually 10% or less, of this test value is then taken as the design loading rate. In effect, these percentages are safety factors derived from successful RI designs which take into account wastewater characteristics, soil variability, and the need for cyclic operations.

As noted earlier, the successful operation of RI systems requires regular drying periods. A drying period permits previously applied wastewater to percolate through the soil matrix, thereby opening the soil pores to air. This supports rapid aerobic biological degradation of the organic components (BOD) in the wastewater. The length of the drying period is determined on the basis of the solids and degradable organics in the wastewater, as well as climatic influences on aerobic reactions. In order to avoid excessive soil clogging the hydraulic loading for primary and similar wastewater effluents should not exceed 1–2 days (USEPA, 1984).

In practice, the ratio of loading to drying periods (wet/dry ratio) within a cycle varies but, to permit adequate aeration of the soil, it is nearly always less than one. For example, for primary effluents with their high BOD, the ratios are often less than 0.2. A long drying period relative to a wet period is necessary to permit adequate aeration for biological removal of the large amount of BOD. If a given RI treatment design also calls for nitrogen removal, then the wet/dry ratios must be adjusted accordingly to satisfy the oxygen requirements of the nitrifying bacteria and the anoxic requirements of the denitrifying bacteria.

Example 11.1: Calculate the application rate and required land area for a rapid infiltration process.

Solution: When a wet/dry cycle is determined it is combined with the hydraulic loading rate to determine the unit application rate. For example, assume a wastewater flow rate of 500 m³/d, a soil conductivity of 5 cm/h, and a 10% adjustment factor. As noted earlier, a 5 cm/h soil conductivity rate is equivalent to a rate of 438 m/y. Ten percent (the adjustment factor) of this value is equal to 43.8 m/y or approximately 44 m/y. Thus the design loading rate will be set at 44 m/y. The equation for determining land area yields

$$A = \frac{0.0365 \times 500 \text{ m}^3/\text{d}}{44 \text{ m/y}} = 0.41 \text{ ha}$$

Assume further that the wet/dry cycle is 9 days; 2 days wet (loading) and 7 days dry. If the summer and winter cycles are the same, 41 cycles would occur over a year and therefore:

$$\text{Wastewater loading} = \frac{44 \text{ m/y}}{41 \text{ cycles/y}}$$

Hence the application rate, R, during a two day wet period would be:

$$R = \frac{1.1 \text{ m/cycle}}{2 \text{ d/cycle}} = 0.55 \text{ m/d}$$

Note that in the example given above, an assumption is made that the wastewater flow and characteristics are uniform throughout the year and the wet/dry cycle is the same winter and summer. Obviously for a given community it will be necessary to design the RI system to accommodate fluctuations in flow and wastewater characteristics. In addition, depending on climatic conditions, it may be necessary to adjust the length of the wet/dry cycle for different periods of the year. For example, the summer cycle might be 2 days wet and 7 days dry, whereas a winter cycle may require 2 days wet and 10 days dry. In such a case the winter cycle would require more land area if the wastewater flow remains the same.

One of the advantages of the RI system is that it can be operated on a year-round basis even in cold climates. Systems have been operated year-round in New York, Michigan, Wisconsin, South Dakota and in other states with northern climates (USEPA, 1984). During sub-freezing weather it is essential to operate the system in such a manner as to prevent the formation of an impermeable ice barrier at the surface, or in the soils near the surface, of the RI basins. There are a number of ways this might be accomplished.

First, a ridge and furrow configuration could be constructed on the bottom of each infiltration basin. During cold weather the ice floats on the surface, providing thermal protection. As the final wastewater infiltrates into the soil at the end of a wet period the ice sheet will come to rest on the ridge tops allowing air spaces to develop between the ice sheet and the soil surface below.

Second, if snow is available, drifting might be induced in the basins. The build-up of a protective snow layer prevents the soil in the basin from freezing and allows for wastewater to be applied to it underneath the snow. The snow adds little hydraulic loading to the basin. By considering thermal and hydraulic conditions, yet another management possibility is to manipulate the wet/dry ratio during critical periods so that the near-surface soil layer never freezes irreversibly. Finally, it should be noted that wastewater storage may be necessary in cold climates if nitrogen removal regulations are particularly stringent.

The RI system is quite effective in the removal of bacteria and viruses, therefore chlorination of wastewater prior to its application to an RI basin is not required. If it contains levels of microbes beyond the limits set for certain uses, as for example, in the irrigation of food crops where the food products are consumed raw, it may be advisable to chlorinate the recovered water (USEPA, 1984).

The RI process is not without limitations. Through nitrification processes, wastewater nitrogen may be converted to nitrate, thereby raising nitrate in groundwater to unacceptably high levels. While an RI system can be designed to remove substantial amounts of nitrogen by manipulating wet/dry cycles, the operation of an RI process is more complicated if nitrogen removal is required. Selection of appropriate wet/dry cycles must also take into account potential problems such as hydrogen sulfide formation from sulfates in anaerobic regions of the infiltration zone.

11.5 OVERLAND FLOW PROCESS

In an overland flow (OF) process, wastewater is applied across the upper reaches of sloped terraces and allowed to flow across vegetated surfaces to runoff collection channels (see Figure 11.3). The primary objectives of the OF process are: (1) treat wastewater to a level comparable to that achieved with secondary or, preferably, tertiary treatment, and (2) produce forage crops or maintain greenbelts and open spaces.

In the right situation this process holds much promise for the treatment of municipal wastewater. There is a potential for treatment quality approaching that of tertiary treatment with construction and operating costs considerably lower than conventional and advanced waste treatment designs. With the emphasis placed on overland flow rather than on percolation, groundwater contamination and the phytotoxic effects of high salt

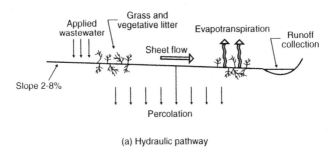

(a) Hydraulic pathway

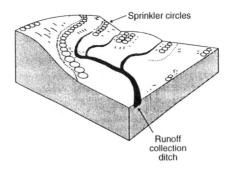

(b) Pictorial view of sprinkler application

Figure 11.3 Overland flow: (a) hydraulic pathway and (b) sprinkler application. (Adapted from U.S. Environmental Protection Agency, Process Design Manual for Land Treatment of Municipal Wastewater, EPA-625/1–81–013m 1981).

concentrations are greatly lessened. In fact, the OF process may be the best choice where soils are of low permeability.

The overland flow process is not unlike a fixed film biological reactor. The rate of wastewater application can be determined on the basis of kinetics and desired treatment levels (Smith and Schroeder, 1985; Witherow and Bledsoe, 1986). Based on results from a research scale project, Tucker and Vivado (1980) reported that the runoff from the OF process contained less than 10 mg BOD/L with application rates of raw wastewater at 10 cm/wk, primary effluent at 15–20 cm/wk, and secondary effluent at 25–40 cm/wk, respectively. Experience with existing systems indicates that application schedules of 6–8 hours on and 16–18 hours off over a 5 to 6 day week provide satisfactory performances (Polprasert, 1989). Table 11.6 describes the application and hydraulic loading rates recommended for OF systems receiving various types of wastewaters. The values shown are based on systems which have been operated successfully.

Example 11.2: Determine the application rate and required land area for an OF system (adapted from USEPA, 1984).

Table 11.6 Suggested Application Rates for Overland Flow Systems

Preapplication Treatment	Application Rate (m^3/h)	Hydraulic Loading Rate (cm/d)
Screening/primary	0.07 – 0.12	2.0 – 7.0
Aerated cell (1 day detention)	0.08 – 0.14	2.0 – 8.5
Wastewater treatment pond[a]	0.09 – 0.15	2.5 – 9.0
Secondary[b]	0.11 – 0.17	3.0 – 10.0

[a]Does not include removal of algae.

[b]Recommended only for upgrading existing secondary treatment.

Source: U.S. Environmental Protection Agency, Process Design Manual for Land Treatment of Municipal Wastewater: Supplement on Rapid Infiltration and Overland Flow, U.S. EPA–625/1–81–013a, U.S. Printing Office, Washington, DC, 1984.

Solution: The design equation is the following:

$$\frac{C-c}{C_o} = A \exp (-kz/q^n),$$

where

C_o = concentration of BOD_5 in applied wastewater, mg/L

C = concentration of BOD_5 at a distance (z) down the terrace, mg/L

c = minimum achievable effluent concentration, mg/L (determined to be 5 mg BOD/L)

A = an empirically determined coefficient which is dependent on the value of q

z = distance down terrace in meters (m)

k = empirically determined rate constant

q = application rate, m^3/h per meter ($m^3/h\cdot m$) of terrace width

n = empirically determined exponent

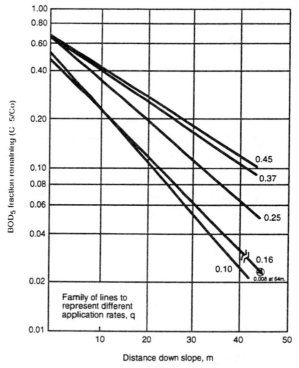

Figure 11.4 Graph used to calculate the application rate and required land area for an overland flow system of land treatment (adapted from Smith, R.G., Development of a Rational Basis for Design and Operation of the Overland Flow Process, EPA 600/9–31–022, in *Proceedings of National Seminar on Overland Flow Technology for Municipal Wastewater*, Dallas, TX, 1981).

Based on empirical studies by Smith (1981), the preceding equation has been used to develop a family of curves in which $(C - c)/C_o$ is plotted against z for different values of q. These curves are displayed in Figure 11.4.

Consider the specific case where the applied wastewater is screened raw municipal sewage; the flow, Q, is 1,000 m³/d; C_o, the influent BOD₅, is 150 mg/L; C, the required effluent BOD₅, is 20 mg/L; and the application rate, q, is 0.37 m³/h·m.

The design calculations are:

(1) Compute the required removal ratio $(C - 5)/C_o$.

$$\frac{C-5}{C_o} = \frac{20-5}{150} = 0.10$$

(2) Determine the required value of the terrace length from Figure 11.4 by noting where the line labeled $q = 0.37$ intersects the horizontal line at 0.10. Thus,

$$z = 41.5 \text{ m}$$

(3) An application period, P, must be chosen. Assume

$$P = 12 \text{ hours per day}$$

(4) Modify q so that a safety factor of 1.5 is incorporated. Thus, if the new value of q is denoted by q_s the result is

$$q_s = \frac{0.37 \text{ m}^3/\text{h} \cdot \text{m}}{1.5} = 0.25 \text{ m}^3/\text{h} \cdot \text{m}$$

(5) Compute the required total area assuming an application frequency of 7 d/wk.

$$Area = \frac{Q\ z}{q_s\ P} = \frac{(1000 \text{ m}^3/\text{d})\ (41.5 \text{ m})}{(0.25 \text{ m}^3/\text{h} \cdot \text{m})\ (12 \text{ h/d})}$$

$$Area = 13800 \text{ m}^2 = 1.38\ ha$$

In the example discussed above, the calculations are based on warm weather conditions. In cold climates wastewater storage capacity would have to be provided for those months when application to OF terraces is not possible.

11.6 CONSTRUCTED WETLANDS

Natural wetlands are broadly recognized as transition zones between terrestrial and aquatic environments. They typically include marshes, swamps, bogs, and other similar land/water systems. A wetland need not be continuously wet, but water is clearly the dominant factor influencing soil development and the plant, animal and microbial communities. Wetlands are very productive because they receive, hold, and recycle nutrients that are washed in from upland sources. They support an abundance of plant growth, which in turn supports diverse invertebrate, fish, bird, and mammal populations.

In addition to their biological productivity, the diverse biota in a wetland are not only capable of removing pollutants from nonpoint sources such as agriculture, but also from wastewater that comes from municipal or industrial sources. The physical and biological structure of a wetland provides an effective filter, sedimentation basin, and a source of biological processes which assist in removal of pollutants. The organic matter and inorganic nitrogen and phosphorus in wastewater are absorbed and biodegraded by plants and their associated microbial populations.

Constructed wetlands are engineered complexes of saturated substrates, emergent, and submergent vegetation designed and maintained for the purpose of treating wastewaters. Constructed wetlands may range from a complex consisting of a largely natural wetland, minimally modified to accommodate wastewater treatment, to a completely engineered site which is designed to simulate the activity of a natural wetland.

Much research on constructed wetlands has focused on the most effective mix of substrate and biota for wastewater treatment. Cattail, bulrush, rush, or giant reed plants seem best able to deal with fluctuating water and nutrient levels and tend to be more tolerant of high pollutant concentrations. According to Hammer and Bastian (1989) five major components of constructed wetlands are substrates (e.g., soil, sand, or gravel) with varying rates of hydraulic conductivity, plants adapted to water-saturated substrates (e.g., cattails), water flowing through or above the substrate, invertebrate and vertebrate animals, and aerobic and anaerobic microbes.

Most wetland treatment systems in the United States have been installed in the southern region, but experts believe there is also good potential for installation in northern areas (Gillette, 1992). Constructed wetlands may be especially advantageous for the treatment of wastewater in small communities since they are, in principle, simple and inexpensive to build and operate, while, at the same time, providing effective wastewater treatment. Such systems may also produce added benefits in the form of green space, wildlife habitats and recreation activities. A recent publication edited by Moshiri (1993) describes numerous examples of constructed wetlands systems research and operation on a wide range of waste streams.

Even though constructed wetlands show considerable promise, especially for small wastewater flows, there are a number of disadvantages. Among them are:

1. Relatively large land area requirements, especially if the most stringent treatment requirements of 30–40 m^2/m^3 wastewater/d (Gillett, 1992) are to be met
2. Design and operating criteria are not as precise as those associated with conventional treatment processes (primary, secondary, and tertiary)
3. The biological and hydrological complexity of wetlands is not yet sufficiently understood
4. Pest problems may arise
5. Several growing seasons may be required to develop a system operating at optimal levels of efficiency

11.7 LAND APPLICATION OF WASTE SOLIDS

Sewage sludge, animal manures, some industrial wastes, and composted MSW have characteristics which make them potentially valuable as fertilizer supplements and soil conditioners. Nitrogen and phosphorus are the two most common plant nutrients found in typical sewage sludge, but other nutrients may also be available and of significant value to agricultural lands. Applied to iron-deficient calcareous soils, sludge may be a valuable source of iron. Sewage sludge also contains zinc and copper, two elements which are becoming deficient in soils that have been cropped for many decades. Because hydrated lime is sometimes used in sludge dewatering, sludge subjected to this treatment may assist in maintaining proper soil pH when applied to soils normally in need of liming.

In addition to adding plant nutrients, the organic matter in biosolids increases soil friability, pore space, organic content, and water-holding capacity. These qualities are especially desirable in the revegetation of disturbed lands, such as strip-mined sites, building and road construction sites, deforested and severely eroded areas, and reclaimed landfills.

Land application, as discussed in the remainder of this chapter, emphasizes relatively low-rate applications of stabilized sewage sludge to agricultural, forested, or reclaimed lands. A stabilized sludge is one that has been processed from a raw state in order to reduce pathogen content and the potential for offensive odors. Two examples of sludge stabilization processes are anaerobic digestion and composting (see Chapter 7). In recent years composting has become an increasingly utilized process.

A number of important factors must be considered when applying sludge to agricultural land on which crops are grown to produce food for animal and/or human consumption. They include the nature of the sludge itself, the agronomic sludge application rate, the annual pollutant loading rate, the cumulative load, the mode of application, the types of crops grown, the soil pH, and the site characteristics.

Biosolids contain three major components that affect the degree of safety in their use on agricultural lands. These include nutrients, potentially toxic components such as mercury, cadmium, and other heavy metals, and pathogens. Each of these is examined in more detail in the paragraphs which follow.

The agronomic rate is the biosolids application rate which provides the correct amount of nitrogen needed by the crop to be grown on the land to which sludge is applied. Adherence to this loading rate minimizes leaching of nitrate to groundwater and avoids the undesirable effects of excessive nitrogen fertilizer on plant growth.

Table 11.7 Summary of Pollutant Limits from U.S. EPA 40 CFR Part 503 Standard

Metal	Ceiling Concentration (mg/kg)	Cumulative Load (kg/ha)	Pollutant Concentration Limit[a] (mg/kg)	APLR[b] (kg/ha/y)
Arsenic	75	41	41	2
Cadmium	85	39	39	1.9
Chromium	3000	3000	1200	150
Copper	4300	1500	1500	75
Lead	840	300	300	15
Mercury	57	17	17	0.85
Molybdenum[c]	75	18	18	0.90
Nickel	420	420	420	21
Selenium	100	100	36	5
Zinc	7500	2800	2800	140

[a]Pollutant concentration limits for biosolids designated as a Class A sludge.

[b]APLR = Annual pollutant loading rate.

[c]Limits for molybdenum stayed by court action in 1994.

Source: Tables 1–4 of 503.13 in *U.S. Federal Register*, 58(32), 9248, U.S. Printing Office, Washington, DC, Feb. 19, 1993.

The annual pollutant loading rate is defined as the maximum amount of a given pollutant contained in sludge that can be applied in a 365-day period (*U.S. Federal Register*, Feb. 19, 1993). For example, the Allowable Pollutant Loading Rate (APLR) for cadmium is 1.9 kg/ha/y (see Table 11.7). If a particular sewage sludge contains 25 mg/kg of cadmium, the annual pollutant loading rate limit per hectare would be reached with the application of approximately 76 tonnes (dry wt. basis) of sludge.

The cumulative load is the maximum amount of an inorganic pollutant (examples are cadmium, lead, and zinc) that may be applied to a given area of land over any period of time. As an example, it can be seen from Table 11.7 that the cumulative load for cadmium is 39 kg/ha.

Biosolids can be applied to land through surface or subsurface processes. While surface application of sludge is the easiest and least expensive method, it increases the potential for runoff of nutrients, such as nitrogen and phosphorus, to surface waters, thereby contributing to eutrophication. Surface-applied solids are also subject to the loss of ammonia nitrogen through volatilization. The injection of sludge below the soil surface avoids these problems but is a more difficult and costly process, yet necessary for public acceptance, especially in areas that are becoming more urban.

Important site characteristics which may place constraints on biosolids applications are soil type, topography, depth to water table, depth to bedrock, and water holding capacity of soil. Other site restrictions include proximity to water supply wells and seasonal limitations resulting from frozen soils.

11.8 PATHOGEN AND VECTOR CONTROL IN BIOSOLIDS

Raw municipal wastewater solids are not permitted on agricultural lands because of their high pathogen content and the potential for nuisance odors. They may also attract vectors, such as rodents, flies, mosquitoes, and other organisms, capable of transporting infectious agents. The recently published Standards for the Use or Disposal of Sewage Sludge (*U.S Federal Register*, February, 1993) contain detailed guidelines for the pathogen standards sludge must meet if it is to be used on lands for food or crop production. The pathogen standards are combined with pollutant limits and vector reduction requirements in formulating two types of sludge designations: Class A and Class B. For a sludge to be designated as a Class A sludge it must meet more stringent requirements than those for a Class B sludge.

CLASS A SLUDGE

A Class A sludge must meet the following standards:

1. Pollutant limits. The biosolids must test for pollutant concentrations at or below those levels listed in the "concentration limit" column of Table 11.7.
2. Pathogen standards. The Class A sludge must test for fecal coliform density of less than 1000 Most Probable Number per gram of solids (dry weight basis) or the concentration of Salmonella species of bacteria must be less than three Most Probable Number per four grams of sludge solids. The density of viruses common to the human intestinal tract (enteric viruses) must be less than one plaque-forming unit per four grams of sludge solids. The density of viable heminth ova must be less than one per four grams of biosolids.
3. Vector reduction. Vector reduction requires that sludge processing remove at least 38% of the raw sludge volatile solids. Volatile solids are defined here as the amount of total solids lost when the sewage sludge is combusted at 550° C in the presence of excess air. A number of methods are available to achieve volatile solids reduction. These include the sludge processing techniques of anaerobic digestion, aerobic digestion, and composting. The requirement for vector reduction can also be met by injecting the sludge below the surface of the soil or by incorporating it into the soil within six hours after surface application.

A wastewater treatment plant may incur substantial sewage sludge processing costs in order to meet Class A sludge standards. However, the very stringent limits on pathogens and the vector and pollutant reduction requirements for Class A sludge make this "safe sludge" attractive for large-scale land application and for packaging in conveniently sized consumer quantities for sale to the general public. Such sludge can be promoted as a valuable resource (Chaney, 1990), encouraging the beneficial recycling of sludge nutrients to land.

CLASS B SLUDGE

The standards for Class B sludge are somewhat less stringent than for Class A sludge. Class B sludge must meet the following standards:

1. Pollutant limits. The maximum limits for pollutants in a sludge intended for land application are those values listed in the column titled "pollutant ceiling concentration" in Table 11.7.
2. Pathogen standards. Pathogen limits for a Class B sludge are somewhat more relaxed than those for a Class A sludge. The density of fecal coliform bacteria should not exceed 2,000,000 Most Probable Number per gram of total solids. The pathogen criterion for Class B sludge can also be met by sludge treatment with "processes to significantly reduce pathogens", (*U.S. Federal Register*, Feb. 19, 1993). Examples of such processes are anaerobic digestion, composting, and drying on sand beds.
3. Vector reduction. The same vector reduction requirements described previously for Class A sludge are imposed when a Class B sludge is applied in bulk quantities to agricultural or forested land, to a public contact site or a reclamation site.
4. Site restrictions. The restrictions encompass sites where animal feed or human food crops are grown, or lands on which animals graze. Class B sludge shall not be applied for a period of 14 months prior to harvest of food crops with harvested parts above the soil surface that come in contact with the sludge/soil mixture. Similarly, Class B sludge shall not be applied for a period of 20 months prior to harvest of food crops with harvested parts below the surface of the soil. Thirty days shall elapse after a sludge application before harvesting food crops, feed crops, and fiber crops. Finally, animals shall not be allowed to graze on land for 30 days after a sludge application.

11.9 LOADING RATES WHEN BIOSOLIDS ARE APPLIED TO AGRICULTURAL LAND

The appropriate rates of application of biosolids to agricultural lands are limited by three loading rate components: (1) the agronomic rate, (2) the annual pollutant loading rate, and (3) the cumulative load. Each of these types of loading rates were defined in Section 11.7. In this section we will provide several examples which illustrate the manner in which these quantities are calculated in typical agricultural settings.

AGRONOMIC LOADING RATE

As noted previously, nitrogen is the key factor in determining the agronomic rate of a sludge application. Municipal wastewater sludge often contains substantial amounts of nitrogen. The application of excessive quantities of nitrogen may have detrimental effects due to the leaching of nitrates into underground aquifers. The phosphorus content of sludge is also significant but it is a lesser environmental hazard than nitrogen because it is not prone to leach, even in sandy soils.

The determination of the agronomic application rate based on nitrogen loading is influenced by several factors. These include the nitrogen requirements of the crop, nitrogen species and content in the biosolids, volatilization of ammonia, mineralization of organic nitrogen in the solids (residual sludge nitrogen), available nitrogen in the soil, and soil type and timing of biosolids application.

The agronomic application rate based on crop nutrient requirements can be calculated with the following steps:

1. Determine crop nitrogen requirements, A (units are kg/ha) (see Table 11.4).
2. Calculate the available NH_4-nitrogen, B, and organic nitrogen, C, in the biosolids (units are kg/tonne):

$$B = \% \ NH_4 - \text{nitrogen} \times 1000 \text{ kg/t sludge}$$

(Note: if nitrogen is applied to the surface, B should be multiplied by one-half to allow for losses of ammonia N from volatilization.)

$$C = \% \text{ organic N} \times 1000 \text{ kg/t sludge} \times \% \text{ mineralization rate}$$

(see Table 11.8)

Table 11.8 Release of Available Nitrogen Per Tonne of Solids During Mineralization of Anaerobically Digested Sludges[a]

Year After Sludge Application	Mineralization Rate (%)	Organic N Content of Sludge[b] (kg/t)				
		2.0%	2.5%	3.0%	3.5%	4.0%
First	20.0	4.0	5.0	6.0	7.0	8.0
Second	10.0	2.0	2.5	3.0	3.5	4.0
Third	5.0	1.0	1.3	1.5	1.8	2.0
4–10 (each year)	3.0	0.6	0.8	0.9	1.0	1.2

[a]The mineralization rates for composted sludges are approximately 1/2 those for digested sludges during years 1–3 and the same for years 4–10.

[b]Expressed as kg of N released/tonne of sludge.

Source: Adapted from U.S. Environmental Protection Agency, Process Design Manual—Land Application of Municipal Sludge, U.S.EPA–625/1–83–016, U.S. Printing Office, Washington, DC, 1983.

3. Determine the residual sludge nitrogen, D, in the soil (units are kg/ha). If the site has received biosolids in the three-year period prior to the time of application, the residual nitrogen can be calculated from Table 11.8.
4. Calculate the sludge application rate in t/ha

$$\text{Application rate (t/ha)} = \frac{A - D}{B + C}$$

Example 11.3: Calculate the agronomic biosolids application rate for an agricultural plot where corn is grown on a soil with a high yield potential.

Assume the nutrient requirements (based on soil test and crop management level) are:

Nitrogen	150 kg/ha
Phosphorus (as P_2O_5)	80 kg/ha
Potassium (as K_2O)	200 kg/ha

Digested sewage sludge with the following nutrient contents is to be applied: NH_4-N, 2%; Organic N, 2.5%; P, 2.0% and K, 0.2%.

Determine the surface application for year 2, if 10 t/ha are applied during year 1.

Solution: The nitrogen required by the crop: A = 150 kg/ha.

The available nitrogen in the biosolids is:

$$B = 0.02(2\% \ NH_4 - N) \times 1000 \times 0.5 \text{ (surface application)}$$

or

$$B = 10 \ kg/\text{t}$$

$$C = 0.025(2.5\% \text{ organic N}) \times 1000 \times 0.2 \text{ (mineralization 1st year; see Table 11.8)}$$

or

$$C = 5 \text{ kg/tonne}$$

Now, determine the residual N (from Table 11.8), for 2.5% organic N sludge.

$$D = 10 \text{ tonnes/ha (biosolids added year 1)} \times 2.5 = 25 \text{ kg/ha}$$

The application rate is

$$\text{Application rate (t/ha)} = \frac{A - D}{B + C} = \frac{150 - 25}{10 + 5}$$
$$= 8.33 \text{ t/ha.}$$

In the example described above, the application rate was based on the premise that sufficient solids would be applied to provide for the nitrogen requirements of the corn crop. Phosphorus (P) and potassium (K) are also essential for high crop productivity, hence these elements must also be supplied if the nitrogen provided by the sludge is to be fully utilized by the crop in question.

The digested biosolids in our example contain 2% phosphorus, therefore the phosphorus contained in 8.33 tonnes of this product is 167 kg. The 8.33 tonnes of sludge needed to supply the annual nitrogen requirement would add phosphorus at a level well above the 200 kg/ha required for the crop. Therefore no commercial fertilizer containing phosphorus is required.

The digested sludge also contains 0.2% potassium; therefore, the total amount of potassium available in 8.33 tonnes of sludge is 16.7 kg. The amount of potassium provided by the sludge would be far below the 200 kg/ha requirement for the corn crop. Thus potassium fertilizer would have to be added in order to obtain the expected yield for the corn crop.

ANNUAL POLLUTANT LOADING RATE

The annual pollutant loading rate refers to the regulatory limits on the amounts of specific pollutants that are permitted on a land site in a given year. Table 11.7 lists the annual pollutant loading rates (APLR) for various heavy metals. This table also lists the maximum metals content allowed for a given sludge. If one knows the metals content of a sludge, it is a simple calculation to determine the maximum amount of sludge that can be applied to a land site before the annual pollutant loading rate would be exceeded.

Ordinarily the agronomic rate will be the limiting factor in determining the amount of sludge to be applied during a given year. In the case of Example 11.3 described earlier it was calculated that 8.33 t/ha of biosolids could be applied to the site during year two in order to conform to the agronomic rate.

Consider what the annual pollutant loading might be for one of the heavy metals, mercury, if the sludge contains the maximum allowable mercury content of 57 mg/kg. The application of 8.33 t/ha of biosolids would add 0.475 kg of mercury/ha, well below the allowable limit of 0.85 kg/ha/y for this metal.

However, suppose the sludge discussed in Example 11.3 had only 0.5% ammonium nitrogen. In the example as described, calculating the "B" component of the agronomic loading rate equation would yield a value of 2.5, and the amount of sludge required to meet the agronomic loading rate would increase to 16.7 t/ha.

$$\frac{A-D}{B+C} = \frac{150-25}{2.5+5} = 16.7 \ t/ha$$

If the sludge has a mercury content of 57 mg/kg, the 16.7 t/ha rate would add 0.95 kg of mercury per hectare, thus exceeding the annual pollutant loading rate for this heavy metal.

CUMULATIVE LOAD

The cumulative sludge load, sometimes called the lifetime of a sludge amended site, refers to the maximum amount of biosolids which can be applied to a site before the cumulative levels of cadmium, nickel, or other nondegradable toxicants are exceeded. Table 11.7 lists the most recently published standards.

The recommendations have evolved from the extensive research over the past several decades on the safety of the application of biosolids to land. The continuing sludge-on-land research and risk assessment analyses have been driven by the need to dispose of increasingly large volumes of biosolids, due to population growth and improved wastewater treatment, in an environmentally acceptable manner. In contrast to incineration, landfilling, and ocean dumping, land application of sewage sludge is viewed by many as the most desirable alternative for many communities, provided sufficient safeguards are followed.

The various heavy metals that are found in sewage sludge have been studied in great detail to determine their impact on plant growth and on animals and humans. Zinc, for example, is an essential element for both plants and animals, but high concentrations of zinc in soil may lead to phytotoxicity. However, the danger for animals and humans is believed to be minimal, since the very high concentrations of zinc that would lead to toxic effects in plants are still within the margin of safety for animals.

Cadmium is one of the heavy metals which has been the subject of much research because of its potential harmful effects. Given the right circumstances, cadmium can accumulate in sufficient quantities in plant tissues to be toxic to animals or humans who consume these plants over an extended time period. The risk of cadmium toxicity to humans is complicated by the fact that plants such as lettuce, chard, and spinach grown on cadmium-rich soils may accumulate cadmium at concentrations exceeding 100 ppm without exhibiting phytotoxic effects.

Potentially toxic organic compounds such as polychlorinated biphenyls (PCBs) added to sludge-treated soils tend to move through the soil air, rather than dissolving in the soil water. Therefore, there is little potential for plant uptake of organic compounds and subsequent transport to plant tissues, such as the grain in corn or soybeans. However, if the sludge is surface-applied, the presence of organic compounds in the soil air permits their transfer to the surface of plant roots or foliage. It has been shown that organic compounds can accumulate on the peel of carrots in this manner.

Example 11.4: A sludge with a cadmium content of 70 mg of Cd per kg of sludge is applied on land at the rate of 10 t/ha/y. Using the cumulative load limit for cadmium in Table 11.7, calculate the lifetime of the site. Assume that the cadmium content and the annual sludge loading rate remain constant.

Solution: This question is answered with the following calculation:

$$\frac{39 \text{ kg/ha (Cd limit from Table 11.7)}}{10 \text{ tonnes/ha/y}} \times \frac{1 \text{ tonne}}{1000 \text{ kg}} \times \frac{1 \text{ kg of sludge}}{0.00007 \text{ kg of Cd}} = 56 \text{ y}$$

The lifetime of the sludge application in the example above is based on a single metal and assumes the cadmium content and sludge loading rates stay the same throughout the 56 years. Obviously, if the wastewater treatment plant were able to reduce the cadmium content of its sludge, the lifetime of a site, based on cadmium considerations alone, could be correspondingly increased.

A comprehensive analysis to determine a site lifetime requires calculations similar to the one above for all the metals listed in Table 11.7. For example, if the lead content of the sludge were 750 mg of lead per kg of sludge, then the lifetime of the site, based on lead considerations alone, is calculated as follows:

$$\frac{300 \text{ kg/ha (Pb limit from Table 11.7)}}{10 \text{ tonnes/ha/y}} \times \frac{1 \text{ tonne}}{1000 \text{ kg}} \times \frac{1 \text{ kg of sludge}}{0.00075 \text{ kg of Pb}} = 40 \text{ y}$$

Therefore the lifetime of the site based on lead limits is 40 years. Even though the cadmium calculation shows a site lifetime of 56 years, it is actually less than this due to limitations imposed by lead. When other metals are considered, the site lifetime may be reduced even further because the actual lifetime is determined by the metal whose cumulative limits are reached first.

It is evident from these examples that a community which produces biosolids containing very low levels of heavy metals and other types of toxic components would be able to apply this "clean sludge" to a given land site for many years. In this way the difficulties and expenses associated with the identification and development of new land areas for biosolids application can be avoided. Thus there is an incentive to remove toxic species at the source and prevent them from contaminating the biosolids in the first place. As an example of an application of this concept, an industry which generates a toxic material, such as cadmium, is required to remove as much of the metal from its wastewater as is technologically feasible before discharging it to a municipal wastewater treatment plant.

11.10 FUTURE PROSPECTS FOR LAND APPLICATION OF SEWAGE SLUDGE AND OTHER TYPES OF WASTE SOLIDS

The recently published sewage sludge use/disposal standards (*U.S. Federal Register*, February 19, 1993), give managers of wastewater treatment facilities two options for adhering to pollutant limits when applying sludge to land. The first is based on Alternative Pollutant Limits (APLs). The second is based on cumulative or lifetime loads. An APL sludge, also referred to as "clean sludge," is a product which has pollutant concentrations below the levels in the "Concentration Limit" column of Table 11.7. If the APL standards are not met it may still be possible to apply sludge to land. However, producers of such sludge will have to conform to detailed and strictly interpreted regulatory requirements. Cumulative load limits for given pollutants may not be exceeded.

It is anticipated that the biosolids regulations will encourage communities to meet the APL standards because the costly paperwork and monitoring associated with regulatory requirements for the land application of poorer quality sludge are greatly reduced. In effect the USEPA is telling wastewater treatment managers that a "clean sludge" with low pollutant and pathogen levels, when properly applied to agricultural and nonagricultural lands, is a fertilizer and not a waste.

While the options for sewage sludge disposal in the United States have been receiving renewed and critical attention in the past decade, nations in the European Union have also been engaged in extensive discussions concerning the deposition of wastes on agricultural lands. Although the guidelines for the application of sewage sludge to land are not uniform among the European nations, they are expected to become more restrictive as the decade of the 1990s unfolds. In addition, in some countries competition exists in finding land for the application of various types of organic wastes. For example, in the Netherlands there is a surplus of organic fertilizers derived from sewage sludge, animal wastes, and composted solid waste. Hence, even sewage sludge that meets the standards for land application will be competing for land space with the other types of organic fertilizers. This problem is especially serious near large urban centers where large quantities of sewage sludge are produced (Kruize, 1988).

According to Santori (1988), the prospects for continuing to apply sewage sludge to land may be dimming in many European countries. This does not bode well for those who would like to see land application become the choice for handling larger quantities of sewage sludge. There are a number of reasons for this outlook. An important factor is a decline in the quality of sludge due to heavy metals and other types of toxic substances. Some deterioration in sludge quality also seems to result from technologies used to improve final effluent quality. Other factors include increases in transportation costs due to greater hauling distances, more stringent controls on sludge quality, more critical public attitudes, and difficulties in determining with reliability the risks imposed by potential pathogens. While the United States has much larger land areas which can serve as potential sites for the application of biosolids to land, the European experience is a reminder that this approach is not a panacea.

DISCUSSION TOPICS AND PROBLEMS

1. When applying biosolids to agricultural land, why is the effect from excessive application of phosphorus less than that from excessive application of nitrogen?

2. When biosolids are immediately incorporated into the soil its ammonia nitrogen content is of greater value for crop utilization than when the biosolids are applied on the surface. Why?

3. Suppose an anaerobically digested sludge with the following characteristics is injected into the soil: Ammonia N = 1.5%, Organic N = 2%, Pb = 400 mg/kg, Zn = 2,000 mg/kg, Cu = 400 mg/kg, Cd = 30 mg/kg, and Ni = 300 mg/kg. The crop to be raised on the application site has a nitrogen requirement of 180 kg/ha/y.

    (a) How much sludge should be applied the first year to meet the nitrogen needs of the crop?

    (b) Assuming the same nitrogen requirement, how much sludge should be applied the second year?

    (c) Use the values in Table 11.7 to determine the number of years sludge can be applied (assuming the sludge characteristics and nitrogen requirements remain constant) before the cumulative loading limit is reached for one of the heavy metals. Which heavy metal limit is reached first?

4. Using the same wastewater volume and soil characteristics as described in the example problem for the Rapid Infiltration (RI) process, calculate the land requirement for differing summer and winter cycles. Assume a nine day wet/dry cycle for the period May 1 to October 31 (184 days) and a fourteen day wet/dry cycle for the period November 1 to April 30 (181 days).

    (a) How many total cycles occur during the year?

    (b) How much land area would be needed for the summer cycle?

    (c) How much land area would be needed for the winter cycle?

5. How much sewage sludge (dry weight basis) is produced by your community annually?

    (a) How is the sludge used or disposed of?

    (b) Which heavy metals are of greatest significance in your community's sewage sludge based on concentration and the impact it would have on the lifetime of a sludge-on-land site?

    (c) Compare the heavy metals content in your community's sewage from 1980 with what it is today.

6. A sludge containing substantial amounts of ammonia nitrogen but no nitrate nitrogen is applied to agricultural crop land. Later, increased nitrate concentrations are discovered in the groundwater below this site. Explain how this could have occurred. Identify various environmental factors which may contribute to this increase in nitrate.

7. What soil conditions favor the conversion of nitrate to nitrogen gas? What is the name commonly applied to this process?

8. What soil conditions favor the formation of hydrogen sulfide?

9. What are the limitations to the use of land as a treatment method in the application of the Slow Rate (SR) process?

10. A community with an average wastewater flow of 400,000 liters per day is planning a constructed wetland for treating wastewater. Calculate the approximate area in hectares required to achieve a high level of waste treatment for this community.

11. What is meant by the term "stabilized sewage sludge"?

12. List examples of processing techniques that are used to stabilize sewage sludge.

REFERENCES

Chaney, R.L., Public health and sludge utilization, *Biocycle*, 10, 31–33, 1990.

Gillette, B., The green revolution in wastewater treatment, *Biocycle*, 12, 44–48, 1992.

Hammer, D.A. and Bastian, R. K., Wetlands ecosystems: Natural water purifiers? in *Constructed Wetlands for Wastewater Treatment: Municipal, Industrial and Agricultural*, Lewis Publishers, Inc., Chelsea, MI, 1989.

Kruize, R.R., Sludge treatment in Amsterdam: Economic, technical and environmental experiences, in *Sewage Sludge Treatment and Use: New Developments, Technological Aspects and Environmental Effects*, (Symposium Proceedings), Elsevier Applied Science, 1988.

Lipták, B.G., *Municipal Waste Disposal in the 1990s*, Chilton Book Co., Radnor, PA, 1991.

Moshiri, G. A., Ed. *Constructed Wetlands for Water Quality Improvement*, Lewis Publishers, Boca Raton, FL, 1993.

Polprasert, C., *Organic Waste Recycling*, J. Wiley and Sons Ltd., New York, NY, 1989.

Santori, M., Evaluation of the conference and closure, in *Sewage Sludge Treatment and Use: New Developments, Technological Aspects and Environmental Effects*, Symposium Proceedings, Elsevier Applied Science, 1988.

Sheikh, B., Cort, R.P., Kirkpatrick, W.R., Jaques, R.S., and Asano, T., Monterey wastewater reclamation study for agriculture, *Res. J. Water Pollut. Con. Fed.*, 62, 216–226, 1990.

Smith, R.G., Development of a Rational Basis for Design and Operation of the Overland Flow Process, EPA 600/9–31–022, in *Proceedings of National Seminar on Overland Flow Technology for Municipal Wastewater*, Dallas, TX, 1981.

Smith, R.G. and Schroeder, E.D., Field studies of the overland flow process for the treatment of raw and primary treated municipal wastewater, *J. Water Pollut. Control Fed.*, 57, 785–793, 1985.

State of California, Wastewater Reclamation Criteria, *Calif. Code of Regulat. Title 22, Div 4.*, Environ. Health Dept. Health Serv., Berkeley, CA, 1978.

Tucker, D.L. and Vivado, N.D., Design of an overland flow system, *J. Water Pollut. Control Fed.*, 52, 559–567, 1980.

U.S. Environmental Protection Agency, Process Design Manual for Land Treatment of Municipal Wastewater, U.S. EPA-625/1–81–013, 1981.

U.S. Environmental Protection Agency Process Design Manual—Land Application of Municipal Sludge, EPA-625/1–83–016, U.S. Printing Office, Washington, DC, 1983.

U.S. Environmental Protection Agency Process Design Manual for Land Treatment of Municipal Wastewater: Supplement on Rapid Infiltration and Overland Flow, U.S. EPA-625/1–81–013a, U.S. Printing Office, Washington, DC, 1984.

U.S. Federal Register, 54(23), 5745–5902, U.S. Printing Office, Washington, DC, Feb. 6, 1989.

U.S. Federal Register, 58(32), 9248–9404, U.S. Printing Office, Washington, DC, Feb. 19, 1993.

Witherow, J.L. and Bledsoe, B.E., Design model for the overland flow process, *J. Water Pollut. Control Fed.*, 58, 381–386, 1986.

CHAPTER 12

Hazardous Waste

12.1 OVERVIEW

The term hazardous waste refers to any solid, semisolid, liquid, or gaseous waste that cannot be handled by routine waste management methods because of chemical, biological, or physical properties that present a significant threat to human health, the health of other living organisms, or the environment. In some cases a hazardous waste may be a mixture of solid, liquid, or gaseous components. Substances that qualify as hazardous include, but are not limited to, poisons, explosives, flammables, oxidizers, irritants, corrosives, pesticides, acids, caustics, pathological waste, and radioactive waste. Radioactive waste is a special waste because the procurement, use, and disposal of radioactive materials are tightly regulated by the U.S. Nuclear Regulatory Commission and are excluded from the federal and state regulations pertaining to the other categories of hazardous waste. For this reason, radioactive waste management disposal is not discussed in this chapter.

Although the title of the legislation fails to convey its significance to hazardous waste management, the Resource Conservation and Recovery Act of 1976 (RCRA) is the foundation for hazardous waste management in the U.S. Subtitle C of RCRA deals with hazardous waste management and Subtitle I addresses the regulation of underground storage tanks, a common source of environmental contamination by hazardous materials.

Information on the generation, management, and disposition of RCRA-regulated hazardous wastes is collected by the USEPA. It is estimated that approximately 180 million tonnes of RCRA hazardous waste was generated by 20,233 large quantity generators during 1989 (USEPA, 1993). A small number of sites accounted for most of these wastes. This is indicated by the fact that the largest 1% of sites generated almost 97% of the hazardous waste.

It is difficult to quantify exactly how much hazardous waste is generated. For example, some hazardous waste is exempt from reporting requirements; a case in point is hazardous waste managed on the same site as it is produced. Further, the RCRA hazardous waste amounts reported above did not include wastes classified as hazardous by the new Toxicity Characteristics Leaching Procedure (TCLP). This procedure, which was adopted in 1990, will be discussed later in this chapter.

Another source of uncertainty in hazardous waste estimates lies with the difficulty in determining whether a particular waste satisfies the hazardous waste criteria. As an example, it has been widely debated whether combustion ash and incinerator ash should be classified as hazardous waste or exempted as a special, but not hazardous, waste. As noted in Chapter 8 (Section 11), a recent Supreme Court ruling clarified this issue. The essence of the ruling was that each facility would have to test its waste and if it qualified as a hazardous waste, it would be subject to hazardous waste regulations. Another example is the significant quantities of soil and other material excavated from a contaminated site and transported to an approved treatment or disposal facility. Should the excavated material be included as newly generated hazardous waste?

Finally, there is a continuing refinement of the classification criteria to determine what constitutes hazardous waste. An example is the change in the classification of sodium sulfate. In USEPA's 1987 Toxic Release Inventory (TRI), (USEPA, 1989) sodium sulfate dwarfed the amounts of all other chemicals, accounting for 54% of the total releases and transfers. Sodium sulfate has since been removed from the TRI reporting because it is generally considered to be nontoxic. The chemicals representing the ten largest releases or transfers in 1989 (USEPA, 1991) and accounting for about 1.6 million tonnes are shown in Figure 12.1.

At the heart of an effective hazardous waste management system is an efficient record-keeping or manifest system to track the transfers of hazardous waste from a registered generator to a licensed hauler and from the hauler to an approved and licensed treatment, storage, or disposal facility. Such a tracking system is commonly called "cradle-to-grave management." The purpose of the manifest system is to prevent illegal dumping by unscrupulous generators and haulers who wish to avoid paying the cost of proper disposal. The USEPA delegates the responsibility for managing hazardous waste and maintaining a manifest system to states which establish management programs at least as stringent as the federal guidelines.

The RCRA regulations established practices for the safe disposal of hazardous waste currently being generated, but they did not address the problem of the many existing sites where hazardous waste was dumped at

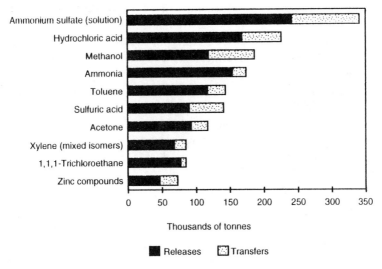

Figure 12.1. The ten chemicals with the largest quantities released and transferred in 1989 according to the USEPA Toxic and Release Inventory (USEPA, 1991).

some time in the past. The highly publicized Love Canal incident in the late 1970s typifies this type of problem. To address the problem of old dump sites, the Comprehensive Environmental Response, Compensation and Liability Act (CERCLA) or "Superfund" was passed in 1980. The purpose of this legislation was to provide funds to identify and remediate dump sites judged to present serious hazards to humans and the environment. The results have been disappointing. There have been numerous site investigations, but only a few site cleanups. Of the 1,077 sites selected for urgent cleanup in 1981, only 30 were completed by 1990 (Lipták, 1991).

An important aspect of CERCLA was to assign the financial responsibility for the cleanup of contaminated disposal sites to the generator. Thus, generators could no longer absolve themselves of the responsibility for proper waste disposal by paying someone to remove and dispose of the waste. To protect themselves from future financial obligations, generators have an important stake in assuring that wastes are disposed properly. Generators also have an incentive for reducing the quantity of hazardous waste produced because proper disposal is often expensive.

Public concern about hazardous waste management continues. The seemingly insurmountable problems associated with the cleanup of contaminated "Superfund sites" are still with us. The cleanup of these sites appears to progress far too slowly in the minds of many, to say nothing about the enormous costs entailed in the cleanup process. On the other hand, the continuing refinement of regulations associated with the management of "new" hazardous wastes appears to be having an impact as manufacturers, institutions, and government regulators are finding effective ways of reducing the volume of hazardous wastes and better ways of storing, treating, and disposing of them. Concern about the proper handling of hazardous waste has even reached individual households and, as a result, many communities have implemented household hazardous waste collection programs to assist with the removal of potentially dangerous materials from residences.

The field of hazardous waste management is a rapidly growing area of study. It includes a complex mix of skills drawn from chemistry, the life sciences, engineering and technology, toxicology, environmental law, and public policy, to name just a few of the many disciplines needed to address the many components involved in hazardous waste issues. Texts by Blackman (1993) and Wagner (1992) are recommended to those who wish to pursue the subject of hazardous waste beyond the level of discussion provided in this chapter.

12.2 CHARACTERIZATION OF HAZARDOUS AND TOXIC WASTE

Whether a particular waste is considered hazardous or not is important to both the regulatory agencies and the waste generators. Regulatory agencies need a clear basis for requiring a generator to dispose of a waste which has been classified as hazardous. The decision to designate a waste as hazardous imposes significant financial obligations on a generator.

The RCRA regulations define hazardous waste as being either: (1) a listed hazardous waste, or (2) a characteristic hazardous waste. A listed waste is one that appears on the USEPA list of substances known to have hazardous and/or toxic properties. If a waste does not appear on the list, it may still be deemed a hazardous waste based on its characteristics and a laboratory test—the Toxicity Characteristics Leaching Procedure (TCLP). In addition, the regulations specify that a solid waste is a hazardous waste if it is a mixture containing a listed hazardous component along with other nonhazardous components or is a waste derived from the treatment, storage, or disposal of a listed hazardous waste. The RCRA regulations pertaining to hazardous waste identification are published in Part 261 of Title 40 of the Code of Federal Regulations, annotated as 40 CFR 261.

LISTED HAZARDOUS WASTE

Tables 12.1 through 12.4 contain selected examples of listed hazardous waste. Each specific type of listed hazardous waste has been assigned a hazardous waste code number by the USEPA. These code numbers are used in the notification and recordkeeping requirements that are part of the hazardous waste storage, transportation, and disposal regulations. The following hazard codes are used in these listings:

Ignitable waste (I)
Corrosive waste (C)
Reactive waste (R)
Acutely hazardous waste (H)
Toxic waste (T)

The hazardous wastes listed in Table 12.1 originate from nonspecific sources. This means that different manufacturing processes and industries produce the same waste. For example, waste xylene solvent (EPA waste number F001) originates in pharmaceutical, paint manufacturing, and cosmetic manufacturing industries.

Some hazardous wastes are generated at specific sources only. For example, waste sludges produced by the wood preserving industry using creosote and/or pentachlorophenol (EPA waste number K001) are produced only by this particular industry. Other selected examples of hazardous wastes from specific sources are listed in Table 12.2.

Some hazardous wastes, identified as acutely hazardous waste, are deemed to be so dangerous that small amounts are regulated as strictly as larger amounts of other hazardous wastes. The EPA waste numbers for these substances are prefaced with the letter "P". Selected examples of acutely hazardous wastes are shown in Table 12.3.

Another category of hazardous wastes, with examples shown in Table 12.4, consists of toxic commercial chemical products and chemical intermediates—compounds used in the manufacture of a final product. Some examples are benzene, chloroform, creosote, sulfuric acid, and pesticides such as kepone.

CHARACTERISTIC HAZARDOUS WASTE

A characteristic hazardous waste is deemed to be hazardous even though it does not appear on the lists described above because it possesses one or more of the following characteristics: toxicity, reactivity,

Table 12.1 Examples of Hazardous Waste from Nonspecific Sources (40 CFR 261.31)

EPA Waste Number	Hazardous Waste	Hazard Code
F001	The following spent halogenated solvents used in degreasing: tetrachloroethylene, trichloroethylene, methylene chloride, 1,1,1-trichloroethane, carbon tetrachloride, and chlorinated fluorocarbons; all mixtures and blends of spent solvents used in degreasing containing, before use, a total of 10% or more by volume, of one or more of the above halogenated solvents or those solvents listed in F002, F004, and F005; and still bottoms from the recovery of these spent solvents and spent solvent mixtures.	(T)
F002	The following spent halogenated solvents: tetrachloroethylene, methylene chloride, trichloroethylene, 1,1,1-trichloroethane, chloroethane, chlorobenzene, 1,1,2-trichloro-1,2,2-trifluoroethane, ortho-dichlorobenzene, trichloro-fluoromethane and 1,1,2-trichloroethane; all mixtures and blends of spent solvents containing, before use, a total of 10% or more by volume, of one or more of the above halogenated solvents or those listed in F001, F004, or F005; and still bottoms from the recovery of these spent solvents and spent solvent mixtures.	(T)
F003	The following spent nonhalogenated solvents: xylene, acetone, ethylacetate, ethyl benzene, ethyl ether, methyl isobutyl ketone, n-butyl alcohol, cyclohexanone and methanol; all mixture and blends of spent solvents containing, before use, only the above spent nonhalogenated solvents; and all spent solvent mixtures/blends containing, before use, one or more of the above nonhalogenated solvents and a total of 10% or more, by volume, of one or more of those solvents listed in F001, F002, F004 and F005; and still bottoms from the recovery of these spent solvents and spent solvent mixtures.	(T)
F004	The following spent nonhalogenated solvents: cresols, cresylic acid, and nitrobenzene; all mixtures and blends of spent solvents containing, before use, a total of 10% or more, by volume, of one or more of the above nonhalogenated solvents or those solvents listed in F001, F002 and F005; and still bottoms from the recovery of these spent solvents and spent solvent mixtures.	(T)
F007	Spent cyanide plating bath solutions from electroplating operations	(R,T)

**Table 12.2 Hazardous Waste from Specific Sources (Selected Examples)
(40 CFR 261.32)**

EPA Waste Number	Hazardous Waste	Hazard Code
Wood Preservation		
K001	Bottom sediment sludge from the treatment of wastewaters from wood preserving processes that use creosote or pentachlorophenol.	(T)
Inorganic Pigments		
K002	Wastewater treatment sludge from the production of chrome yellow and orange pigments.	(T)
K003	Wastewater treatment sludge from the production of molybdate orange pigments.	(T)
K004	Wastewater treatment sludge from the production of zinc yellow pigments.	(T)
Organic Chemicals		
K009	Distillation bottoms from the production of acetaldehyde from ethylene.	(T)
K010	Distillation side cuts from the production of acetaldehyde from ethylene.	(T)
K013	Bottom stream from the acetonitrile column in the production of acrylonitrile.	(R,T)
Pesticides		
K032	Wastewater treatment sludge from the production of chlordane.	(T)
K033	Wastewater and scrub water from the chlorination of cyclopentadiene in the production of chlordane.	(T)
K043	2,6-Dichlorophenol waste from the production of 2,4-D.	(T)
K099	Untreated wastewater from the production of 2,4-D.	(T)
Explosives		
K044	Wastewater treatment sludges from the manufacturing and processing of explosives.	(R)
K047	Pink or red water from TNT operations.	(R)
Petroleum Refining		
K048	Dissolved air flotation float from the petroleum refining industry.	(T)
K049	Slop oil emulsion solids from petroleum refining.	(T)

Table 12.3 Selected Examples of Acutely Hazardous Commercial Chemical Products and Manufacturing Chemical Intermediates (40 CFR 261.33)

EPA Waste Number	Substance	Hazard Code
P023	Acetaldehyde, chloro	(H)
P002	Acetamide	(H)
P070	Aldicarb	(H)
P004	Aldrin	(H)
P010	Arsenic acid	(H)
P015	Beryllium	(H)
P021	Calcium cyanide	(H)
P037	Dieldrin	(H)
P051	Endrin and metabolites	(H)
P056	Fluorine	(H)
P063	Hydrogen cyanide	(H)
P071	Methyl parathion	(H)
P095	Phosgene	(H)
P105	Sodium azide	(H)
P123	Toxaphene	(H)

Table 12.4 Toxic Commercial Chemical Products and Manufacturing Chemical Intermediates (Selected Examples) (40 CFR 261.33)

EPA Waste Number	Substance
U001	Acetaldehyde
U002	Acetone
U011	Amitrole
U019	Benzene
U225	Bromine cyanide
U044	Chloroform
U228	Trichloroethylene
U043	Vinyl chloride

ignitability, corrosivity, and infectivity. Descriptions of these characteristics follow.

Toxicity

Toxicity refers to the capacity of a substance to produce personal injury or illness to humans through ingestion, inhalation, or absorption through any body surface. Exposure to such a waste can result in acute or chronic health damage. Examples of toxic wastes include heavy metals, such as arsenic, cadmium, lead, and mercury, and many synthetic organic compounds, such as pesticides, PCBs, and solvents.

A waste may be defined as toxic because of its characteristics, or on the basis of a test, the Toxicity Characteristics Leaching Procedure (TCLP). In this test a sample of waste is treated with an acidic solution to simulate leaching activity in a landfill. After 24 hours, the extract is analyzed to determine if the concentrations of toxic constituents exceed the levels in Table 12.5. If they do, the waste is considered hazardous.

The terms hazardous waste and toxic waste are sometimes used interchangeably, but RCRA guidelines make a clear distinction between them. According to the RCRA guidelines, certain waste streams are designated as hazardous if they satisfy any one of several characteristics. Toxicity is one of these characteristics, but there are others as well. Therefore toxic waste is, in effect, a subset of hazardous waste that is regulated on the basis of human toxicity.

Reactivity

Reactivity refers to the tendency of a substance to react vigorously or violently with air, water, or other substances, resulting in the generation of harmful vapors or fumes or precipitating an explosion. Examples of reactive hazardous wastes are obsolete explosives and munitions, elemental alkali metals, such as sodium and potassium, and certain wastes from the explosive and chemical industries.

Ignitability

Ignitable wastes present a fire hazard because of their tendency to easily ignite or undergo spontaneous combustion at low temperatures. Such materials may also carry a risk of explosion and pose threats by dispersing toxic particulates and gases. Examples of ignitable wastes include organic solvents such as benzene and toluene, oils, plasticizers, and paint and varnish removers. Liquids with a flash point of less than 60°C

Table 12.5 Maximum Concentration of
Contaminants by TCLP Extraction,
Effective September 1, 1990

Contaminant	Regulatory Level (mg/L)
Arsenic	5.0
Barium	100.0
Benzene	0.5
Cadmium	1.0
Carbon tetrachloride	0.5
Chlordane	0.03
Chlorobenzene	100.0
Chloroform	6.0
Chromium	5.0
o-Cresol	200.0[a]
m-Cresol	200.0[a]
p-Cresol	200.0[a]
Cresol	200.0[a]
2,4-D	10.0
1,4-Dichlorobenzene	7.5
1,5-Dichloromethane	0.5
2,4-Dinitrotoluene	0.13[b]
Endrin	0.02
Heptachlor (and its epoxide)	0.008
Hexachlorobenzene	0.13[b]
Hexachlorobutadiene	0.05
Hexachloroethane	3.0
Isobutanol	1000.0
Lead	5.0
Lindane	0.4
Mercury	0.2
Methoxychlor	10.0
Methyl ethyl ketone	200.0
Nitrobenzene	2.0
Pentachlorophenol	100.0
Pyridine	5.0
Selenium	1.0
Silver	5.0
Tetrachloroethylene	0.7
Toxaphene	0.5
Trichloroethylene	0.5
2,4,5-Trichlorophenol	400.0
2,4,6-Trichlorophenol	2.0
2,4,5-TP (Silvex)	1.0
Vinyl chloride	0.2

[a]If o-, m-, and p-cresol concentrations cannot be differentiated, the total cresol concentration of 200 mg/L is used.

[b]If the quantitation limit is greater than the calculated regulatory level, the quantitation limit becomes the regulatory level.

Source: 40 CFR 261.24

(140°F) as determined by standardized tests are considered ignitable. Solids that spontaneously combust as a result of friction or through absorption or loss of moisture also are described as ignitable.

Corrosivity

An aqueous waste is corrosive if it is strongly acidic with a pH of 2 or less, or strongly alkaline with pH of 12.5 or more. Also considered corrosive is a liquid waste capable of corroding steel at a rate of 6.35 mm per year as determined by the National Association of Corrosive Engineers. Examples of corrosive substances are highly alkaline compounds such as caustic soda or strong acids such as sulfuric and nitric acids.

Infectivity

Infectious waste is capable of producing disease. For this to happen a number of factors are necessary, including (1) presence of a pathogen of sufficient virulence, (2) dose (quantity) of a pathogen, (3) route of a pathogen's entry, and (4) resistance of a host.

The infectious waste category includes pathological waste from hospitals, clinics, and certain laboratories, and raw sewage sludge. Special attention must be given to human and animal tissues, used hypodermic needles, and various viral and other types of microbiological wastes.

12.3 HAZARDOUS WASTE GENERATORS

Hazardous waste regulations in the United States place the primary accountability and liability for the proper disposal of hazardous waste on the producer of the waste. The physical removal of hazardous waste from a site constitutes generation. Generators are subject to the regulations contained in the 40 CFR 262—Generator Standards. The USEPA monitors generator activity by assigning a unique EPA ID number to each generator, hauler, and to each treatment, storage, and disposal facility. The generator may only transfer hazardous waste to a licensed hauler for transport to a licensed facility.

The tendency is to think of industries as the only hazardous waste generators, but commercial firms such as dry cleaners, exterminators, service stations, metal plating shops, printers, and others, also handle and dispose of chemicals subject to hazardous waste regulations. Households, schools, university and research laboratories, and hospitals are also generators. The quantity of hazardous waste generated by these different sources varies greatly. Consequently, RCRA Subtitle C regulations

Table 12.6 Requirements for Hazardous Waste Generators

	Requirements		
	Small-Quantity Generators	**Medium-Quantity Generators**	**Large-Quantity Generators**
HW quantity limits[a]	<100 kg/mo	100–1000 kg/mo	>1000 kg/mo
Manifest	Not required	Required	Required
Biennial report	Not required	Not required	Required
Personnel training	Not required	Basic training required	Required
Contingency plan	Not required	Basic plan required	Full plan required
EPA ID number	Not required	Required	Required
On-site storage limits	May accumulate up to 999 kg for 180 or 270 days if waste is to be transported over 200 miles	May accumulate up to 6000 kg for 90 days	May accumulate any quantity
Storage requirements	None	Basic requirements of 40 CFR 265	Full compliance with 40 CFR 265

[a]To qualify as a small or medium generator, the amount of acutely hazardous waste produced must be less than 1 kg/mo.

Source: Adapted with permission from Wagner, T.P., *The Hazardous Waste Q & A*, Van Nostrand Reinhold, New York, NY, 1992, p.53.

recognize three categories of generators, each differentiated by the quantity of hazardous waste produced: the small quantity generator (SQG); the medium-quantity (MQG); and the large quantity generator (LQG). These categories are defined on the basis of three factors: (1) the amount of hazardous waste generated per calendar month; (2) the amount of hazardous waste accumulated on-site at any one time; and (3) whether the waste is acutely hazardous. The regulations which apply to the various sized generators of hazardous waste are summarized in Table 12.6.

Some wastes are excluded from Subtitle C requirements because they do not fit the statutory definition of solid waste, are not intended by Congress to be regulated under Subtitle C, or are subject to other EPA regulations. Examples of these wastes include domestic sewage, mining overburden returned to the mine site, and household waste. Excluded wastes are listed under Section 261.4 of Subtitle C.

The on-site storage of hazardous wastes is regulated so that they are handled in a safe manner during the storage period. Wastes can be accumulated in containers or aboveground tanks provided they meet the technical standards of 40 CFR 265. Containers or tanks must be constructed of a material which is compatible with the hazardous waste to be stored, maintained in good condition, inspected weekly to ensure they are not leaking, and clearly labeled "Hazardous Waste."

12.4 TRANSPORTING HAZARDOUS WASTE

Hazardous waste may only be transported by a licensed hazardous waste transporter. Shipments of hazardous waste must comply with U.S. Department of Transportation (USDOT) regulations for packaging and shipping—found in 49 CFR Parts 170-177—and the USEPA manifest requirements for tracking the shipment. Detailed shipping requirements are contained in the Hazardous Materials Table (49 CFR 172.10). The hazardous waste must be properly packaged with packaging materials that are compatible with the waste being shipped. All hazardous waste shipping containers must be properly marked with the generator's name and address, USDOT shipping name and ID number, manifest numbers and the words:

> Hazardous Waste—state and federal law prohibits improper disposal. If found please contact the nearest police department, Division of Emergency Government or Department of Natural Resources.

In addition to this labeling, the generator must provide emergency response information, including a 24-hour emergency telephone number. The containers must have a hazard label that identifies the type of hazard presented by the waste so that those responding to an emergency can assess the hazards from a safe distance.

Transporters of hazardous materials must also display the appropriate placards on their vehicles. These are generally larger versions of the hazard labels that are affixed to the containers and tanks containing hazardous materials.

12.5 TREATMENT OF HAZARDOUS WASTE

Some hazardous waste substances may be treated for the purpose of eliminating, or at least greatly reducing, their hazardous characteristics. Often familiar chemical, physical, or biological treatments are capable of accomplishing this objective. The following selected processes illustrate some common treatment possibilities.

CHEMICAL AND PHYSICAL TREATMENTS

Neutralization

Acidic and alkaline (basic) liquid wastes can be treated by adjusting the pH to 7 by adding alkaline or acid solutions, respectively, as appropriate. This may be a simple process if an industry produces both acidic and alkaline wastes. Otherwise it may be necessary to purchase appropriate acids or bases for on-site treatment, or procure the appropriate chemicals from another company. Organizations called Waste Exchanges exist for the purpose of arranging swaps and sales of wastes.

Precipitation

Liquids with dissolved heavy metals can often be treated by raising the pH using lime. The hydroxide ion (OH^-) reacts with the metal cations forming insoluble compounds that precipitate or form flocs. The process is then followed by solidification or another process to immobilize the metal ions.

Oxidation and Reduction

The toxicity of some chemical species depends on the chemical oxidation state. When the oxidation state is increased by the removal of electrons or decreased by the addition of electrons, the resulting form may be less toxic or, perhaps, more easily removed from a solution. For example, hexavalent chromium (+6 oxidation state) is a highly toxic component of metal plating waste. But when the chromium ions are reduced to the comparatively innocuous trivalent form (+3 oxidation state) by the addition of lime, the chromium can be precipitated from the solution as chromium hydroxide. The representative reactions using sulfurous acid to reduce the oxidation state of chromium (Wentz, 1989) are the following:

$$2CrO_3 + 3H_2SO_3 \rightarrow Cr_2(SO_4) + 3H_2O$$

$$Cr_2(SO_4)_3 + 3Ca(OH)_2 \rightarrow 2Cr(OH)_3 + 3CaSO_4$$

As another example, cyanide solutions produced in metal-finishing processes can be effectively treated by oxidation to first produce the less toxic cyanate, CNO^-. Carbon dioxide and nitrogen can then be obtained from the cyanate. An oxidizing agent, such as a hypochlorite solution, hydrogen peroxide, chlorine, or ozone, can be used in this process. Typical reactions using chlorine as the oxidizing agent (Wentz, 1989) are:

$$NaCN + Cl_2 + 2NaOH \rightarrow NaCNO + 2NaCl + H_2O$$

$$2NaCNO + 3Cl_2 + 4NaOH \rightarrow 2CO_2 + N_2 + 6NaCl + 2H_2O$$

Activated Carbon Adsorption

Activated carbon may be used to remove organic substances from an aqueous waste. The fine particles of carbon present a large surface area, 600–1000 m^2/g, onto which organic molecules may be adsorbed. The spent carbon may be regenerated by heating in an oven or by passing steam through it.

Solidification or Encapsulation

Solidification is a method of immobilizing substances, such as heavy metal precipitates, by mixing them with a mixture of a pozzolanic material, such as cement or fly ash, adding water, and permitting the mixture to set into a concrete. Encapsulation of the waste into a glass is another possibility.

Distillation

Many organic solvents are disposed after they become contaminated with dirt and dissolved grease. Spent solvents can be recovered by distillation. Waste vehicle oil is commonly collected and reprocessed by distillation.

BIOLOGICAL TREATMENT

Leaking underground storage tanks and other types of uncontrolled spills result in large volumes of contaminated soils and groundwater. The cleanup of such contaminated environments has placed great stress on available technologies. Biological treatment, often called **bioremediation**, uses the same basic biological processes discussed in Chapter 7. A variety of soil bioremediation techniques are in use, including in situ, biopiles, landfarming, and engineered reactors.

In situ bioremediation is accomplished by introducing nutrients and oxygen or other electron acceptors to the contaminated site so that microbes can actively degrade the contaminants in place; that is, without excavating and removing contaminated soil. An alternative is to excavate the site and place the soil in a "pile" with the objective of maintaining proper moisture and pH conditions. Various amendments such as microbes, nutrients (e.g., nitrogen and phosphorus), and oxygen, may be included in the "pile" to facilitate the degradation process. Landfarming consists of spreading relatively thin layers of contaminated soil on the ground and then periodically tilling it with common farm equipment to maintain a well aerated environment. Again, nutrient levels, moisture, and pH conditions are adjusted as needed. Finally, the engineered reactors used for biological treatment of hazardous waste are based on design principles and operations described previously in Chapters 6 and 7.

Many variables influence biological degradation of toxic organic compounds in a complex environment such as soil. Prior to treatment, it is necessary to demonstrate the ability of the microbes to effectively degrade the toxic, without the microbes succumbing to the toxic substances. It is also essential to identify and eliminate potentially toxic intermediates produced during the biological breakdown of the parent compound(s).

Some organic compounds are not amenable to biological breakdown except under very specific conditions. For example, polychlorobiphenyl compounds (PCBs) are resistant to aerobic degradation but may be readily dechlorinated in a reducing environment under anaerobic conditions, after which the dechlorinated organic structures become much more susceptible to breakdown by aerobically metabolizing microbes.

An important question in biological treatment is whether the microbes are able to degrade toxic organic substances to sufficiently low concentrations. For some toxic substances, reductions to parts per billion (ppb) or even parts per trillion (ppt) may be necessary to achieve the desired level of cleanup so that humans and the environment are protected. Another consideration is that bioremediation is an inherently slow process, possibly requiring several months. This may be unacceptably slow for contractors adhering to a stringent work schedule, even if the cost of biological treatment is less than that for an alternative cleanup method. Costs vary greatly and tend to be very site-specific. Torpy et al. (1989) reported conventional land treatment costs at $65 to $105/m^3 and composting fees averaging $130/m^3 of soil. Incineration with a mobile rotary kiln can cost $290 to $1450/m^3 of soil (USCOTA, 1989b). The removal of contaminated soil to a sanitary landfill may cost $260-$390/m^3 (USEPA, 1988).

It is evident from the above discussion that the application of biological methods to environmental cleanup of hazardous waste is complex and there are many uncertainties, but the USEPA sees bioremediation as a source of promising technologies for solving hazardous waste problems (USEPA, 1992). Bioremediation is attractive because it is a natural process generating harmless residues such as carbon dioxide and water. A major advantage of bioremediation over other hazardous waste disposal options, such as incineration or landfilling, is that the biological approach degrades the target chemical, whereas the other two options may merely transfer the contaminant (or emissions from the contaminant) from one medium to another.

12.6 DISPOSAL OF HAZARDOUS WASTE

Practically speaking, there are only two methods for the ultimate disposal of hazardous waste—incineration and land disposal. Other methods have been tried and tested. These include ocean dumping, ocean incineration, underground disposal, underground injection, and land spreading. However, these methods have since been prohibited or severely restricted.

Hazardous waste disposal in a landfill, even in a well-engineered one, poses a long-term risk of a future financial liability, a risk waste generators are increasingly reluctant to assume. As a consequence many generators are looking for safe ultimate disposal methods such as incineration.

LAND DISPOSAL

Landfills designed to accept municipal waste are not adequate sites for the disposal of hazardous waste, but some types of hazardous waste can be landfilled in specially-designed sites known as secure landfills. Such landfills are equipped with double liners, leakage detection equipment, leachate monitoring and collection, and groundwater monitoring equipment. Synthetic liners with a minimum thickness of 0.76 mm (30 mil) are mandated. Improvements in technology now allow liners to be put in place in very large sections, thus minimizing the number of joints. Special on-site welding machines make it possible to create leakproof joints between individual sections of liner material. Some states allow the use of natural clays for liners, provided the hydraulic conductivity is less than 10^{-6} cm/s and a leachate collection system is in place.

INCINERATION

Incineration in a facility with appropriate emission control equipment is the preferred method for destroying organic compounds, particularly halogenated organic compounds. The predominant combustion products are CO_2 and H_2O with small amounts of HCl or other halogen compounds. The latter must be captured from the stack gases through reactions with lime or other alkaline compounds to produce nonhazardous salts.

Incineration does not destroy heavy metals, but can alter their form from sulfides or cyanides to oxides. The heavy metal compounds appear in the bottom ash, the fly ash, and the scrubber sludge. All of these solids require stabilization using processes such as the solidification or encapsulation methods described in the previous section. The scrubber liquor discharge from cleanup of such stack emissions may be defined as a hazardous waste if the original waste stream contained substantial amounts of heavy metal elements (Whitworth and Waterland, 1992).

Kilns used in the production of portland cement kiln are widely used for destroying certain types of wastes. Cement production is an energy intensive process using fossil fuels such as coal, gas, and oil. With appropriate kiln management and choice of wastes, some of these fuels can be replaced with wastes. Over 1.2 million tonnes (1.3 million tons) of hazardous waste were combusted in cement kilns in 1991 (Hansen, 1992). Kleppinger (1993) estimates that 70% of liquid and solid hazardous wastes commercially incinerated in the U.S. are being burned in cement kilns.

The cement kiln is well suited for destruction of wastes as a result of the high temperatures, oxidizing conditions, alkaline environment, and long residence times. A kiln is, in effect, a large furnace approximately four meters in diameter and over 100 meters in length. The raw materials for cement production enter at one end and are heated to approximately 870°C during their passage through the first one-third of the kiln. The temperature increases to 1,400°C in the last one-third of the kiln. The process requires 3–4 hours, which is a sufficient period of time to effectively destroy wastes.

Although many organic hazardous wastes may be effectively destroyed in a cement kiln, the fate and impact of heavy metals is uncertain. According to Kleppinger (1993), the concentrations of the heavy metals in the cement produced by these kilns are likely to increase. He recommends the development of a new standard for cement containing hazardous or other types of wastes.

12.7 MINIMIZATION OF HAZARDOUS
WASTE QUANTITIES

An important aspect of managing hazardous waste in an industry or institution is the intensive and ongoing examination of practices which hold potential for minimizing its generation. Figure 12.2 provides an overview of waste minimization techniques. Actions that can be initiated to implement these techniques are:

1. Segregate hazardous waste—Mixing hazardous waste with nonhazardous waste results in the total amount being considered hazardous.
2. Monitor inventory—Avoid purchasing excess amounts of chemicals to reduce surpluses. Advertise and share materials between departments within a large organization. A surplus chemical reagent from one department might be used in another. It is advantageous for an organization to set up a system of communication so that key individuals are aware of such "exchange" possibilities and of the fact that disposal of unused materials defined as hazardous may often cost more than the original purchase of the material.
3. Investigate material substitution—Substituting a nonhazardous material for a hazardous one in a manufacturing process may make it possible to completely avoid the hazard and cost of disposal associated with some materials. For example, benzene was widely used as an industrial solvent in the past but the identification of its high level of hazard has led to its replacement by other materials in many manufacturing processes.
4. Modify processes—Process changes may lead to production efficiencies which generate less waste. If members of the production staff are aware of the great cost associated with hazardous

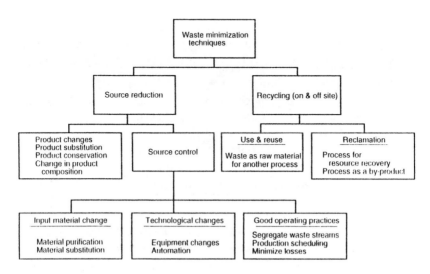

Figure 12.2 Minimizing hazardous waste generation.

waste generation they may be able to find new ways to maintain productivity with less waste generation. The decision to purchase a new machine or adopt a new technology should include waste generation as a consideration. For example, ultrasonic cleaning with aqueous detergents has been shown to be as effective as the previously used chlorinated solvents in removing rust-preventive oil, lapping oil, and machine coolant from aluminum, brass, and iron and steel parts (Vaccari, 1993).

5. Change product design—Encourage research and development and marketing staff to design products with an emphasis on minimizing waste during production, use, and disposal.

6. Investigate waste exchanges—Waste exchange is becoming an increasingly effective way for manufacturers and others to exchange excess inventory, off-specification materials and hazardous wastes. The waste solvent of one company might be acceptable to another as a raw material. Such exchange services deal with nonhazardous as well as hazardous materials.

7. Investigate recycling opportunities—Some waste disposal costs might be avoided by recycling. For example, a solvent contaminated with dissolved greases and suspended solids might be reclaimed by distillation. In addition, organic solvents can be burned to achieve energy recovery.

Hirschhorn (1989) describes waste reduction in terms of four increasingly more complex and costly levels of activity: (1) common sense waste reduction; (2) information-driven waste reduction; (3) audit dependent waste reduction; and (4) research and development waste reduction.

Common sense waste reduction refers to low cost, low risk, readily visible, perhaps obvious, approaches to waste reduction. There is little need for new technology in achieving this level of waste reduction. It may be possible to walk through a production process and spot waste reduction opportunities that can be implemented easily and almost immediately. These efforts may not even require much in the way of engineering analysis or investment. For example, the use of water in cleaning operations may be reduced significantly with the aid of certain mechanical techniques. Commingling of hazardous and nonhazardous wastes should be avoided or greatly minimized. Volatile chemicals can be contained in closed vessels. Closed loop recycling of water may be feasible.

The second level of waste reduction, information-driven, is based on more detailed information about the waste and its production origins. Production people need to understand how a waste is generated and its connection to the raw material and product quality in order to determine potential waste reduction opportunities. For example, it may be helpful to know whether a raw material change can provide a means for reducing waste, such as replacing an organic solvent with a water-based product or a highly toxic solvent with a much less toxic solvent. The information required in this stage of waste reduction may be available from many sources: in-house, literature, conferences, workshops, discussions with other production people, and, increasingly, from government waste reduction programs.

Hirschhorn's third level of waste reduction, audit dependent, is necessary if additional reduction is to occur after the benefits from the first two stages have been achieved. It now becomes necessary to have an audit program to identify the reduction possibilities at a technological and economically feasible level. At this stage technological barriers may be identified, resulting in uncertainties about economic payback. More capital investment is often necessary and financial risks are increased. Payback periods may be long. Testing and development in the laboratory and in production becomes necessary. The impact on product quality requires careful study. Finally, new wastes may be created that could negate attempts at overall hazardous waste reduction.

The fourth stage of waste reduction, research and development (R&D), implies substantial research and development efforts related to process technology and equipment, as well as product composition and design. A potential payback for these efforts might result if a marketable process is developed.

USEPA'S 33/50 PROGRAM

In 1991 USEPA initiated a program aimed at over 600 U.S. companies asking them to voluntarily reduce the release and transfer of 17 high-priority toxic chemicals (USEPA, 1991). The 33/50 Program calls for a stepwise reduction in the release and transfer of these 17 target chemicals from 33% by the end of 1992 to 50% by the end of 1995. Reduction will be measured against the 1988 data on these chemicals. The 33/50 Program chemicals were selected on the basis of criteria which included health and environmental effects, possibility of exposure, volume of production and release, and the potential for reduced pollution. The compounds or classes of compounds identified under the 33/50 Program are listed in Table 12.7 and the relative amounts of 33/50 chemicals for 1989 are depicted in Figure 12.3. The program overlaps with the Clean Air Act Amendments of 1990 because over 70% of the releases and transfers of these target chemicals are in the form of air emissions. As a consequence, USEPA sees the two programs working in concert to promote pollution reduction in advance of statutory timetables.

Table 12.7　Chemicals in the 33/50 Program

Cadmium and compounds
Chromium and compounds
Lead and compounds
Mercury and compounds
Nickel and compounds
Benzene
Methyl ethyl ketone
Methyl isobutyl ketone
Toluene
Xylene
Carbon tetrachloride
Chloroform (trichloromethane)
Methyl chloride (dichloromethane)
Tetrachloroethylene (perchloroethylene)
1,1,1-Trichloroethane (methyl chloroform)
Trichloroethylene
Cyanides

Source: U.S. Environmental Protection Agency, Toxics in the Community: National and Local Perspectives, Office of Toxic Substances, Economics and Technology Division, EPA/560/4-91-014, U.S. Printing Office, Washington, DC, 1991.

12.8　CLEANUP OF CONTAMINATED WASTE SITES (SUPERFUND)

Any hazardous waste disposal facility remaining in active operation, or newly constructed, after November 19, 1980 is required to comply with RCRA regulations. As a result, many generators, treaters, and disposers elected to close their existing facilities. But RCRA failed to address the environmental degradation created by these and other abandoned contaminated sites; hence, the Comprehensive Environmental Response, Compensation and Liability Act (CERCLA) was passed. CERCLA established the Hazardous Waste Trust Fund, nicknamed "Superfund," with an original budget of $1.6 billion.

The Superfund program included two primary components: (1) an emergency response, providing for the immediate removal of spills or abandoned materials posing an imminent danger to humans or the envi-

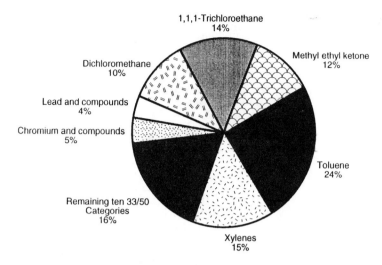

Figure 12.3 USEPA Toxic Release Inventory releases and transfers of 33/50 chemicals in 1989. (USEPA, 1991).

ronment, and (2) provision for the long-term cleanup and restoration of abandoned toxic waste sites which are identified as "priority" sites based on the criterion that releases of hazardous substances had occurred or might occur in the future. Sites identified for cleanup are placed on the National Priorities List (NPL).

An important aspect of the program is the allocation of financial responsibility for the remediation activities to the responsible parties, among whom are the producers of the wastes deposited at the site, provided they can be identified. If responsible parties cannot be identified, then federal and state monies fund the cleanup costs. In its first decade of operation, the EPA reached settlements with 372 responsible parties and received $619 million for cleanup costs (Lipták, 1991). However, the cleanup process is slow, requiring an average of five years or more to complete cleanup of a site. Only 14 sites had been cleaned up by 1986, and only 30 by 1990. The U.S. Congress demanded more and faster action. This led to the Superfund Amendments and Reauthorization Act (SARA) which was signed into law on October 17, 1986.

SARA was a 5-year extension of the Superfund program to clean up hazardous releases from abandoned or uncontrolled hazardous waste sites. SARA was funded at a much higher level, $8.5 billion, and included a separate $0.5 billion fund for cleanup of leaking underground petroleum storage tanks. SARA provisions also called for more stringent remedial

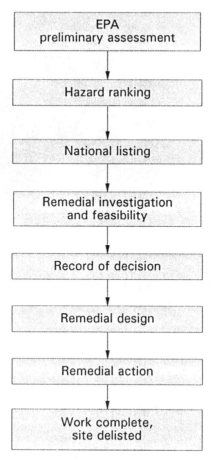

Figure 12.4. Flow chart of steps in the process for identifying sites for the National Priorities List.

standards. Overall support for Superfund cleanup comes from taxes on petroleum and chemicals, an environmental tax on corporations, $1.25 billion from general appropriations, costs recovered from responsible parties, punitive damages and penalties, and earned interest.

It is important to note that NPL sites are considered the most hazardous of the many thousands of contaminated sites known to exist. An NPL site is ranked as such on the basis of an extensive assessment. The hazard identification and eventual cleanup process is depicted in the form of a flow chart in Figure 12.4.

The government (USEPA) becomes aware of a potential NPL site from citizen complaints or through nominations by states. A preliminary site assessment is made by the USEPA to determine how the site was originally used and what wastes were placed there. At this point, many sites are judged sufficiently harmless so that no further attention is received from the EPA. However, if evidence indicates potential for significant harm, the site enters the formal hazard ranking process. The Hazard Ranking System (HRS) evaluates five pathways: air, groundwater, surface water, fire and explosion, and direct contact. This process results in a numerical score where a value beyond a certain number puts the site on the NPL list.

Once a site is listed, a Remedial Investigation/Feasibility Study (RI/FS) is conducted. The site is examined in terms of its characteristics and the quantity and types of wastes deposited there. Cleanup alternatives and their costs are identified. After the RI/FS, the regional EPA administrator issues a record of decision (ROD) after inviting comments from the community and state and local government officials. At this point remediation begins, and when it is completed the site is eligible for "delisting" from NPL.

12.9 INTERNATIONAL ASPECTS OF HAZARDOUS WASTE MANAGEMENT

The nations of western Europe are perceived by many to be generally ahead of the United States in their approach to efficient and effective hazardous waste management. In Europe, there is a particular emphasis on pollution prevention, in contrast to the American tendency to use the end-of-the-pipe treatment approach in dealing with wastes (Davis et al., 1987). The European emphasis on pollution prevention is attributed to four basic factors: (1) scarcity of land for land disposal, (2) great concern for groundwater contamination, (3) a conservation ethic derived from the post-World War II material scarcity and (4) a tradition of a high level of cooperation between government and industry in all areas of technology and industrial policy. There seems to be general agreement that the heavily industrialized nations of western Europe have been quicker to adopt hazardous waste reduction strategies, such as source separation, process modification, end-product substitution, and recycling.

The sophistication of European countries in matters of hazardous waste disposal may be overstated in some instances. For example, the British have deliberately mixed hazardous wastes with conventional MSW in permeable landfills. The belief is that the toxic chemicals will be converted to nontoxic forms through leaching and biological degradation (Piasecki and Brooks, 1987). Many in the field of waste management in

Europe and the United States disagree with this co-disposal approach to hazardous waste management.

Comparing nations with respect to their approaches in dealing with complex environmental issues such as hazardous waste management is difficult at best. There is no agreement on which chemical substances are toxic. In addition, a unanimous agreement on the best approach to treatment and ultimate disposal does not exist. European countries and Japan use incineration for disposal extensively, while the United States continues to emphasize the land disposal option.

The high cost of hazardous waste management encourages international businesses to locate manufacturing plants that produce hazardous waste in countries with less stringent disposal regulations. This issue was raised prior to the passage of the North American Free Trade Agreement (NAFTA) by the United States, Canada, and Mexico in 1994. The financial motivations for generators to export hazardous waste to third world countries which lack sufficient management regulations are substantial. To address this problem, on March 22, 1989 in Basel, Switzerland, over 100 nations adopted a treaty prohibiting shipments of hazardous waste across international boundaries without obtaining permission of the receiving country prior to shipment (Lipták, 1991).

12.10 HOUSEHOLD HAZARDOUS WASTE

More than 100 substances listed as RCRA hazardous wastes are present in household products such as cleaners, solvents, pesticides, paints, and household and automotive maintenance products. Selected examples of such hazardous materials and the products in which they may be found are shown in Table 12.8. The disposal of these products is exempt from hazardous waste regulations only because of the small quantities generated by each household.

The U.S. Congress Office of Technology Assessment (1989a) summarized data from several studies which determined the amounts of household hazardous waste (HHW) generated. The studies yielded similar conclusions; namely, HHW accounts for 0.2 to 0.4% of the residential portion of the MSW stream. However, even using the conservative value of 0.2% as the HHW generation rate implies over 220,000 tonnes of HHW are discarded each year. Many hazardous household products are emptied directly into sewer systems, and would not be reported in these studies. In fact, it is worth noting that regardless of whether the HHW is discarded into the sewer system or is landfilled and transformed into the collected leachate at a state-of-the-art landfill and subsequently treated at a municipal sewage treatment plant, a significant amount of non-biodegradable HHW may eventually be released to surface waters by way

Table 12.8 Selected Examples of Hazardous Ingredients Found in Common Household Products[a]

Ingredient	Types of Products Found In
Acrylic acid	Adhesives
Aniline	Cosmetics(perfume), wood stain
Arsenic(III) oxide	Paint (nonlatex anti-algae)
Benzene	Household cleaners,(spot remover, oven cleaner) stain, varnish, adhesives cosmetics (nail polish remover)
Cadmium	Ni-Cd batteries, paints, photographic chemicals
Chlordane	Pets (flea powders)
Chlorinated phenols	Latex paint
Chlorobenzene	Household cleaners (degreaser)
Hexachloroethane	Insect repellents
Lead	Stain/varnish, auto batteries, paint
Mercury	Batteries, paint (nonlatex anti-algae), fluorescent lamps
Methylene chloride	Household cleaners, paint strippers, adhesives
Nitrobenzene	Polish (shoe)
Silver	Batteries, photographic chemicals
Warfarin	Rodent control
Xylene	Transmission fluid, engine treatment (degreaser), paint (latex, nonlatex, lacquer thinners), adhesives, microfilm, fabric, cosmetics (nail polish)

[a]Hazard based on 40 Code of Federal Regulations 261.

Source: Selected examples from Table 3.5 in U.S. Congress Office of Technology Assessment, Facing America's Trash: What Next for Municipal Solid Waste?, OTA-O-424, U.S. Printing Office, Washington DC, October, 1989.

of municipal treatment plant effluent or to land by the land application of sewage sludge.

One can argue that the environmental impact of disposing of HHW is small since the HHW is spread among thousands of locations, but there is concern and interest in removing those materials which present the greatest risks. By 1993, 37 states banned the disposal of vehicle batteries in landfills and 17 banned the disposal of motor oil in landfills (see Appendix E for a listing by state).

An increasing number of communities have organized periodic HHW collection days. Fewer than 100 programs were in existence in 1984 but by 1989 this number had grown to more than 600 (Duxbury, 1990). The impact of the periodic HHW collections seems to be mixed. While some HHW is indeed diverted from the usual MSW stream, the participation in such programs tends to be low, often less than 1%, and the costs of the special HHW collection programs are high, often several thousand dollars per tonne collected. In addition, occasional programs may not always produce the desired objective of keeping hazardous waste out of the MSW stream. As noted in Chapter 1, the University of Arizona Garbage Project (Rathje and Murphy, 1992) found in a 1986 study that more HHW appeared in the MSW stream following a special HHW collection day than before. It was speculated that the public awareness resulting from the HHW media campaign stimulated people to gather their household hazardous waste, but when, for one reason or another, they missed the special collection day they proceeded afterwards to discard their HHW along with the other household wastes.

In an effort to increase the participation in their HHW collection programs, some communities have established permanent facilities with regular operating hours. There were fewer than 10 such facilities in 1985, but the number has grown rapidly to over 125 in 1993.

Support for HHW collection programs continues. Many people believe there is educational value in these efforts, especially as a means of promoting the use of alternative, less toxic products. The efforts may also encourage manufacturers of household products to reduce the hazardous material content of their products. For example, a major manufacturer of alkaline batteries notes on a battery package, "No mercury added for a safer and cleaner environment—contains only naturally occurring trace elements."

DISCUSSION TOPICS AND PROBLEMS

1. Prepare a list of items in your residence that qualify for classification as hazardous waste.
2. Distinguish between a Listed Hazardous Waste and a Characteristic Hazardous Waste.
3. What is meant by the term "cradle-to-grave management" when applied to hazardous waste? Why is it an important concept in hazardous waste management?
4. What is the definition of an acutely hazardous waste?

5. A waste is analyzed using the TCLP procedure. The following concentrations of potentially hazardous materials are identified:

Arsenic	4 mg/L
Cadmium	0.8 mg/L
Lead	6 mg/L
Mercury	0.1 mg/L

Is this mixed waste considered hazardous? Justify your answer.
6. What are the objectives of the 33/50 program? Which companies are affected by this program?
7. Select an industry in your community or region and describe how it handles its hazardous waste.
8. Does your community have a household hazardous waste (HHW) collection program? If it does,

   (a) How often is the HHW collected?
   (b) How is HHW collected?
   (c) How much HHW is collected annually?
   (d) What is the average amount of HHW collected annually per person?
   (e) What is the cost per tonne of disposing of the HHW?
   (f) From your perspective, what appear to be the most significant weaknesses of the current collection program?

   If your community does not have a HHW collection program, should one be established? Provide an analysis to justify your answer.
9. Select a hazardous waste generator in your community and describe what it has done in the past five years to reduce the volume of hazardous waste generated.
10. What programs does the school system in your community or your college campus have: (a) to minimize the use of hazardous materials in classrooms and laboratories and the operation of the schools in general, and (b) to collect and dispose of hazardous materials?

REFERENCES

Blackman, W.C., Jr., *Basic Hazardous Waste*, Lewis Publishers, Boca Raton, FL, 1993.

Davis, G., Huisingh, D., and Piasecki, B., Waste reduction strategies: European practice and American prospects, in *American's Future in Toxic Waste Management*, by Piasecki, B., and Davis, G., Quorum Books, New York, NY, 1987.

Duxbury, D., Emerging prominence for HHW, *Waste Age*, 21(6), 37–38, 40, June, 1990.

Hansen, E.R., Treatment and Destruction of Hazardous Waste in Portland Cement Kilns, Fifth Conference on Toxic Substances, Montreal, Canada, 1992.

Hirschhorn, J.S., Hazardous waste reduction, in *Management of Hazardous Materials and Wastes: Treatment Minimization and Environmental Impacts*, Pennsylvania Academy of Science, Easton, PA, 1989, Chapter 9 .

Kleppinger, E.E., Cement clinker: An environmental sink for residues from hazardous waste treatment in cement kilns, *Waste Management*, 13(8), 553–572, 1993.

Lipták, B.G., *Municipal Waste Disposal in the 1990s*, Chilton Book Company, Radnor, PA, 1991.

Piasecki, B. and Brooks, J., Government's aid: The role of citizen and environmental groups in Europe, in *In American's Future in Toxic Waste Management*, by Piasecki, B., and Davis, G., Quorum Books, New York, NY, 1987.

Rathje, W. and Murphy, C., *Rubbish! The Archaeology of Garbage*, Harper Collins Publishers, New York, NY, 1992.

Torpy, M.F., Stroo, H.F., and Brubaker, G., Biological treatment of hazardous waste, *Pollut. Engineer*, 21, 80, 1989.

U.S. Congress Office of Technology Assessment (USCOTA), Facing America's Trash: What Next for Municipal Solid Waste?, OTA-O-424, U.S. Printing Office, Washington DC, October, 1989a.

U.S. Congress Office of Technology Assessment (USCOTA), Coming Clean: Superfund's Problems Can Be Solved, OTA-ITE-433, U.S. Printing Office, Washington, DC, October, 1989b.

U.S. Environmental Protection Agency, Survey of Household Hazardous Wastes and Related Collection Programs, Office of Solid Waste and Emergency Response, EPA/530-SW-86-054, U.S. Printing Office, Washington, DC, October, 1986.

U.S. Environmental Protection Agency, Cleanup of Releases from Petroleum USTs: Selected Technologies, Office of Underground Storage Tanks, EPA/530/UST-88/001, U.S. Printing Office, Washington, DC, 1988.

U.S. Environmental Protection Agency, The Toxics Release Inventory: A National Perspective, Office of Toxic Substances, Economics and Technology Division, EPA 560/4-89-005, U.S. Printing Office, Washington, DC, 1989.

U.S. Environmental Protection Agency, Toxics in the Community: National and Local Perspectives, Office of Toxic Substances, Economics and Technology Division, EPA/560/4-91-014, U.S. Printing Office, Washington, DC, 1991.

U.S. Environmental Protection Agency, Bioremediation of Hazardous Wastes, Biosystems Technology Development Program, EPA/600/R-92126, U.S. Printing Office, Washington, DC, 1992.

U.S. Environmental Protection Agency, National Biennial RCRA Hazardous Waste Report, EPA/530-R-92-027, U.S. Printing Office, Washington, DC, 1993.

Vaccari, J., Ultrasonic Cleaning with Detergents, *American Machinist*, 41–42, 1993.

Wagner, T. P., *The Hazardous Waste Q&A: An In-depth Guide to the Resource Conservation and Recovery Act and The Hazardous Materials Transportation Act*, Van Nostrand Reinhold, New York, NY, 1992.

Wentz, C.A., *Hazardous Waste Management*, McGraw-Hill Publishing Company, New York, NY, 1989.

Whitworth, W.E., and Waterland, L.R., Pilot-Scale Incineration of PCB-Contaminated Sediments from the New Bedford Harbor Hot Spot Superfund Site, EPA/600/SR-92/068, U.S. Printing Office, Washington, DC, 1992.

CHAPTER 13

Costs and Management of Waste Facilities and Systems

13.1 OVERVIEW

In the previous chapters primary attention was given to the scientific and engineering aspects of waste management systems. An understanding of the technical factors which underlie the generation, collection and transport, and processing of waste and wastewater is indispensable for identifying and implementing good waste management practices. In addition, the evaluation of alternative means for the disposal of solid residuals and sludges, and an understanding of the environmental impacts of all elements of a waste management system require scientific and technical expertise. While technical analyses provide necessary and important information concerning the feasibility and suitability of pursuing specific waste management alternatives, this information is not sufficient. Good estimates of costs are needed so that wise decisions can be made on the investment of scarce economic resources.

Waste management planning requires the forecasting of future costs over a project lifetime—typically 15 or 20 years—for various alternatives under consideration. This task is often difficult because many assumptions must be made about the appropriate sizes of facilities, interest and inflation rates, the costs of labor and maintenance, availability of markets for the sale of products, and the revenues that will be received during the lifetime of a project. If major waste generators join or leave a system, the unit cost for treatment or processing and for disposal of the residues may change significantly. Other events, such as a change in regulations, may also severely impact costs. In response, major modifications to processes, buildings or equipment, or the abandonment of facilities may be required.

In spite of these uncertainties, planning and decision-making must proceed.

In this chapter we describe and discuss the types of costs associated with the facilities which are part of waste management systems. Methods for estimating future costs and procedures for updating costs associated with comparable facilities which have been incurred in the past will also be discussed. Revenues must be raised to cover costs associated with waste management systems. Alternative types of fee policies, financing arrangements for raising the required revenues, ownership of facilities, and managerial arrangements are addressed in the concluding sections.

13.2 CAPITAL AND OPERATING COSTS OF FACILITIES

There are two distinct types of costs attributed to a waste management facility: the capital or fixed costs of buildings and equipment and the operating or variable costs associated with the daily plant activities as waste is processed or handled.

CAPITAL COSTS

Capital costs for facilities are those incurred in the planning and construction phases of a project and the cost of the equipment needed for waste processing or handling. These costs are directly related to the type and size of the project. Capital costs are generally too large to be paid in a single year, so they are amortized over some period of time, often the project lifetime.

The capital cost is strongly dependent upon the size or capacity of a facility; however, it becomes a financial obligation whether the facility is used to its capacity or not. When the capital costs are allocated to the waste throughput—a normal practice—it is easy to recognize that the capital cost per unit of waste processed increases when the facility receives less waste than its design capacity. The facility planner tries to select a capacity that will adequately handle current waste quantities and their seasonal fluctuations, but one that will also allow for some future growth. Allowance for future growth must be balanced against an unreasonable excess capacity that raises the capital cost component of unit costs to an unacceptably high level.

One method for obtaining a preliminary estimate of the capital cost of a proposed facility is to base it on published studies of waste management system costs. Such estimates are often expressed as a simple "rule-of-thumb" number. For example, the capital cost of a field erected incinerator has been estimated to be $150,000–$175,000 (1992 dollars) per tonne of daily design capacity (Beck, 1992). Using these figures, an estimate of the capital cost of a 400 t/d incinerator would be 60–70 million dollars.

A second method for estimating the capital cost of a facility to be built today or at some future date is by updating the costs of a similar existing facility built at an earlier time. In this case, the costs from the earlier year must be adjusted to reflect changes in construction and equipment costs resulting from higher wages and material prices. These costs can be adjusted using the Construction Cost Index (CCI). Table 13.1 shows the Construction Cost Index values for the years 1965 through 1993. These values are referenced to a value of 100 for the year 1913, which is the base year for the Construction Cost Index. The capital cost of a facility built in an earlier year can be updated to the current year by the simple calculation

$$\text{Present cost} = \text{past cost} \times \frac{\text{present year's CCI value}}{\text{past year's CCI value}}$$

If cost data are available for a similar type facility, but of different capacity, then a procedure is needed to scale the cost to correspond to the different size. Several authors (Miller, 1973; Grant & Cooper, 1982; Wilson, 1981) postulate a relationship between capital cost, C, and plant capacity, W, as follows:

$$C = a\, W^s$$

Table 13.1 Annual Average Values of the Construction Cost Index

Year	Index Value	Year	Index Value
1913	100	1979	3003
1965	971	1980	3237
1966	1019	1981	3535
1967	1074	1982	3825
1968	1155	1983	4066
1969	1269	1984	4146
1970	1381	1985	4195
1971	1581	1986	4295
1972	1753	1987	4406
1973	1895	1988	4519
1974	2020	1989	4615
1975	2212	1990	4732
1976	2401	1991	4835
1977	2576	1992	4985
1978	2776	1993	5210

Source: *Engineering News Record*, McGraw-Hill Publishers, New York, NY, March, 1994.

where a is a constant of proportionality and s is a **scale factor**. If the scale factor is known for a particular type of plant, the relationship between the costs C_A and C_B of two plants with capacities W_A and W_B, respectively, is given by the equation

$$C_A / C_B = (W_A / W_B)^s$$

If the value of the scale factor is 1, then cost is a linear function of capacity. A value of s less than 1 indicates that an economy of scale exists, which means that the capital costs do not increase as rapidly as the design capacity does. Wilson (1981) states that economies of scale are expected in the capital costs of most engineering plants, but that evidence for the existence of economies of scale is lacking. However, several authors, including Wilson, have investigated this issue and the scale factors, s, derived from their studies of different waste processing and disposal plants are displayed Table 13.2.

Table 13.2 Scale Factors, s, for Waste Processing and Disposal Plants

Process	Midwest Research Institute (1973)	Schultz (1976)	Bridgewater (1977)	Wilson (1981)	Rhyner & Wenger (1986)
Landfill	0.93	–	0.93	0.8	–
Transfer station	0.93	–	–	0.85	–
Pulverization	–	–	–	0.85	–
Baling	–	–	–	0.9	–
Incineration					
High capacity	0.78	0.65	0.9–0.95	0.95	–
Low capacity	–	–	0.78	0.85	–
Heat recovery	–	–	–	–	1.01
Modular incineration			–	1.0	0.85
Composting					
High capacity	0.65	–	0.95	0.95	–
Low capacity	0.65	0.60	0.65	0.65	–
RDF production	0.70	0.60	0.70	0.8	1.24
Suspension firing RDF	–	–	0.9	–	–
Incinerator turbine	0.70	–	0.9	–	–
Wet pulping	0.72	0.75	–	0.9	–
Pyrolysis	0.78	0.60–0.80	–	0.70–0.80	–
Acid hydrolysis	–	–	–	0.8	–
Anaerobic digestion	–	0.60	–	0.9	–

Sources: Wilson, D.C., *Waste Management, Planning, Evaluation, Technology*, Clarendon Press, Oxford, U.K., 1981. Rhyner, C.R., and Wenger, R.B., Capital costs of resource recovery facilities in the U.S.A., *Waste Management and Research*, 4, 321, 1986.

Finally, the most accurate method of estimating capital costs is through a detailed analysis in which all construction and equipment items are delineated and actual prices used to determine costs. As a project progresses and reaches its conclusion, a detailed cost analysis of this type is a required component of the planning process.

Capital costs are typically amortized or paid over the useful lifetime of a facility or equipment item. The lifetime of a major waste treatment facility is typically assumed to be 20 or 25 years, whereas the lifetime of an equipment item is usually considerably shorter. For example, the lifetime of a truck is usually considered to be about 5 years. The cost to be amortized is the initial capital cost less the value of the facility or equipment item at the end of its projected lifetime (i.e., the salvage value). Usually, the cost is paid in equal annual installments over the projected lifetime of the facility. In this case, the annual payments required to pay off the principal and interest on a loan or a bond issue is calculated in a manner similar to that for a mortgage on a house or an automobile loan. The formula for computing the payments for a borrowed amount P (principal) at an interest rate i per period (expressed as a decimal) in N payments is:

$$\text{Payment} = Pi/[1 - (1 + i)^{-N}]$$

Under this formula, payments can be easily calculated using the built-in @PMT(P,i,N) function available on most computer spreadsheets.

Example 13.1: The capital cost of a materials recovery facility is $3 million dollars. The money is borrowed at an annual interest rate of 9%. For the purposes of amortization, the lifetime of the facility is considered to be 20 years. (a) What is the annual payment to amortize the capital cost? (b) If the facility is to handle 12,000 tonnes of recyclable materials annually, what is the capital cost per tonne of material?

Solution: (a) The amount to be amortized is $3,000,000. Using the formula

$$\text{Annual Payment} = Pi/[1 - (1 + i)^{-N}]$$

yields

$$\text{Annual payment} = (\$3,000,000)(0.09)/[1 - (1 + 0.09)^{-20}]$$

or

$$\text{Annual payment} = \$328,639.43.$$

(b) The cost per tonne = $328,639.43/12,000 t = $27.39 /t

OPERATING COSTS

Operating, or variable costs, are those that arise from the daily operation of a facility as waste is processed or handled. They include labor, maintenance, utilities, supplies, and administrative costs. Within limits these costs can be adjusted for varying throughput quantities. For example, if larger than expected quantities are to be handled at a materials recovery facility, its managers may decide to operate the plant over longer hours, resulting in a need for additional labor, utilities, and maintenance on equipment. Obviously, these additional inputs will increase operating costs. However, the unit cost on a per tonne basis may actually decline because of improved efficiencies.

As in the case of capital costs, operating costs for a facility can be estimated from published information on waste management systems, from the operating costs of comparable facilities, or by direct calculation of the cost of each item on a detailed list of labor, utilities, supplies, maintenance, and administrative needs. Other costs which need to be taken into account are insurance, taxes, and licenses.

13.3 LIFECYCLE COST ANALYSIS

In almost all cases, cost is a major influence in determining whether decision-makers elect to construct a particular facility or pursue another alternative. A common procedure for comparing alternatives is to conduct a lifecycle cost analysis. Under this procedure, expenditures incurred over the lifetime of a facility are calculated. This total is adjusted based on revenues received from the sale of materials or energy. The result is then divided by the total quantity of waste handled during this time period to arrive at a unit cost.

In an actual application, the calculation is not as straightforward as described above. Not all expenditures occur at the same time, nor are all revenues received at the same time. For example, in the construction of an incinerator almost all of the capital expenditures are incurred prior to the opening of the facility for operation. In contrast, only the first sequence of landfill cells needs to be readied before operation begins. The construction costs for subsequent sequences of cells are incurred at a later time, after the cells in the initial set have been filled.

Future costs and revenues are difficult to forecast because of inflation, changing market values of materials and energy, and unanticipated technical problems. Despite these difficulties, it is important that serious attempts be made to obtain good estimates. To account for inflation a certain annual inflation rate, expressed as a percentage, is usually assumed. This approach was used in discussing landfill costs in Chapter 9. The annual inflation rate can then be used to estimate changes in expected expenditures or revenues when they occur in a future year.

Once good estimates are in hand, money flows occurring in a future year must be discounted to the present. This must be done because a given sum of money is worth more if it is available today than if it were to become available in the future, say five years from now. This concept is understood by anyone who invests a sum of money in a savings account and leaves it untouched for five years while the original amount grows through the accumulation of interest. Discounting is necessary in order to convert future money flows to present values so that alternatives can be compared on an equal footing.

Calculations to determine the **present value** of future money flows are based on a discount rate. While several different methods are used in choosing a discount rate (Wilson, 1981), a common one is based on the rate at which money could be profitably invested in an alternative venture. This rate would not be less than the cost of borrowed money and, in general, is usually somewhat higher. An amount S_N to be paid N years from now, with an assumed discount rate r, has a present value, PV, determined by the formula:

$$PV = S_N / (1 + r)^N$$

For example, if one assumes a discount rate of 12% (i.e., $r = 0.12$), a money flow of $1000 occurring 5 years from now has a present value of $567.43. The calculations are as follows:

$$PV = \$1000 / (1 + 0.12)^5$$

$$PV = \$567.43$$

When all costs for a facility are converted to present values, the costs can be added to obtain a single present value cost. Similarly, future revenues can be discounted to the present, totaled, and subtracted from the total cost to obtain a net present value cost figure. It is this number which can be compared with those for other alternatives to assist in decision-making.

A number of standard formulas are available which can be used to convert money flows of various types to a present value. For example, if uniform flows occur each year over a period of years, as is the case when annual payments are required to amortize capital costs, the formula which applies is the following:

$$PV = \frac{A[(1+r)^N - 1]}{r(1+r)^N}$$

where A is the annual payment. The number derived with this formula is often referred to as the **capital recovery factor**.

Example 13.2: Suppose an incinerator facility is under consideration and annual expenditures are estimated to be $625,000 over the 20 year lifetime of the project. If the discount rate is 12%, what is the present value of this cost flow?

Solution: Substituting in the capital recovery formula we have

$$PV = \frac{\$625,000[(1+.12)^{20}-1]}{0.12[1+.12]^{20}}$$

or

$$PV = \$4,668,402.27$$

Calculations to determine payments to amortize capital costs over a chosen time period, present values of a list of payments made at regular intervals, and the like, are routinely done using computer spreadsheet programs.

As an example of how a unit cost can be determined, consider the materials recovery facility discussed in Example 13.1 with a capital cost of $3,000,000 to be paid over its projected useful life of 20 years. With an 8% interest on this debt, the annual payments are $305,557. Let us also assume that the facility will process 7000 tonnes of recyclables each year and that the operating and maintenance costs are initially $475,000 per year, but these costs will rise with inflation at a rate of 4% per year, and that the revenues received from the recovered materials will be $268,000 per year in the first year and then increasing at the rate of 4% per year. A discount rate of 12% is assumed. Table 13.3 summarizes the costs for the 20-year period. The total cost of the facility in 1992 dollars is $4,795,948, during which time 140,000 tonnes of material are processed. This results in a unit processing cost of $34 per tonne.

The value of the life-cycle analysis becomes apparent when there are alternative systems that need to be compared, perhaps with several proposals and financing arrangements submitted by different vendors. We note again, as we did earlier, that there are significant uncertainties in forecasting future expenditures, revenues, and inflation rates. A prudent course of action is to perform an analysis to determine the sensitivity of the results to changes in the assumptions. For example, one might perform the calculations for reduced or greater revenues from the sale of the recovered materials, or for lesser or greater tonnages processed (with corresponding changes in operating costs and revenues). The value of using

Table 13.3 Example of a Calculation of the Life Cycle Costs of a Materials Recovery Facility

Principal	$3,000,000	Discount rate	0.12
Interest	0.08	Inflation rate	0.04
Periods	20		
Payment	$305,557		

Year	Capital	Operating	Revenues	Net Cost	PV Factor	PV Cost	t/y	Cost/t
1	$305,557	$475,000	($268,000)	$512,557	1.0000	$512,557	7000	$73
2	$305,557	$494,000	($278,720)	$520,837	1.1200	$465,033	7000	$66
3	$305,557	$513,760	($289,869)	$529,448	1.2544	$422,073	7000	$60
4	$305,557	$534,310	($301,464)	$538,403	1.4049	$383,225	7000	$55
5	$305,557	$555,683	($313,522)	$547,717	1.5735	$348,084	7000	$50
6	$305,557	$577,910	($326,063)	$557,404	1.7623	$316,286	7000	$45
7	$305,557	$601,027	($339,105)	$567,478	1.9738	$287,502	7000	$41
8	$305,557	$625,068	($352,670)	$577,955	2.2107	$261,437	7000	$37
9	$305,557	$650,070	($366,777)	$588,850	2.4760	$237,827	7000	$34
10	$305,557	$676,073	($381,448)	$600,182	2.7731	$216,432	7000	$31
11	$305,557	$703,116	($396,705)	$611,967	3.1058	$197,037	7000	$28
12	$305,557	$731,241	($412,574)	$624,224	3.4785	$179,449	7000	$26
13	$305,557	$760,490	($429,077)	$636,970	3.8960	$163,494	7000	$23
14	$305,557	$790,910	($446,240)	$650,227	4.3635	$149,015	7000	$21
15	$305,557	$822,546	($464,089)	$664,014	4.8871	$135,870	7000	$19
16	$305,557	$855,448	($482,653)	$678,352	5.4736	$123,932	7000	$18
17	$305,557	$889,666	($501,959)	$693,264	6.1304	$113,086	7000	$16
18	$305,557	$925,253	($522,037)	$708,772	6.8660	$103,229	7000	$15
19	$305,557	$962,263	($542,919)	$724,901	7.6900	$ 94,266	7000	$13
20	$305,557	$1,000,753	($564,636)	$741,674	8.6128	$ 86,113	7000	$12
					Totals	$4,795,948	140,000	$34

a computer spreadsheet becomes evident when these additional calculations are made.

13.4 FEE POLICIES

The cost of waste processing and disposal has increased significantly as communities and industries have been required to construct and operate facilities to conform to more stringent environmental standards mandated by federal, state, and local regulations. These costs will continue to increase as landfills are constructed and operated to meet the requirements of the Subtitle D regulations. The providers of the solid waste collection, solid waste disposal, and wastewater treatment services must recover their costs and, in the case of private companies, generate a profit to remain in operation. Fee policies are required to raise the necessary revenues to pay the costs involved.

In the United States, most households, businesses, and industries pay for solid waste collection and disposal services and wastewater treatment directly through a fixed weekly, monthly, or quarterly service charge or indirectly through their property taxes. Many communities choose to use property taxes as the means to charge for the services because it is consistent with how they charge for many other services (e.g., schools, police and fire protection, library, transportation), and no separate rate billing procedure is needed. The major disadvantage of this approach is that the costs are often allocated solely on the value of a property and not the quantity of waste produced therein. This method also has the disadvantage of isolating the waste generator from the costs of waste management. Thus, the waste generator has little incentive to reduce waste quantities.

Increasingly, however, communities are implementing user charges based on some form of unit pricing. In the case of sewage treatment services, it is not economical to meter household sewage as it is for large industrial users. A surrogate measurement that is sometimes employed is the volume of water used by a household, the assumption being that this volume of water will approximate the volume of sewage generated. Industrial wastewater generators are commonly charged based on the concentrations of wastewater constituents such as BOD or COD, as well as dissolved chemical species that require removal.

The cost of solid waste disposal at a landfill or incinerator is commonly recovered by a tipping fee based on the weight or volume of wastes received. The generator is insulated from this cost because of intervening collection and transportation charges, which in many communities are significantly larger than the tipping fees.

Unit pricing based on the the size or number of containers in which the waste is collected is commonly used by private waste handlers for commercial and industrial accounts, but it is not widely used for residential collection by either private haulers or by municipalities. Possible types of rate structures based on unit pricing are: a fixed per-container charge, a declining per-container charge, an increasing per-container charge, and a weight-based charge. Studies and reports addressing the effects of these unit pricing approaches are limited. In one report concerning this issue (USEPA, 1990), case studies are discussed for the suburban community of Perkasie, Pennsylvania, the semi-rural community of Ilion, New York and the city of Seattle, Washington. All of these communities based charges on the number and size of bags or cans picked up under their curbside collection programs.

In Perkasie and Ilion, the evidence indicates that householders reduced their total waste generation by 10% or more, more than doubled the amount of material recycled, and reduced the amount of mixed waste by 30% or more. Less dramatic effects were observed in Seattle even though there were significant unit price increases. In that city, some citizens decided to reduce their level of service, some increased the compaction of the refuse in the containers, and, overall, slightly increased their recycling rate. A statistical analysis of data collected by the Seattle Solid Waste Utility between 1970 and 1987 (before curbside recycling) showed a relationship between how much customers pay for refuse service, the level of service they demand, and the amount of refuse they generate. The Seattle data showed that for every 1% increase in the price of the services, there was a 0.07% decrease in the amount of refuse generated and a 0.14% decrease in the quantity of refuse services demanded as measured by the number and size of cans used (Gale, 1990).

The evidence, meager as it is, suggests that there may be positive benefits from fee structures based on unit pricing. In theory, this should be the case because the user of the waste collection service is in a much better position to make choices based on a knowledge of costs. A concern which has been raised is that unit pricing for residential waste might encourage clandestine dumping of waste along roadsides, in commercial dumpsters, or in vacant lots to avoid having to pay a user fee. What happens to the waste that no longer appears at curbside when a unit pricing policy is implemented? Have citizens become more careful with their purchasing habits to reduce waste and more diligent in their recycling efforts, or have they tossed more waste into a gutter or back alley? The likely answer is that some of each occurs but little is known concerning the degree of each.

13.5 OWNERSHIP AND OPERATION OF WASTE FACILITIES

The ownership of a wastewater treatment plant is usually not a major issue. Municipalities typically have their own sewage treatment plant or are members of a sewerage district or commission formed to serve a larger metropolitan area. However, there are some communities with privately operated sewage treatment plants. Municipal facilities may accept industrial wastewater provided there is sufficient capacity and the wastewater is amenable to treatment by the processes used in the facility. High volume industrial wastewater generators, or those with special wastewater characteristics or treatment needs, generally will operate their own treatment facilities.

In the United States, solid waste management is a 30-billion dollar per year enterprise. Municipalities account for about 45% of this amount, publicly-held companies, such as WMX (formerly Waste Management, Inc.) and Browning-Ferris Industries (BFI), account for 30%, and independent local operators the remaining 25% (Schweich, 1992). Approximately 17% of MSW landfills are privately-owned, 78% are owned by local governments, 4% by the federal government, and 1% by state governments (USEPA, 1988). The private sector has a major involvement in collection, hauling, and recycling. Even in communities where there is a municipal collection service, a large portion of commercial and industrial wastes are handled by private contractors.

The decisions that a municipality makes concerning the ownership, operation, and financing of processing and disposal facilities not only determine the waste management costs, but also the risks it assumes, its control of the operation, and its flexibility for making future changes. Municipalities may choose from several ownership and operation alternatives, including public ownership and operation, public ownership and private operation, or private ownership and operation. The factors that may influence the decision to choose a public or private operation of a chosen disposal alternative are management expertise needed to provide efficient and effective service, degree of financial and technical risk, and the source and type of financing.

Public ownership and operation provides the greatest control and flexibility for a municipality. Matters included under this control are siting of a facility, procurement and construction, maintenance, disposal fee structures, and facility modifications to respond to changing technologies or regulatory requirements. If the facility is to serve industrial waste producers, the municipality can negotiate a sharing of costs and risks with them.

The combination of public ownership and private operation provides the opportunity to minimize the cost of financing through the use of tax-exempt municipal bonds. This arrangement also has the advantage of incorporating efficiencies of private enterprise and of eliminating the need to hire additional public employees.

Private ownership and operation allows a municipality to shift much of the development and performance risk to a private contractor. The municipality's role may be to select the most qualified vendor to own and operate a facility and to negotiate a contract to deliver a specified quantity of waste to the facility for a given length of time in return for a favorable disposal fee. The primary disadvantage to a municipality under the private ownership and operation scheme is a loss of overall control. This becomes especially acute if the service provided is unsatisfactory or the operation of the facility is unacceptable to the municipality or to a regulatory agency. If a facility fails to satisfy regulatory requirements the facility could be closed down with the municipality having no feasible alternatives to replace it.

13.6 FINANCING OF FACILITIES

Municipalities finance large projects by issuing bonds that are to be repaid from tax revenues or user fees. Two common types of municipal bonds are **general obligation bonds** and **revenue bonds.** General obligation bonds usually have a lower interest rate because they present a lower risk to the purchaser. The bonds are backed by the taxing power of the municipality. If the project should fail, the municipality provides the money from property taxes, sales taxes, or other tax revenues to repay the incurred obligations. Debt incurred under general obligation bonds is counted toward a community's debt limit. This may be a concern if a community is financing a number of additional projects through bonds.

Revenue bonds are issued to finance a specific project. User fees and other revenues from the sale of recovered materials and energy are used for the repayment of revenue bonds. The interest rates are typically higher than those for general obligation bonds because repayment is dependent upon the success of the specific project and therefore the risk of nonpayment is usually higher. Revenue bonds offer an advantage to a municipality because the cost does not add to its debt limit.

Private companies finance large projects from profits generated in other enterprises, borrowed money, or stock sales. They recover costs and earn a profit by charging users a service fee. This fee may be based on the weight or volume of the waste handled, or on the estimated cost for a given time period.

Other forms of financing include installment purchases, leases, lease purchase agreements, and leveraged lease agreements. These approaches provide sophisticated means to reduce the amount of up-front public money required for a project or, perhaps, allow a change in payment amounts during different periods of the project's lifetime. The private developer or participating financial institution receives tax benefits, as well as a repayment on its investment. For additional details on different types of financial instruments, Robinson (1986) is an excellent source. It should be noted that tax laws relating to different financial approaches frequently undergo changes and vary from one state to another.

13.7 MANAGEMENT RESPONSIBILITIES

The scope and magnitude of waste management duties have grown as federal and state regulations for wastewater discharges and solids disposal have become more restrictive and complex. Concomitantly, the requirements for reporting and supplying operating data to the regulatory agencies have increased. Many communities and waste collection companies have also initiated recycling programs with special collection equipment, facilities for material separation and sorting, and educational programs. Consequently, the number of people involved in waste management has increased significantly during the past decade.

Many communities with solid waste collection and disposal programs, or which own and operate sewage treatment plants, assign the operational responsibilities to a sanitation, an engineering, or a solid waste department. Departmental managers implement policies and ordinances passed by a county, city, or other governmental body whose members also approve budgets and expenditures. The council represents the public, who elects its members.

The increased complexity and cost of waste management programs have caused some communities to consider and, perhaps, become part of intercommunity, county, or even regional waste management organizations. One reason is economics. As was pointed out in the discussion on landfills in Chapter 9, the trend is toward fewer, but larger, landfills because the siting, approval processes, environmental controls, and financial requirements are formidable and expensive. In addition, many of the operational and long-term care requirements are also costly. A second reason for the rise in inter-governmental waste management organizations is to enable municipalities to gain access to facilities or resources. Urban communities may need access to rural land for landfill space. Rural communities may need access to materials recovery facilities, a landfill, or an incinerator.

The implementation of cooperative arrangements among municipalities may require legislative authorization at the state level to define the scope of the organization's activities and its powers. Agreements, contracts or resolutions by public bodies defining the management or administrative structure, duties, powers, procedures, and compensations are important considerations in establishing a successful cooperative arrangement. These agreements may result in the formation of entities such as waste boards, authorities, districts, or nonprofit public corporations.

An authority is a corporate body which usually has a charter and is authorized and approved by a state legislature. Authorities are established to free waste management activities from the influence of municipal bureaucracies and to enable the establishment of financing arrangements that do not add to constitutionally or legislatively imposed debt limits. Authorities typically cannot finance their facilities from general taxes but must rely on revenue bonds.

A nonprofit public corporation could also be formed to operate in a manner similar to an authority. The creation of a corporation requires approval of articles of incorporation by the state in which it is to be located. In order to have a tax-exempt status, the corporation must satisfy specific criteria of the U.S. Internal Revenue Service regarding the control and approval of projects, assignment of assets, and the composition of its board of directors. Corporations must satisfy local, state, and federal regulatory requirements.

DISCUSSION TOPICS AND PROBLEMS

1. An incinerator built in 1977 cost 25.8 million dollars. What would be the cost of a comparable facility in 1993 dollars?
2. The capital cost of a 500 tonne per day landfill was 3.3 million dollars in 1988. Using a scale factor of 0.9, compute the expected cost of a 2000 tonne per day landfill in 1993.
3. (a) Calculate the capital cost per tonne of waste for a 500 tonne per day incinerator that cost 80 million dollars if it is assumed: (1) the incinerator operates at capacity; (2) the incinerator has a 25 year lifetime; and (3) the incinerator is financed over its lifetime at an interest rate of 5%.

   (b) What would be the cost per ton if the interest rate were 7%?
4. Compare the advantages and disadvantages of having a publicly-owned disposal facility to handle a community's waste to those of a privately-owned facility. Is the disposal site for your community publicly or privately owned?
5. What are the annual costs of sewage treatment, refuse collection, and solid waste disposal in your community? What is the cost of waste treatment and disposal as a percentage of the annual municipal budget? What is the per capita cost of waste management?

6. Compare the advantages and disadvantages of financing projects using general obligation bonds and revenue bonds.

7. What is the cost of MSW disposal, excluding collection, in your community? Is the disposal cost disproportionately high or low compared to surrounding communities? If so, can you explain the differences?

8. User charges based on some form of unit pricing are now being used by some communities to recover the costs of solid waste collection and disposal services. What are the advantages and disadvantages of this approach?

REFERENCES

Beck, R.W. and Associates, *Brown County Waste-to-Energy Study*, prepared for Brown County, Wisconsin, 1992.

Bridgewater, A.V., Technological economics applied to waste recovery and treatment processes, *Effluent Water Treatment Journal*, 17, 223, 1977.

Engineering News Record, McGraw-Hill Publishers, New York, NY, March, 1994.

Gale, D., personal communication, December 14, 1990, Seattle Solid Waste Utility, to Paul Curl, Washington Utilities and Transportation Committee.

Grant, K.D. and Cooper, D.W., Landfill Disposal Costs are Based on Daily Tonnage, *1982 Sanitation Industry Yearbook*, Solid Wastes Management Magazine, Atlanta, GA, 1982, pp. 24–26.

Midwest Research Institute, *Resource Recovery: The State of Technology*, Report for the President's Council on Environmental Quality, Report PB 214 149, National Technical Information Service, U.S. Department of Commerce, Washington, DC, 1973.

Miller, C.A., Current concepts in capital cost, forecasting, *Chemical Engineering Progress*, 69, 77–83, 1973.

Robinson, W.D., *The Solid Waste Handbook: A Practical Guide*, John Wiley & Sons, New York, NY, 1986.

Rhyner, C.R. and Wenger, R.B., Capital costs of resource recovery facilities in the U.S.A., *Waste Management and Research*, 4, 321–326, 1986.

Schweich, S.L., The solid waste industry: 1992 review and 1993 outlook, *Waste Age*, 23(12), 47,50,52,54,56–57, Dec., 1992.

Schulz, H.W., Benziger, J.B., Bortz, B.J., Neamatalla, M., Szostak, R.M., Tong, G., and Westerhoff, R.P., Resource Recovery Technology for Urban Decisionmakers, National Technical Information Service Report PB 252 458, prepared for the National Science Foundation by Urban Technology Center, Columbia University, NY, 1976.

U.S. Environmental Protection Agency (USEPA), Report to Congress: Solid Waste Disposal in the United States, EPA Publication 530-SW-88-011, U.S. Printing Office, Washington, DC, 1988.

U.S. Environmental Protection Agency (USEPA), Charging Households for Waste Collection and Disposal: The Effects of Weight or Volume-Based Pricing on Solid Waste Management, Environmental Protection Publication 530-SW-90-047, U.S. Printing Office, Washington, DC, 1990.

Wilson, D.C., *Waste Management, Planning, Evaluation, Technology*, Clarendon Press, Oxford, U.K., 1981.

Glossary

Acetogens – anaerobic bacteria which metabolize various organic acids to acetic acid in methanogenic environments.

Actinomycetes – a large group of primarily filamentous bacteria in soil. They contribute to the stabilization of solid waste (composting) and sewage.

Activated carbon adsorption – refers to the use of activated carbon to remove potentially toxic or undesirable substances from aqueous waste.

Activated sludge – the biological solids produced in an activated sludge basin (secondary wastewater treatment).

Active site life – the period from the initial receipt of waste at a landfill until the site ceases to accept waste and the site undergoes final closure.

Acute effect – an adverse effect on the receptor organism, with symptoms of severity coming quickly to a crisis.

Acutely hazardous waste – waste that is deemed so hazardous that small amounts are regulated by the USEPA as strictly as larger amounts of other hazardous waste. Acutely hazardous wastes are listed in 40 CFR 261.31 and designated with the symbol (H), and in addition those wastes listed in 40 CFR 261.33(e) with EPA Waste Numbers beginning with the letter "P".

Adsorption – the attraction and adhesion of ions or molecules from an aqueous solution onto a solid particle or surface.

Advanced wastewater treatment – *See* Tertiary Treatment.

Aeration – the process of exposing something to air or oxygen.

Aerobic – the biological state of living and growing in the presence of oxygen.

Aerobic decomposition – the natural decay and breakdown of organic matter by bacteria that utilize oxygen in respiration.

Aerosols – minute droplets of liquid that can remain suspended in air for long periods of time.

Afterburner – a device used to burn or oxidize the combustible constituents remaining in the exhaust gases produced by prior combustion processes.

Agitated bed composting system – an in-vessel composter which uses augers or other mechanical mixing to promote aeration and good mixing of the compost mass.

Agronomic sludge application rate – the amount of sludge applied to satisfy the annual nutrient needs of a crop; usually based on the nitrogen content of the sludge.

Air, combustion (excess) – air supplied in excess of the theoretical quantity of air needed for complete combustion.

Air, combustion (primary) – air supplied to a combustion system for the initial oxidation of the fuel or waste, often through the fuel bed.

Air combustion (secondary) – air introduced above or beyond the fuel bed by natural, induced, or forced draft. It is often called overfire air when it is supplied above the fuel bed.

Air, combustion (stoichiometric) – the theoretical quantity of air required for complete combustion calculated from the chemical composition of the waste.

Air knife – a device for separating light materials from heavier materials by passing a mixture of materials through a plane of moving air.

Air pollution – the presence of contaminants in the air.

Alcohol fermentation – the process whereby microbes (usually yeasts) convert simple sugars to alcohol and carbon dioxide.

Algae – plants found in sunlit situations on land as well as in fresh and salt water over a wide range of latitude. They grow as individual cells, small clumps, or large masses.

Alkalinity – a quantitative measure of the capacity of liquids or suspensions to neutralize strong acids. Alkalinity results from the presence of bicarbonates, carbonates, hydroxides, silicates and phosphates, and some other substances. Numerically it is reported in terms of the concentration of calcium carbonates that would have an equivalent capacity to neutralize strong acids.

Ammonia stripping – the removal of ammonia from high pH wastewater by the passage of air through the water.

Ammonia volatilization – the loss of ammonia gas to the atmosphere.

Amu – atomic mass unit; 1 amu = 1.66×10^{-27} kg.

Anaerobic contact reactor – an anaerobic system which is completely mixed and uses recycle of the solids to maintain high microbial populations in the reactor.

Anaerobic decomposition – the natural decay and breakdown of organic matter by bacteria that do not require oxygen for respiration.

Anaerobic digestion (sludges) – the process by which sewage sludges are degraded by anaerobic bacteria to form digested (stabilized) sludge and methane gas.

Anaerobic reactor – a reactor in which microbes degrade organic matter in the absence of oxygen using a variety of fermentation processes.

Anion – a negatively charged ion.

Annual loading pollutant rate – the maximum amount of a given pollutant that can be applied to a sludge land application site in a 365 day period.

Anoxic conditions – refers to environments lacking molecular oxygen. In wastewater treatment it may also be used to denote the presence of an electron acceptor such as nitrate available for bacterial respiration.

Aquifer – underground water-bearing geologic formation capable of yielding a significant amount of groundwater to wells or springs.

Aquifer, perched – a region in the unsaturated zone where there is a local region of saturation because of an underlying confining layer of finite extent.

Aquifer, unconfined – an aquifer that has no confining layers between the zone of saturation and the surface.

Artesian – the occurrence of groundwater under greater than atmospheric pressure.

Ash – the noncombustible solid by-products or residue from incineration processes.

Audit dependent waste reduction – refers to the use of a program (audit) to identify the potential for waste reduction at technological and economically feasible levels in an organization.

Average linear velocity – the rate a fluid moves through a porous medium.

Avoided cost – costs a utility may pay for electric power purchased from a waste-to-energy facility, based on how much it would have cost the utility to generate the power itself; or, costs not incurred because of diversion of waste from a landfill (e.g., disposal, environmental, and opportunity costs).

BACT – best available control technology.

Bacteria – single-celled microscopic organisms, simple in structure. Some types are capable of causing human, animal, or plant diseases. Others are important in sewage or refuse stabilization.

Bacteria, aerobic – bacteria which require the presence of dissolved or molecular oxygen for their metabolic processes.

Bacteria, anaerobic – bacteria that do not require oxygen for metabolism, and in fact, their growth is often hindered by the presence of oxygen.

Bacteria, facultative – bacteria which can exist and reproduce under either aerobic or anaerobic conditions.

Baffle – any refractory construction intended to change the direction of flow of the products of combustion.

Baffle chamber – a device designed to promote the settling of fly ash and/or coarse particulate matter by changing the direction and/or reducing the velocity of the gases produced by combustion.

Bagasse – an agricultural waste material consisting of the dry pulp residue remaining from the processing of sugar cane after the juice has been extracted.

Baghouse (or fabric filter) – emission control device; an array of cylindrical bags used to trap solid particles and dust.

Baler – a machine used to reduce the volume of materials by compression.

Bar screen – a system of parallel bars placed in a waterway to remove debris.

Base flow – the portion of the flow, in surface streams, that has been discharged to the stream as influent groundwater.

Basic Oxygen Furnace (BOF) – a steel-making process using pure oxygen to oxidize a molten mixture of pig iron and steel scrap. The oxygen process can use a maximum of 35% steel scrap.

Bedding (animal) – material, commonly organic in nature, which is placed on the floor surface of livestock buildings for animal comfort and cleanliness.

Bedrock – solid rock underlying soils.

Belt filter press – a continuous feed device for dewatering sludge by passing it between opposing belts with increasing pressures along its length.

Beneficiation – the concentration, enhancement or upgrading of waste materials in a resource-recovery processing system so that they may be more readily marketed and reused.

Berm – a constructed ridge of earth.

Bio-pile – refers to placement of contaminated soils in "piles" to create conditions for rapid degradation of contaminants.

Bioaccumulation – the retention and concentration of a substance by an organism.

Biochemical Oxygen Demand (BOD) – a determination of the amount of organic matter in a water sample based on measuring the microbial oxygen consumption necessary to decompose the organic material. $(BOD)_5$ is the BOD measured in a five-day test.

Biodegradability – the degree to which a substance may be broken down or degraded by the enzymatic activities of microbes.

Biodegradable plastic – a plastic that can be broken down by microorganisms such as bacteria and fungi; as generally used, the term does not necessarily mean complete degradation into carbon dioxide and water.

Biodegradable waste – organic material which is capable of being decomposed by microorganisms.

Biodegradation – decomposition of a substance into more elementary compounds by the action of microorganisms such as bacteria.

Bioremediation – refers to the use of biological processes to degrade contaminants such as oil or pesticides in water or soil.

Biosolids – the biological cell mass generated during wastewater treatment. Sewage sludge is commonly referred to as biosolids.

Blue baby disease – a condition induced in young infants by nitrate interfering with the oxygen-carrying ability of hemoglobin.

Bottom ash – the relatively coarse uncombusted or partially combusted residue of incineration that accumulates on the grate of a furnace.

Broke – paper that has been discarded anywhere in the manufacturing process. It is usually returned to a repulping unit for reprocessing.

Btu (British thermal unit) – the quantity of heat required to increase the temperature of one pound of water from $59.5°F$ to $60.5°F$.

Bulky waste – large items of refuse, such as household appliances, and furniture.

Burner, secondary (afterburner) – a burner installed in the secondary combustion chamber of an incinerator to maintain a minimum temperature and complete the combustion process.

Buyback – a facility that pays individuals for recyclable materials and further processes them for market.

Calcareous soils – soils that are rich in calcium carbonate.

Capacity utilization – ratio of quantity of production to total capacity of production facilities.

Capillary attraction – the force of attraction between a solid and a liquid. Capillary attraction accounts for the tendency of water to move into the spaces between soil particles, regardless of gravity.

Capture rate – amount of recyclables collected, divided by total amount of MSW generated by participating households or commercial establishments.

Carbon dioxide (CO_2) – an odorless, colorless, and nonpoisonous gas. Carbon dioxide is a major by-product of combustion and of aerobic and/or anaerobic microbial decomposition. It is slightly soluble in water, forming carbonic acid.

Carbon monoxide (CO) – a colorless, odorless, and very toxic gas associated with incomplete combustion of organic materials.

Carbon-nitrogen ratio (C/N ratio) – the ratio of carbon to nitrogen. This number is important in the composting of wastes.

Carcinogen – an agent that has the potential to induce the abnormal, excessive, and uncoordinated proliferation of certain cell types, or the abnormal division of cells (i.e., a material that causes cancer cells to develop and proliferate).

Casing (well casing) – a pipe used to keep a well open in unconsolidated materials or unstable rock.

Cast-iron – an iron-carbon alloy with a high carbon content, usually above 2.5%, and substantial amounts of cementite (Fe_3C) or graphite, making it unsuitable for working.

Catalytic conversion – using a catalyst to promote the burning or oxidizing of hydrocarbons, odorous contaminants, and other undesirable gases in the exhaust gas stream.

Cation – a positively charged ion.

Cation-exchange capacity – the ability of a particular soil to absorb cations.

Cell, landfill – a basic building unit of a landfill in which waste is surrounded by compacted soil.

Centrifuge – a device to remove solids from a liquid phase; in wastewater treatment it is used to thicken or dewater sludges.

CERCLA – the Comprehensive Environmental Response Compensation and Liability Act of 1980. The program implementing this act is commonly called Superfund.

CFCs – chlorofluorocarbon compounds, such as Freon-12 (CCl_2F_2), used chiefly as refrigerants and as propellants for aerosol cans. The use of CFCs as propellants was prohibited in 1979 because of their depleting effect on ozone in the atmosphere.

CFR – the Code of Federal Regulations.

Characteristic hazardous waste – any substance or waste that is shown to be hazardous on the basis of characteristics and a laboratory test, the Toxicity Characteristics Leaching Procedure (TCLP).

Charging chute – a channel or passage through which waste materials enter an incinerator from above by gravity.

Charging ram – a reciprocating device to meter and move refuse into a furnace.

Chemical oxygen demand (COD) – a determination of the amount of organic matter in a water sample based on measuring the amount of chemical oxidants necessary to oxidize the organic material.

Chemical precipitation – in wastewater treatment, the use of chemicals to precipitate dissolved and suspended matter.

Chipper – a size-reduction device using the shearing, cutting, or chipping action produced by sharp-edged blades attached to a rotating shaft which shaves or chips off pieces of objects.

Chronic effect – an adverse effect on a receptor organism, with symptoms that develop slowly over a long period of time or that recur frequently.

Class A sludges – sludges which must meet the most stringent guidelines mandated by the USEPA for pathogen and pollution content (e.g., a fecal coliform density of less than 1000 Most Probable Number per gram of dry solids).

Class B sludges – sludges that do not meet the strict guidelines for Class A sludges but must nonetheless meet certain standards as defined by the USEPA.

Classification, material – to separate or sort waste materials into uniform categories or classes, based on physical characteristics such as size, weight, density, color.

Clay – impermeable soil with particle size less than 0.002 mm in diameter.

Clean sludge pollutant limits – sludges which contain less than EPA mandated "Clean Sludge" concentrations of pollutants such as heavy metals.

Climate – long-term manifestations of weather conditions.

Closed composting – generally refers to "reactor" systems where the waste is completely enclosed in a reactor in which conditions of moisture, aeration, pH, and temperature can be closely controlled. Sometimes referred to as closed mechanical composting or in-vessel composting.

Co-collection system – recyclables and solid wastes collected simultaneously in the same truck either mixed together in the same compartment or placed in separate compartments.

COD – *See* Chemical oxygen demand.

Codisposal – disposal in one area of two or more types of solid waste; for example, unprocessed MSW and incinerator ash in a landfill.

Cogeneration – production of both electricity and steam at one facility, from the same primary fuel source.

Collection – picking up or gathering materials for processing or disposal.

Collection time (pick-up time) – the elapsed or cumulative time spent by the refuse collection crew in collecting refuse from a collection stop, including travel time between collection stops on the route.

Colloids – very fine suspended solids which do not settle by gravity, but may be removed by chemical precipitation or membrane filtration.

Combined ash – mixture of bottom ash and fly ash.

Combustion – *See* Incineration.

Combustion chamber (primary) – the chamber in an incinerator where ignition and burning of the waste occurs.

Combustion chamber (secondary) – the chamber in an incinerator where combustible solids, vapors, and gases from the primary chamber are burned and settling of fly ash occurs.

Combustion, complete – the complete oxidation of the fuel, with either the theoretical amount of air or excess air.

Commerce Clause – a constitutional clause granting Congress the power to regulate all commerce; the "dormant commerce clause" makes it explicit that state lines cannot be made barriers to the free flow of commerce.

Commercial waste – waste produced by commercial buildings, including wholesale warehouses, retail stores, and other service organizations.

Commingled recyclables – mixed recyclable materials separated from MSW at the point of generation. Further separation into individual components occurs at the collection vehicle or a centralized materials processing facility (MRF).

Commodity – a useful or valuable material.

Common sense waste reduction – a method of achieving low cost waste reduction by walking through a production process with waste reduction in mind, looking for easy-to-spot waste reduction opportunities.

Communicable disease – an illness due to an infectious agent or its toxic products which is transmitted directly or indirectly to a well person from an infected person or animal, or through the agency or an intermediate host, vector, or inanimate environment.

Compactor (steel wheel) – a vehicle equipped with steel wheels to provide good compaction of waste and soil.

Compactor collection trucks – solid waste pick-up vehicles with hydraulic equipment for compacting the waste in order to increase the payload.

Composite liner – a liner system composed of an engineered soil layer overlain by a synthetic flexible membrane liner.

Composting – a controlled microbial degradation of organic waste yielding a nuisance-free product of potential value as a soil conditioner.

Confined aquifer – an aquifer bounded above and below by impermeable beds or by beds of distinctly lower permeability than that of the aquifer itself; an aquifer containing confined groundwater.

Constructed wetlands – engineered complexes of saturated substrates and vegetation for the purpose of treating wastewater.

Construction and demolition waste – waste building materials and rubble resulting from construction, remodeling, repair, and demolition operations on houses, commercial buildings, pavements, and other structures.

Conventional fertilizers – mineral based fertilizers widely used in agriculture to maintain or increase soil fertility.

Conveyor – a device that transports material by a moving belt.

Core – a cylindrical sample of an underground formation, cut and raised by a rotary drill with a hollow bit.

Corrosive hazardous waste – refers to strongly alkaline or acid material capable of corroding steel at the rate of 6.35 mm/year.

Cover material – granular material, generally soil, that is used to cover compacted solid waste in a sanitary landfill.

Cracking – a process in which relatively heavy hydrocarbons are broken into lighter products by heat.

Cradle-to-grave hazardous waste management – the tracking system based on the use of manifests to track transfers of hazardous waste from generators to haulers to approved storage or disposal facilities.

Crane, bridge – a crane consisting of a lifting bucket that hangs from, and can travel along, a movable horizontal rail that rides between two parallel, horizontal rails.

Crawler – a bulldozer or other vehicle mounted on a pair of roller-chain tracks.

Crusher – a mechanical device used to break up waste material into smaller sized pieces by a pounding action by hammers or beaters (e.g., hammermill).

Cullet – scrap glass, usually broken into small, uniform pieces.

Cumulative load (pollutants) – the maximum amount of sludge which can be applied to a land application site over its lifetime. Usually based on specific pollutants such as heavy metals.

Curbside collection – collection at individual households or commercial buildings by municipal or private haulers, for subsequent transport to management facility.

Curing of compost – the gradual changes which take place in compost after the initial composting activity has occurred. Cured compost is stable, can be stored for long periods, and is relatively odorless.

Cyclone separator – a mechanical separator which uses a rotating air flow to sort materials according to mass or weight.

Dechlorination – the removal of chlorine from water after it has been chlorinated.

Demand-limited materials – secondary materials for which buyers are relatively scarce even though supplies may be available.

Demolition waste – waste materials resulting from the destruction of buildings, roads, and other structures. The major components are usually concrete, wood, bricks, plaster, glass, and other miscellaneous materials.

Denitrification – the bacterial conversion of nitrate to nitrogen gases under anaerobic conditions.

Density – the mass or quantity of material per unit volume.

Destructive distillation (pyrolysis) – the heating of organic matter in the absence of oxygen resulting in the evolution of volatile matter and leaving solid char consisting of fixed carbon and ash.

Detention time – the average length of time that a waste remains in a reactor or treatment basin.

Dewatering – the removal of water by filtration, centrifugation, pressing, open-air drying, or other methods of reducing the water content of a waste.

Diffusion – the process by which molecules and ions in solution move from regions of high concentration to regions of lower concentration.

Digested sludge – sludge that has been stabilized, usually by means of anaerobic digestion.

Dioxin – a group of polychlorinated compounds that are characterized by a chemical structure with two benzene rings linked by two oxygen bridges in the presence of specific chlorine atom arrangements.

Discharge – *See* specific discharge.

Disinfection – any process which kills or inactivates most disease causing microorganisms and viruses.

Dispersion – the phenomenon by which concentrations of dissolved substances decrease as a result of mixing with uncontaminated water.

Dissolved air flotation (DAF) – a process whereby air released under pressure in a wastewater treatment basin forms minute bubbles which carry suspended microbial flocs to the surface where they are skimmed.

Distillers dried grain – the dried yeast cells and remains of grains after the brewing of beer. The product is an excellent animal feed rich in protein and other nutrients. Also called stillage.

Dose – the quantity of a chemical or radiation that is absorbed by an organism.

Drainage – provision for guiding the surface water resulting from precipitation or overland flow in a planned way.

Dredge – to dig under water, commonly to remove sediment.

Drop-off centers – locations in a community where residents deliver their recyclables or solid waste.

Dross – the scum that forms on the surface of molten metal.

Dry injection – injection of a dry reagent such as lime powder into an incinerator boiler or the original MSW, to aid in control of acid gases.

Economies of scale – increases in processing or disposal capacity that reduce the average cost per ton of output.

Eddy current separator – a device in which a changing magnetic field induces currents in nonferrous metals, which in turn produce a repulsive force and change the path of motion of the metals.

Effluent – the liquid, partially or completely treated, which leaves a treatment stage, reservoir or plant. For example, primary effluent is the partially treated wastewater which leaves a primary treatment basin.

Electrostatic precipitator – a device for collecting particulates from a gas stream by placing an electrical charge on the particles in a strong electric field, then removing the particles attracted to the electrodes.

Emergent vegetation – plants growing in water but whose stems and leaves are above the surface of the water.

Encapsulation – placement of a toxic substance into glass or concrete to isolate it from the environment.

Enclosed static pile composter – same as static pile except that the pile is enclosed in a building.

Energy recovery – the retrieval of energy from the combustion or oxidation of organic materials or methane gas from landfills.

Enteric viruses – viruses common to the human intestinal tract.

EPA – the United States Environmental Protection Agency, whose main responsibility is to oversee environmental regulations.

EPA hazardous waste number – the number assigned by EPA to each hazardous waste listed in Subpart D of 40 CFR 261, and to each characteristic waste identified in Subpart C of 40 CFR 261.

EPA identification number – the number assigned by EPA to each generator, transporter, and treatment, storage, or disposal facility.

EPA Region – the states and territories found in any one of the 10 standard federal regions of the U.S.

EPCRA – the Emergency Planning and Community Right-to-Know Act is Title III of the Superfund Amendments and Reauthorization Act (SARA) of 1986. EPCRA requires notification of local and state authorities and the public in the event of accidental releases of hazardous substances.

Equipotential surface – a three-dimensional surface such that the total hydraulic head is the same everywhere on the surface.

Erosion, soil – the wearing away of the land surface, normally by wind or running water.

Essential nutrients (plants) – mineral elements such as nitrogen, phosphorus, potassium, iron, and others that are essential for the growth of plants.

Eutrophication – the term which describes nutrient enrichment in aquatic environments due to the addition of substances such as ammonia, nitrate, phosphate, and organic matter.

Evapotranspiration – water loss by both evaporation and uptake by vegetation.

Explosives – materials capable of producing a large-scale violent reaction or expansion.

Exothermic reactions – chemical reactions which release energy.

Fabric filter – *See* Baghouse.

Facility – the structures, land, and other improvements on the land, used for treating, storing, or disposing of waste. A facility may consist of several treatment, storage, or disposal operational units (e.g., one or more landfills, surface impoundments, or combinations of them).

Facultative organisms – organisms capable of growing under aerobic and anaerobic conditions, in contrast to "obligate" aerobes which can grow only in the presence of oxygen.

Fermentation – any microbial energy yielding process in which organic compounds serve as both primary electron donors and ultimate electron acceptors.

Ferric chloride – a widely used chemical precipitant in sludge conditioning.

Ferrous metals – iron and steel. They have the distinguishing property of being attracted to a magnet.

Field capacity – the maximum quantity of water held by saturated compacted solid waste, such that any additional water will cause water to be released by gravity.

Final closure – the activities required to terminate the use of a facility such as a landfill.

Fixed costs – costs that do not vary with level of output of a production facility (e.g., administrative costs, building rent, mortgage payments).

Fixed film processes – biological wastewater treatment processes, trickling filters, or rotating biological contactors are examples, where microbes form thin films on media such as rock or plastic.

Flash point – the lowest temperature at which vapors above a volatile combustible substance ignite in air when exposed to a flame.

Flint glass – clear glass.

Floc – in wastewater treatment, small masses of solids formed with the aid of polyelectrolytes or other chemical precipitants. Floc is also the name applied to the biological solids formed in the activated sludge process.

Flocculation – in water and wastewater treatment, the treatments which aggregate suspended matter to form a rapidly settling floc.

Flow control ordinances – ordinances that require delivery of collected MSW to specific management facilities.

Fly ash – particles that are carried off an incinerator grate by turbulence or volatilization.

Food chain crops – crops grown for human consumption or crops grown for feed for animals whose products are consumed by humans.

Fungi – unicellular and multicellular microorganisms that are nonphotosynthetic, requiring organic compounds for growth. They are important agents of decay, with many able to degrade complex organic compounds such as lignin and cellulose.

Furans – a group of polychlorinated compounds that are characterized by a chemical structure with two benzene rings linked by one oxygen bridge in the presence of specific chlorine atom arrangements.

Furnish – the pulp used as raw material in a paper mill.

Gaylord container – a large reusable corrugated container used for shipping materials. The volume is approximately 1 m^3. The approximate dimensions are 1.0 m $\times$ 1.2 m $\times$ 0.93 m (40 $\times$ 48 $\times$ 37 in.).

General obligation bonds – municipal bonds with repayment guaranteed by community taxpayers.

Geographic Information System (GIS) – a computer software system which can store, manage, analyze, and display spatially referenced data.

Gibbs function – the expression used to denote the energy changes when compounds are metabolized. A negative value signifies that a given chemical reaction releases energy to do useful work such as permitting an organism to grow.

Glasphalt – an asphalt product that uses crushed glass as a partial substitute for aggregate in the mix.

Gradient (groundwater or water table) – the slope of the water table or potentiometric surface. A positive slope is referred to as "up-gradient"; a negative slope is called "down-gradient."

Grate – the devices used to support the burning material in an incinerator.

Gravel – rock fragments from 2 mm to 64 mm in diameter, possibly mixed with sand and boulders.

Gravity separation (flotation, heavy media) – the collection of substances immersed in a liquid by taking advantage of differences in specific gravities.

Grinder – a mechanical device used to pulverize waste material into small particles using friction.

Grit – dense, mineral, suspended matter such as sand or silt in wastewater.

Groundwater – water below the land surface in a zone of saturation.

Groundwater flow – movement of groundwater in an aquifer.

Groundwood pulp – *See* Mechanical pulp.

Grout – a mixture of cement, water, and possibly some filler.

Half-life – the time required for the concentration, strength, quantity, emission rate, etc.) of a substance to decrease, due to natural decay processes, to half of the original value.

Halogenated organic compounds – compounds which contain halogens such as chlorine, bromine, or fluorine as part of their structure. Polychlorinated biphenyls (PCBs), trichloroethylene (TCE) are examples.

Hammermill – a size reducing machine that operates by impacting material with freely-swinging heavy metal hammers pinned to a horizontal or vertical shaft rotating at a high angular velocity.

Hazard codes – descriptions used to identify the hazardous qualities of a particular hazardous material; for example, ignitability (I), corrosivity (C), reactivity (R), acutely hazardous waste (H), Toxic (T).

Hazard labels – refers to labels placed on hazardous waste containers and transporting vehicles to identify the type of hazard present.

Hazard Ranking System (HRS) – the method used by the USEPA to evaluate the potential hazard of an abandoned toxic waste site to determine whether it should be on the National Priorities List.

Hazardous Materials Table – the guide (49 CFR 172.10) which specifies requirements to be met in transporting HW.

Hazardous waste (HW) – any solid, liquid, or gaseous waste or mixture of wastes that presents a significant threat to human health and/or the environment. From a regulatory standpoint HW is defined on the basis of regulations in the Resource Conservation and Recovery Act administered by the USEPA.

Hazardous waste generator – any business, industry, or institution which generates HW beyond a USEPA specified amount. *See* the categories Small, Medium and Large quantity HW generators.

Hazardous waste transporter – refers to transporters who are licensed to transport HW in compliance with U.S. Dept. of Transportation regulations 49 CFR Parts 170-177.

Head, total hydraulic – the sum of the elevation head, the pressure head, and the velocity head at a given point in the aquifer.

Hearth, drying – a surface upon which waste material with high moisture content is placed for drying or burning.

Heat balance – an accounting of the distribution of the energy inputs and outputs of an incinerator.

Heat exchanger – any device designed to transfer heat from one fluid to another.

Heat value, high (HHV) – the heat liberated per unit mass of refuse when burned completely and the products of combustion are cooled to the initial temperature, as measured in a calorimeter.

Heat value, low (LHV) – the high heat value (HHV) less the latent heat of vaporization of the water formed by burning the hydrogen in the fuel.

Heavy metals – metals of high atomic weight and density, such as mercury, lead, and cadmium, that are toxic to living organisms.

Helminth ova – eggs produced by intestinal worms (Helminths).

Heterogeneous – a mixture of dissimilar material.

High rate digester – *See* Stirred tank reactor.

Home scrap – waste produced and reused inside a production facility.

Homogeneous – the characteristic of having uniform properties everywhere.

Household hazardous waste – waste derived from products used at residences which contain substances which if present in larger quantities would be regulated under RCRA.

Humus – the stabilized organic material that remains after microbial degradation of plant and animal matter.

Hydraulic conductivity (soil) – the amount of water that can move through a cross-section of soil per unit time.

Hydraulic gradient – change in the hydraulic head per unit horizontal distance.

Hydrogen sulfide (H_2S) – a highly toxic gas, often produced in the decomposition of materials when sulfur is present. It has the recognizable odor of rotten eggs in low concentrations as small as parts per billion.

Hydrology – the science dealing with the study of the distribution and flow of water on or in the earth.

Hydrolysis, cellulose – a process to chemically convert cellulose (paper) to yield a dilute sugar solution which can then be made into ethyl alcohol, or other products.

Hydrolyzing fermentative organisms – in methanogenesis, the microbes which hydrolyze starches, proteins, and other complex organic molecules fermenting the products to form alcohols, organic acids, carbon dioxide, and hydrogen.

Hydropulper – a large mechanical processor used in the paper industry to pulp wastepaper and separate foreign matter. Hydropulpers have been used for wet-processing of solid waste.

Ignitable hazardous waste – hazardous waste that presents a fire hazard because of the tendency to easily ignite or undergo spontaneous combustion.

Imperforate basket centrifuge – a batch mode centrifuge used to thicken or dewater sludge.

Impermeable – not permitting fluids to pass.

In situ bioremediation – degrading contaminants in place (in soil or ground water) by enhancement of microbial degradation processes.

In-vessel composting – *See* Closed composting.

Incineration – the controlled combustion of solid, liquid, or gaseous wastes resulting in by-product gases and a residue containing little combustible material.

Incinerator – a facility used for burning waste under controlled conditions.

Industrial waste – waste produced in processing raw materials and manufacturing products.

Infective hazardous wastes – refers to pathological wastes from hospitals, clinics, certain laboratories, and raw sewage sludges.

Infiltration – the downward flow of water through the upper surface of a landfill or ground surface.

Information driven waste reduction – refers to an approach to waste reduction based on the collection of information on the nature of waste and how it is generated in an industrial process.

Institutional waste – waste produced by schools, hospitals, research institutions, and public buildings.

Integrated collection strategy – refers to the design of a solid waste collection system with goals that include service levels which meet political, health, and regulatory requirements, are flexible, and support waste reduction/diversion — all at the lowest possible cost — and, where appropriate, with partnerships between the private and public sectors.

Integrated waste management – coordinated use of a hierarchy of management methods, including recycling, composting, incineration, and landfilling.

Invertebrates – small animals lacking a spinal column.

Investment tax credit – a tax credit that allows businesses to subtract a portion of the cost of qualifying capital purchases from their federal or state tax liability, thus reducing the net after-tax cost of capital.

Ion – a charged atom or molecule.

Ion exchange – a chemical process where there is a reversible exchange of ions between a liquid and a solid.

Isotropic – having physical properties that are the same regardless of the direction of measurement.

Junk – a collection of secondary materials, sorted but not processed.

Kjeldahl method – a standard method for determining total nitrogen content of a substance.

LAER – lowest achievable emission rate.

Lagoon sludge dewatering – a method of dewatering based on the slow settling of solids and evaporation in large earthen basins.

Laminar flow – fluid flow in which the water particles follow smooth streamlines.

Land application – applying wastewater or sludges onto or into soil.

Land farming – refers to placement of contaminated soils in thin layers where they are tilled to promote aeration and enhance biological removal of contaminants.

Landfill cell – a discrete unit of volume in a landfill which uses a liner to provide isolation of wastes from adjacent cells or wastes.

Landfill, sanitary – *See* Sanitary landfill.

Landfilling – disposing of solid waste on land in a series of compacted layers and covering it, usually daily, with soil or other materials.

Large quantity HW generator – any organization which generates more than 1000 kg of hazardous waste per month.

Leachate – the chemically and biologically contaminated liquid that has percolated through or drained from waste.

Leachate collection and removal system – a system of pipes used to collect leachate that settles on a liner and prevent it from migrating into groundwater.

Lift – a layer of landfill cells in a designated area of a sanitary landfill at approximately the same elevation.

Lightweighting – reducing the overall weight of a container by using less material.

Lime – specifically calcium oxide (CaO), but more generally any of the chemical forms of quicklime, hydrated lime, and hydraulic lime. Lime is commonly used as a chemical precipitant in wastewater treatment.

Liner – a protective layer, made of soil or synthetic materials, installed along the bottom or sides of a landfill to reduce migration of leachate into groundwater beneath the site or laterally away from the site.

Listed hazardous waste – any substance on the USEPA's list of substances known to have hazardous and/or toxic properties.

Litter – post-consumer waste that is carelessly discarded outside of the regular disposal system.

LOAEL – the lowest observed adverse effect level; the lowest dose in an experiment that produces an observable adverse effect.

Loam – a broad grouping of soil texture classes ranging from coarse-textured to fine-textured, containing sand, salt, and clay.

Loan guarantee – government-funded insurance that protects lenders against the failure of a project to pay back the principal and interest on a loan.

Low-interest loans – government subsidy that allows loans for specific purposes to be offered at below market interest rates.

Lysimeter – a collection device for determining the quantity and characteristics of water that has percolated through waste, either in the field (i.e., in situ) or in the laboratory.

Macro-routing – the process of determining well-balanced collection routes.

Magnetic separator – a device that uses a magnet to sort ferrous materials from other waste materials.

Manifest – the shipping document EPA Form 8700-22 and, if necessary, EPA Form 8700-22A, originated and signed by a generator of hazardous waste in accordance with the instructions included in the Appendix to 4-0 CFR 262.

Manure – The fecal and urinary defecations of livestock and poultry.

Mass burn incinerator – incinerator capable of burning MSW without prior sorting or processing, generally under conditions of excess air; built on-site.

Material substitution – a concept in hazardous waste management which encourages the use of less toxic substances in a manufacturing process to reduce the volume or toxicity of hazardous waste.

Materials management – a MSW management approach that would: 1) coordinate product manufacturing with different management methods (e.g., design products for recyclability); and 2) manage MSW on a material-by-material basis, by diverting discarded materials to the most appropriate management method based on their physical and chemical characteristics.

Materials recovery – retrieval of materials from MSW.

Materials Recovery Facility (MRF) – a building with equipment to sort recyclable materials and convert them to a salable form.

Mechanical pulp — pulp produced by grinding wood into fibers. The primary use is for newsprint.

Medium quantity hazardous waste generator – any organization which generates 100–1000 kg of hazardous waste and no more than 1 kg of acutely hazardous waste in a calendar month.

Membrane filtration – filtration using thin membrane-like filter media containing numerous small pores which can remove colloid size particles or smaller.

Mesophilic microbes – organisms which are best adapted to a temperature range of 25–45°C.

Methane (CH_4) – an odorless, colorless, nonpoisonous explosive gas generated during anaerobic microbial decomposition of organic materials.

Methanogenesis – pertaining to the formation of methane by anaerobic microbes.

Methanogens – obligatory anaerobic bacteria which metabolize carbon dioxide, hydrogen, acetic acid, and certain other select one or two carbon compounds to form methane.

Micro-routing – the process of designing waste collection routes so that the number of routes and duplicated travel over specific streets is minimized.

Microorganisms – microscopic living things, including the bacteria, actinomycetes, yeasts, fungi, some algae, spirochaetes, slime molds, protozoans, and some multicellular organisms.

Mixed liquor – the mixture of microbes, other suspended matter, and water in a given stage of wastewater treatment.

Mixed MSW – refuse that is not sorted into categories of materials.

Modular incinerator – incineration without prior sorting or processing of MSW, in a relatively small two-chamber combustion system; usually fabricated elsewhere and then delivered to the incineration site.

Moisture content – the amount of water in a material, typically expressed as a percentage of the total mass.

Mole – the quantity of a substance that contains an Avogadro's number (6.023×10^{23}) of units. The atomic weight of an element expressed in grams contains an Avogadro's number of atoms. The molecular weight of a molecule, equal to the sum of the atomic weights of the atoms in the molecule, expressed in grams contains an Avogadro's number of molecules.

Monitoring well – a well used to measure groundwater levels, or to obtain water samples for water quality analysis.

Monofill – a sanitary landfill for one type of waste only.

MRF – a materials recovery facility.

MSW – *See* Municipal solid waste.

Multiple chamber incinerator – an incinerator designed with two or more refractory-lined chambers, interconnected by gas-passage ports or ducts and designed in such a manner as to provide for complete combustion of the material to be burned.

Municipal solid waste (MSW) – solid waste generated at residences, commercial establishments, and institutions; as used here. MSW excludes construction or demolition debris and automobile scrap.

Mutagen – an agent that causes a permanent genetic change in a cell other than that which occurs during normal genetic recombination (i.e., causes mutations).

National Priorities List (NPL) sites – abandoned toxic waste sites which have been designated as Superfund sites for cleanup.

NCP – the National Contingency Plan is the basis for remedial actions taken under Superfund.

NESHAPS – National Emission Standards for Hazardous Air Pollutants, mandated by the Clean Air Act, and promulgated by EPA.

Neutralization – treatments which change the pH of highly acidic or highly basic substances to neutral (pH = 7).

Newsprint – paper made from groundwood pulp.

Nitrification – the bacterial conversion of ammonia to nitrate.

Nitrobacter – the genus of bacteria which convert nitrite to nitrate.

Nitrogen loading (soil) – the amount of nitrogen applied to soil as ammonia and/or nitrate.

Nitrogen oxides (NO$_x$) – gases formed from atmospheric nitrogen and oxygen at high combustion temperatures.

Nitrosomonas – the genus of bacteria which convert ammonia to nitrite.

Nonspecific sources – listed hazardous waste that comes from a variety of manufacturing and industrial processes.

Nonferrous metals – all metals except iron (steel), nickel, and cobalt. The predominant nonferrous metals in waste are aluminum, copper, brass, and lead.

Nonreturnable container (bottle or can) – a bottle or can that is designed for a single use after which it will be discarded for recycling or landfilling.

Observation well – a well used to observe the elevation of the water table or potentiometric surface. An observation well generally has a larger diameter than a piezometer and typically is screened or slotted over a larger distance.

Open burning – the combustion of any material without the control of combustion conditions.

Open dumping – disposal of waste in a site that has little or no management and no engineered design to minimize pollution or nuisance conditions.

Open-hearth process – a steel-making process using chemicals and combustible gas to oxidize a molten mixture of pig iron and steel scrap. The open-hearth process can use a maximum of 50% steel scrap.

Opportunity cost – the cost of foregoing alternative uses of a resource.

Organic – being, containing, or relating to compounds that contain carbon in combination with one or more elements, whether derived from living organisms or not.

Organic acid – a product of biochemical activity containing the carboxyl group (-COOH) which readily reacts with other compounds.

Overland Flow Process – a treatment method in which wastewater is applied across the upper reaches of sloped terraces and allowed to flow across vegetated surfaces to runoff collection channels.

Oxidation – a chemical reaction which involves combining a substance with oxygen or in which there is an increase in oxidation number resulting from a loss of electrons.

Oxidation and reduction – treatment of hazardous waste to change the oxidation state of a substance in a manner such that the waste will be converted to a less toxic form and/or make it possible to remove it from solution.

Ozone – a highly reactive form of oxygen which is an effective disinfectant.

Paper converting operations – manufacturing facilities that transform paper into products such as envelopes or boxes.

Paperboard – a broad category of paper products that are heavier, thicker, and more rigid than ordinary paper. Also called cardboard.

Paperstock – a general term used to designate waste papers which have been sorted or segregated at the source into various standard grades.

Participation rate – portion of a population participating in a recycling program.

Particulates – any aerosols, dusts, fumes, mists, smokes, and sprays, or matter suspended in gases discharged to the atmosphere.

Pathogen – any infective agent or organism (e.g., virus, bacterium, protozoan) capable of producing disease.

Pathological waste – wastes that may contain various microbes or viruses capable of causing disease.

Peat (humus) – a soft, light swamp soil consisting mostly of decayed vegetation.

Perched water table – underground water lying above dry soil, but sealed from the dry soil by a low permeability layer such as clay or bedrock.

Percolation – a qualitative term applying to the downward movement of water through unsaturated soil, rock, or waste in a landfill by the pull of gravity.

Permeability (qualitative) – a term used to describe the ease with which a fluid can flow through a solid material.

PHB – Poly-beta-hydroxybutyrate, an energy reserve compound produced in many bacterial cells.

Phosphate – Orthophosphate, PO_4^{3-}, is the most common form of phosphorus in wastewater.

Photodegradable plastic – a plastic that will break down in the presence of ultraviolet (UV) radiation.

Phytotoxic effects – showing evidence of plant toxicity.

Piezometer – a nonpumping well used to measure the elevation of the water table or potentiometric surface. The piezometer has a short screened area at its base.

Pig iron – crude iron that is the direct product of a blast furnace.

Pollutants – any solid, liquid or gaseous matter in an effluent which tends to degrade the environment.

Polyelectrolytes – organic compounds possessing numerous ionizable subunits; very effective in inducing the formation of settleable flocs in wastewater treatment.

Polymers – *See* Polyelectrolytes.

Polyvinyl chloride (PVC) – a plastic material (CH_2=CHCl) commonly used in construction materials, in water and drainage pipes, and in a membrane form.

Porosity – the ratio of the void space in a porous material such as a soil to the total volume occupied.

Post-consumer waste – waste generated by consumers.

Potentiometric surface – the surface defined by the levels to which groundwater will rise in tightly cased wells that tap an artesian aquifer.

Pozzolanic materials — materials capable of reacting with lime and water to form concrete.

Pre-consumer waste – waste generated in processing materials or manufacturing them into final products.

Preapplication treatment – the treatment, usually preliminary and primary, that is applied to wastewater prior to its application to land.

Precipitation – the process that occurs when dissolved or suspended matter in water aggregates to form solids which separate from the liquid phase.

Preliminary treatment – in wastewater treatment this refers to the operations such as screening and grit removal that prepare wastewater for primary treatment.

Pretreatment – treatment given domestic sewage that contains industrial waste components that would, if not removed or treated, interfere with (or pass through untreated) the treatment processes in a wastewater treatment plant.

Primary material – a commercial material produced from virgin materials.

Primary sludge – the settled solids removed from a primary sedimentation tank or basin.

Primary treatment – the treatment of wastewater by sedimentation to remove much of the suspended solids.

Processing – preparing individual or mixed MSW materials for subsequent management, using processes such as baling, magnetic removal, shredding.

Procurement – the purchase of materials and services, usually, in the case of government procurement, through awarding contracts to low bidders.

Product fee – a tax or fee on materials or products that is designed to add the cost of their disposal to the purchase price.

Productivity models – mathematical models used in solid waste management to identify factors which offer the greatest productivity opportunities in waste collection.

Prompt industrial scrap – waste produced in an intermediate stage of processing and returned to the basic production facility for reuse.

Pulp – the fiber produced by chemical or mechanical means (or a combination of both) from fibrous cellulosic raw material, such as wood pulp, waste paper, or rags.

Pulpwood – soft woods, such as spruce, aspen, or pine, used in making paper.

Pulverization – reduction in the size of brittle materials by crushing.

Putrescible – capable of being rapidly decomposed by microorganisms so as to cause nuisances from odors and gases.

Putrescible components – constituents of waste that have a tendency to be changed to highly odorous products as raw waste decomposes.

Pyrolysis – thermal treatment of material in a closed chamber in the absence of oxygen. The heat evaporates the moisture and decomposes the material into various hydrocarbon gases and liquids and a carbon-like residue called char.

Quantity reduction – changing the design of a product so that less MSW is generated when the product or its residuals are discarded, or so that the product is more durable or repairable.

Quarry – an open-pit mining operation for sand or rock.

Rapid Infiltration Process – a method of wastewater treatment by placement in earthen basins containing highly permeable soils.

Raw municipal sewage – wastewater, before treatment, that is largely generated by residences and small businesses.

RCRA – the acronym refers both to a law and a program. The Resource Conservation and Recovery Act, as amended—4-2 U.S.C. Sec. 6901 *et seq.*, and the program and regulations that implement the Act.

Reactivity – refers to the tendency of a substance to react vigorously or explosively with air, water, or other substances and generate harmful vapors or fume.

Reclaimed water – wastewater treated sufficiently to be reused, as for example, in crop irrigation.

Recycling – separating and processing a given waste material (e.g., glass) from the waste stream for reuse or processing so as to be suitable for use as a raw material for manufacturing.

Recycling/recovery/diversion rate – the amount of recyclables collected and processed into new products, divided by total amount of MSW generated.

Reduction – a chemical reaction in which there is a decrease in oxidation number as a result of gaining electrons.

Refuse – same as solid waste.

Refuse-derived fuel (RDF) – fuel produced from MSW that has undergone processing; fuel can be in shredded, fluff, or densified pellet forms.

Remedial Investigation/Feasibility Study (RI/FS) – the means by which a NPL Superfund site is examined in terms of what remedies exist for cleanup, what project costs might be, and what the resulting risk would include.

Renovated water – wastewater which has been sufficiently treated to be used for irrigation or to recharge groundwater.

Residual sludge nitrogen – the nitrogen which is carried over from one year's sludge application to land to succeeding years.

Residue – all solid by-products of processing.

Resource recovery – a general term used to describe the retrieval of materials or energy from MSW, for purposes of recycling or reuse.

Responsible parties – the companies or organizations which generated the waste placed in a Superfund site or improperly managed the site. Responsible parties are liable for the cost of cleanup of a site.

Returnable bottle – a bottle that is returned to the bottler after the beverage in it is consumed and refilled.

Reuse – taking a component of MSW and, possibly, with some slight modification (e.g., cleaning, repair), using it again for its original purpose (e.g.. refillable beverage bottles).

Revenue bonds – bonds sold to finance a specific project. Revenues received from facility charges and tipping fees are used to repay the bonds.

Reverse osmosis – a filtration process in which water passes through a membrane under high pressure leaving dissolved matter behind. Reverse osmosis removes particles in the size range of 0.1 to 15 nm.

RFP – acronym for "request for proposal"; a process for inviting qualified firms or individuals to supply goods, services, or construct facilities.

Rotating biological contactor – a method of aerobic wastewater treatment based on the formation of microbial films on thin plastic disks which are alternately rotated through a waste stream and air.

Rubbish – nonbulky domestic and commercial solid waste exclusive of garbage.

Run-off – any rainwater, leachate, or other liquid that drains over land from any part of a facility.

Run-on – any rainwater, leachate, or other liquid that drains over land onto any part of a facility.

Salmonella – a genus of bacteria which includes a number of human pathogens; for example, typhoid fever.

Salvaging – the retrieval of reusable materials.

Sand – coarse soil particles ranging from 0.05 to 2.0 mm in diameter.

Sanitary landfill – an engineered disposal site. The site is designed and operated so as to compact the waste to the smallest possible volume, to cover the waste with soil at the end of each day, and to minimize adverse effects to groundwater, control landfill gases, and to prevent litter, odor, and other nuisance conditions.

Sanitation – the control of environmental factors which affect hygiene and health.

SARA Title III – Title III of the Superfund Amendments and Reauthorization Act of 1986 includes the Emergency Planning and Community Right-to-Know Act.

Saturated zone – *See* Zone of saturation.

Scavenging – the uncontrolled or unauthorized picking of materials from a waste stream.

Scrap – recovered metal, often applied to iron and steel.

Screening – separating pulverized waste material into various sizes by using one or more sieve-like devices.

Scrubber – an air pollution control device that removes certain gases by passing exhaust gases through an alkaline liquid spray that reacts with and neutralizes acid gases.

Secondary materials – materials recovered from the solid waste stream that may be used as raw materials.

Secondary treatment – the biological wastewater treatment process which follows primary treatment.

Secure landfill – specially designed landfills for hazardous waste. They are equipped with double liners and/or other containment features to minimize leachate and emissions from the landfill.

Sedimentation – the removal of suspended solids by gravity in a tank or basin; also called settling or clarification.

Sedimentation basin – a basin or tank in which wastewater is held with minimum turbulence to remove suspended solids by gravity; also called clarifier or settling tank.

Seepage – the movement of water through soil without formation of definite channels.

Sensitivity analysis – determination of the changes occurring in the value in the objective function when deviations from the optimal value take place.

Separation – to manually or mechanically sort and segregate into groups and subgroups of similar materials, such as newsprint, paper, glass (by color), plastics, food wastes, and metals.

Separator ballistic – a device that separates materials based upon density or mass by dropping mixed material onto a high-speed rotary impeller. Objects of different masses are propelled at different velocities and collected in separate bins placed at various distances.

Separator, inertial – a material separation device that relies on ballistic or gravity separation of materials having different masses and densities.

Separator, magnetic – any separating device that removes metals by means of magnets.

Sequential reactors – wastewater treatment basins operated in sequential fashion and in which oxygen is alternately introduced or excluded in order to achieve biological removal of BOD, nitrogen, and phosphorus.

Set-out locations – refers to the locations where MSW will be picked up by collection vehicles (e.g., curbside or alley).

Settling chamber – any chamber designed to reduce the velocity of the exhaust gas to promote the settling of particulates.

Sewage sludge – a semiliquid substance consisting of suspended sewage solids combined with water and dissolved material in varying amounts.

Shears – a size-reduction machine that cuts material by using large blades.

Shredder – a mechanical device used to break up waste materials into smaller pieces. The pieces are usually in the form of irregularly shaped strips.

SIC code – code numbers used to identify economic activities for statistical purposes. The SIC codes are published in the U.S. Standard Industrial Classification Manual (1987).

Silo composters – compost reactors in which waste is fed in at the top and compost removed at the bottom (plug flow). Air is forced through the mixture in a controlled manner for optimum aeration and temperature control.

Silt – a fine soil that is intermediate between clay and sand, with particles 0.05 to 0.002 mm in diameter.

Slag – a glassy substance formed from fusion of materials at furnace temperatures in an incinerator.

Slope – the ratio of the change in height per horizontal distance traveled, usually expressed in percent or by the angle (expressed in degrees) by which the surface is inclined from horizontal.

Slow Rate (SR) land treatment – the application of wastewater to crop land at rates sufficient to satisfy the needs of the crop (irrigation rate).

Sludge – any solid, semi-solid, or liquid waste generated from a municipal, commercial, or industrial wastewater treatment plant or air pollution control facility.

Sludge cake – dewatered sludge that usually has a solids content of 20% or more.

Sludge conditioning – the biological, chemical, or physical treatment of raw sludges to improve their dewatering characteristics.

Sludge dewatering – removal of water from sludge using centrifugation, vacuum filtration, belt presses, or drying beds.

Sludge drying beds – a dewatering technique in which thin layers of sludge are spread over a large area of sand or other porous media. The sludge dewaters by drainage and evaporation.

Sludge thickening – the process of increasing the solids content of a sludge.

Slurry – a fluid mixture of water and fine insoluble particles.

Small quantity hazardous waste generator – an organization which generates less than 100 kg of hazardous waste, and no more than 1 kg of acutely hazardous waste in a calendar month.

Smoke – an aerosol consisting of all the dispersible particulate products from the incomplete combustion of carbonaceous materials entrained in flue gas as gaseous medium.

Soil – the upper layer of the earth in which plants grow.

Soil cation exchange capacity – a measure of the capability of soil to remove cations from leachate by attachment to the soil particles.

Soil conditioner – substances such as compost which improve the aeration, water holding capacity, and other soil properties important to good plant growth.

Soil erosion – the movement of soil from the land surface by wind or water.

Soil matrix – the array of materials of which soil is composed.

Soil permeability – the term used to describe the ease with which water can pass through a given soil.

Solid bowl decanter centrifuge – a continuously fed centrifuge capable of very high throughput, but sensitive to abrasive materials.

Solid waste – garbage, refuse, sludge from a waste treatment plant, water supply treatment plant, or air pollution control facility and other discarded material, including solid, liquid, semisolid, or contained gaseous material resulting from industrial, commercial, mining, and agricultural operations, and from community activities.

Solid waste management – planning and implementation of systems to handle solid waste.

Solidification, hazardous waste – refers to methods used to immobilize toxic substances using lime, cement, or fly ash with water to set into concrete.

Source separation – separation at a household or commercial establishment of MSW into different recyclable components.

Source-separated recyclables – recyclable materials separated from each other and from mixed waste at the point of generation.

Sparger – a device used to inject air in the vicinity of a turbine in an activated sludge basin to create an aerobic environment.

Specific discharge – an apparent velocity calculated from Darcy's law which represents the rate at which water would flow if the water were flowing in an open conduit rather than one filled with a porous material.

Specific hazardous waste sources – listed hazardous wastes that come from a specific source or industry, as for example, penta-chlorophenol waste from wood preserving plants.

Spoil – soil or rock that has been removed from its original location.

Stabilized sludge – a sludge that has been processed to reduce pathogens and the potential for offensive odors.

Static pile composting – a method of composting in which a pile of waste is aerated in place with the use of aeration pipes placed in the waste pile.

Steel – iron containing various amounts of carbon.

Sterilization – destruction of all microorganisms.

Stillage – the remains of grains and yeast cells after the brewing of beer. Referred to as distillers dried grain after it is dewatered.

Stirred tank reactor (anaerobic) – a reactor design used in the anaerobic digestion of sewage sludges. Also referred to as a high rate digester.

Stockpile – material to be available for future use.

Stoichiometric quantities – the amounts or proportions of reactants or products that satisfy a balanced chemical reaction.

STP – standard temperature and pressure; STP conditions are 0°C and 1 atmosphere pressure.

Submergent vegetation – plants growing submerged in water.

Subsidy – direct or indirect payment from government to businesses, citizens, or institutions to encourage a desired activity.

Suction lysimeter – a device for obtaining water samples from the unsaturated zone by applying a vacuum to a porous ceramic cup.

Sulfur oxides – compounds of sulfur combined with oxygen.

Sump – a pit, tank, or reservoir in which water is collected or stored.

Superfund – the program that implements the Comprehensive Environmental Response, Compensation, and Liability Act (CERCLA).

Superfund Amendments and Reauthorization Act (SARA) – a federal act signed into law October 1986 extending the Superfund program for another five years.

Superfund sites – sites which have been identified as hazardous to humans and the environment due to release of toxic emissions from improperly disposed hazardous materials in the past.

Supply-limited materials – secondary materials that are not collected in sufficient amounts or are too highly contaminated for current manufacturing processes.

Surface water – standing or flowing water whose top surface is exposed to the atmosphere.

Suspended solids – in wastewater the particulate matter which is in suspension and much of which may be removed by sedimentation, coagulation, or filtration processes.

Synthetic clay liner – a synthetic landfill liner manufactured by sandwiching a uniform layer of bentonite between two geotextiles which swells when the liner is in contact with water.

Tailings – low grade minerals or waste material separated from richer mineral ore during screening or processing.

Teratogen – a physical or chemical agent which is capable of causing nonhereditary birth defects in offspring.

Terrestrial – referring to land, as distinct from air and water.

Tertiary treatment – the use of chemical, physical, or biological processes to improve the quality of effluent beyond secondary wastewater treatment.

Thermal sludge conditioning – the application of high temperature and pressure to improve the dewatering characteristics of sewage sludge.

Thermal treatment – the treatment of waste in a process that uses elevated temperatures as the primary means to change the chemical, physical, or biological character or composition of a waste. Examples of thermal treatment processes are incineration, calcination, wet air oxidation, and pyrolysis.

Thermophils – bacteria or other microorganisms with optimum growing temperatures of roughly 45°C to 60°C.

Threshold – the lowest dose of a chemical at which a specified measurable effect is observed.

Tin cans – cans made from thin sheet steel coated with a layer of tin to provide protection against corrosion.

Tipping fee – a fee charged for delivering MSW to a landfill, incinerator, or recycling facility; usually expressed in dollars per ton.

Tipping floor – the unloading floor for vehicles that are delivering refuse to an incinerator or other processing plant.

Topographic map – a map indicating surface elevations and slopes.

Toxic Release Inventory (TRI) – the USEPA program which monitors the release and transfer of hazardous wastes by industries.

Toxicity – refers to the capacity of a substance to produce personal injury or illness through ingestion, inhalation, or absorption through any body surface.

Toxicity Characteristics Leaching Procedure (TCLP) – a test applied to potentially hazardous waste that uses an acidic solution to simulate leaching activity in a landfill. If the TCLP test results in concentrations of toxic constituents above USEPA specified limits, the tested waste is declared hazardous on the basis of its toxic characteristics.

Toxicity reduction – elimination or reduction of toxic substances contained in products.

Transfer station – a facility used for removing refuse from collection trucks and placing it in long-haul vehicles.

Transfer trailers – vehicles which move waste from transfer stations to landfills, incinerators, or other waste processing facilities.

Trash – nongarbage portion of solid waste.

Treatment – any process or technique designed to change the physical, chemical, or biological character or composition of a waste so as to reduce objectionable characteristics, recover components or energy, or convert the waste into a more useful form.

Trickling filter – a biological treatment process where wastewater is passed over a bed of coarse media such as rock, or plastic media. Microbial films form on the media and degrade the organic components of the wastewater.

Trommel – a rotating inclined drum with holes or screens used to separate particles by size.

Tunnel composters – horizontal reactors loaded with waste at one end and with removal of compost at the other (plug flow). Aeration is by "headers" along the length of the tunnel.

Ultrafiltration – a membrane filtration system designed to remove material in the size range of 5–100 nanometers.

Upflow anaerobic sludge blanket reactor (UASB) – an anaerobic reactor design which uses a gas/solids separator device to aid in retaining the microbial cell mass in the reactor.

USEPA's 33/50 Program – a program initiated in 1991 asking for voluntary reduction in the release and transfer of 17 high priority toxic chemicals. The objective was to reduce the release and transfer of these chemicals from 33% by the end of 1992 to 50% by the end of 1995.

Utility – a company providing service under a government license or monopoly franchise.

Vacuum filter – a method of sludge dewatering consisting of a large rotating drum covered with a metal or cloth belt. It is partially submerged and as it rotates, the application of a vacuum during part of the rotation produces a sludge cake.

Vacuum flotation – a process in which a vacuum is applied to mixed liquor to separate the biological solids from the liquid phase.

Vector – a living insect or other arthropod, or animal (not human) which transmits infectious diseases from one person or animal to another.

Vector reduction (sludges) – treatments such as anaerobic digestion and composting which make the waste unattractive for vectors.

Vegetative cover – referring to plants or plant growth in soil.

Vertebrates – animals possessing a segmented spinal column.

Virgin material – any raw material for industrial processes which has not previously been used.

Viruses – small noncellular parasitic particles; many viruses cause disease in humans, animals, plants, and other organisms.

Volatile matter – material capable of being vaporized or evaporated quickly.

Waste – a material perceived to have little or no value.

Waste activated sludge (WAS) – the excess biological solids produced during the activated sludge process. It is sometimes referred to as secondary sludge.

Waste exchanges – organizations which encourage the exchange of materials between industries. One company's waste solvent might be acceptable as a process chemical by another.

Waste paper utilization rate – ratio of waste paper consumption to total production of paper and paperboard.

Waste stabilization – changes in the molecular structure of organic wastes that reduce the offensiveness of the wastes in terms of odor, pathogen content and attractiveness to vectors.

Waste stabilization ponds (lagoons) – large earthen basins used in the treatment of wastewater. Mechanical aeration is often required to allow naturally occurring microbes to degrade organic matter (BOD) effectively.

Waste-to-Energy (WTE) – processing waste in incinerators designed for recovery of energy as steam or electricity.

Wastewater treatment – a combination of physical, biological, and chemical processes used to remove suspended solids, pathogenic organisms, and dissolved chemical species.

Water table – the level below which the ground is saturated with water.

Waterwall incinerator – incinerator constructed for capturing energy by heating water or steam passing through tubes embedded in the combustion chamber walls; capable of burning MSW without prior sorting or processing; built on-site.

Well – a hole bored into the earth to extract liquids or gases.

Well-balanced collection routes – solid waste collection routes in which workloads are balanced, overtime is modest, and there is relatively little nonproductive time.

Well casing – *See* Casing.

Well nest – a group of groundwater monitoring wells drilled to different depths at the same location.

Wet/dry cycle – in rapid infiltration wastewater treatment, a period of waste application followed by a period with no application, after which the cycle is repeated.

Wetlands – transition zones between terrestrial and aquatic environments, but where water is the dominant element.

White goods – large, metal household appliances (e.g., stoves. dryers, refrigerators, etc.)

Windrow – a long row of heaped material left on the ground or in a special area.

Windrow composting – a composting method in which waste is placed in long rows which are periodically "turned" to promote aeration.

Wood pulp – the mixture of cellulosic fibers and water from which paper is made.

Zone of saturation – that region of the earth's crust in which all voids are filled with water.

Appendix A

FACTORS FOR CONVERSION OF UNITS

To Convert From	To	Multiply By
acres	hectares	0.4047
acres	sq feet	43,560.0
acres	sq meters	4047
acres	sq miles	0.001562
acre-feet	cu feet	43,560.0
acre-feet	cu meters	1.233×10^3
acre-feet	gallons	3.259×10^5
acre-feet	hectare-meters	0.1233
acre-feet	liters	1.234×10^6
BTU	joules	1054.8
BTU	kilowatt-hrs	2.928×10^{-4}
BTU/cu foot	joules/cu meter	3.724×10^4
BTU/gal	kilojoule/liter	0.2787
BTU/lb	kilojoule/kg	2.326
BTU/min	horsepower	0.02356
BTU/min	kilowatts	0.01757
BTU/min	watts	17.57
BTU/ton·mi	kilojoule/tonne·km	0.7225
Celsius	Fahrenheit	$C^\circ \times 1.8 + 32$
cubic feet	cu meters	0.02832
cubic feet	cu yards	0.03704
cubic feet	gallons (U.S liq.)	7.481
cubic feet	liters	28.32
cubic feet/hr	gallons/min	0.1247
cubic feet/min	gallons/sec	0.1247
cubic feet/min	liters/sec	0.4720
cubic feet/sec	cu yards/min	2.222
cubic feet/sec	gallons/min	448.8
cubic meters	cu feet	35.31
cubic meters	cu yards	1.308
cubic meters	gallons (U.S liq.)	264.2

To Convert From	To	Multiply By
cubic meters	liters	1000
cubic yards	cu feet	27.00
cubic yards	cu meters	0.7646
cubic yards	gallons (U.S. liq.)	202.0
cubic yards/min	cu feet/sec	.4500
cubic yards/min	gallons/sec	3.367
cubic yards/min	liters/sec	12.74
Fahrenheit	Celsius	$0.556 \times (F^\circ - 32)$
feet	meters	0.3048
feet/min	feet/sec	0.01667
feet/min	kms/hr	0.01829
feet/min	meters/min	0.3048
feet/min	miles/hr	0.01136
feet/sec	kms/hr	1.097
feet/sec	meters/min	18.29
feet/sec	miles/hr	0.6818
gallons (U.S. liq.)	acre-feet	3.068×10^{-6}
gallons	cu feet	0.1337
gallons	cu meters	3.785×10^{-3}
gallons	cu yards	4.951×10^{-3}
gallons	liters	3.785
gallons/min	cu ft/sec	2.228×10^{-3}
gallons/min	liters/sec	0.06308
gallons/min	cu feet/hr	8.021
gallons/sec	cu ft/min	8.021
gallons/sec	liters/sec	3.785
gallons/sec	cu yd/min	0.2971
hectares	acres	2.471
hectares	sq feet	1.076×10^5
hectares	sq miles	3.860×10^{-3}
hectares	sq yards	1.196×10^4
hectare-meters	acre-ft	8.110
hectare-meters	gallons	2.642×10^5
hectare-meters	liters	1.000×10^6
hectometers	meters	100.0
horsepower	BTU/min	42.44
horsepower	kilowatts	0.7457
horsepower	watts	745.7
joules	BTU	9.480×10^{-4}
joules	kcal	2.389×10^{-4}
joules	watt-hours	2.778×10^{-4}
joules/cu meter	BTU/cu foot	2.685×10^{-5}
joules/cu meter	kcal/cu foot	6.766×10^{-6}
kilocalories	joules	4186
kcal/cu meter	joules/cu meter	4186
kilograms	pounds	2.205

To Convert From	To	Multiply By
kilograms	tons (short)	1.102×10^{-3}
kilograms/cu meter	pounds/cu ft	0.06243
kilograms/cu meter	pounds/cu yard	1.686
kilograms/sq meter	pounds/sq foot	0.2048
kilojoules/kg	BTU/lb	0.4300
kilojoules/kg	kcal/kg	0.2389
kilojoules/liter	BTU/gal	3.588
kilojoule/tonne·km	BTU/ton·mi	1.384
kilometers	meters	1000.0
kilometers	miles	0.6214
kilometers	yards	1094
kilometers/hr	feet/min	54.68
kilometers/hr	feet/sec	0.9113
kilometers/hr	meters/min	16.67
kilometers/liter	miles/gal	2.352
kilowatts	BTU/min	56.92
kilowatts	horsepower	1.341
kilowatts	watts	1000.0
kilowatt-hrs	BTU	3413
kilowatt-hrs	horsepower-hrs	1.341
liters	acre-feet	8.104×10^{-7}
liters	cu feet	0.03531
liters	cu meters	0.001
liters	cu yards	1.308×10^{-3}
liters	gallons (U.S. liquid)	0.2642
liters	hectare-meters	1.000×10^{-7}
liters/min	cu ft/sec	5.886×10^{-4}
liters/min	gals/sec	4.403×10^{-3}
liters/sec	cu ft/min	2.119
liters/sec	gallons/min	15.85
liters/sec	cu yds/min	0.0785
meters	feet	3.281
meters	yards	1.094
meters/min	feet/min	3.281
meters/min	feet/sec	0.05468
meters/min	miles/hr	0.03728
meters/sec	feet/min	196.8
meters/sec	feet/sec	3.281
meters/sec	kilometers/hr	3.600
miles	kilometers	1.609
miles/gal	kilometers/liter	0.4251
miles/hr	feet/min	88.02
miles/hr	feet/sec	1.467
miles/hr	kilometers/hr	1.609
miles/hr	kilometers/min	0.02682
miles/hr	meters/min	26.82
miles/hr	miles/min	0.1667
pascals (i.e., N/m^2)	pounds/sq inch	1.450×10^{-4}

To Convert From	To	Multiply By
pounds	kilograms	0.4536
pounds	tons (short)	5.000×10^{-4}
pounds	metric tons (tonnes)	4.536×10^{-4}
pounds/cu foot	kgs/cu meter	16.02
pounds/cu yard	kgs/cu meter	0.5931
pounds/cu yard	pounds/cu foot	0.0370
pounds/sq foot	kgs/sq meter	4.8818
pounds/sq inch	pascals (i.e., N/m^2)	6.896×10^3
sq feet	acres	2.296×10^{-5}
sq feet	sq meters	0.09290
sq feet	sq miles	3.587×10^{-8}
sq feet	sq yards	0.1111
sq kilometers	acres	247.1
sq kilometers	sq feet	10.76×10^6
sq kilometers	sq meters	1.000×10^6
sq kilometers	sq miles	0.3861
sq kilometers	sq yards	1.196×10^6
sq meters	acres	2.471×10^{-4}
sq meters	sq feet	10.76
sq meters	sq miles	3.861×10^{-7}
sq meters	sq yards	1.196
sq miles	acres	640
sq miles	hectares	259.0
sq miles	sq feet	27.88×10^6
sq miles	sq kilometers	2.590
sq miles	sq meters	2.590×10^6
sq miles	sq yards	3.098×10^6
sq yards	acres	2.066×10^{-4}
sq yards	sq feet	9.000
sq yards	sq meters	0.8361
sq yards	sq mile	3.228×10^{-7}
tons (metric)	kilograms	1000.
tons (metric)	pounds	2205.
tons (metric)	tons (short)	1.102
tons (short)	kilograms	907.2
tons (short)	pounds	2000.
tons (short)	tons (metric)	0.9072
watts	BTU/hr	3.4192
watts	BTU/min	0.05688
watts	horsepower	1.341×10^{-3}
yards	kilometers	9.144×10^{-4}
yards	meters	0.9144

Appendix B

TYPICAL DENSITIES OF RECYCLABLE MATERIALS

Material	Density (kg/m^3)	Density (lb/yd^3)
Aluminum cans, whole	30–44	50–74
Aluminum cans, flattened	150	
Corrugated cardboard, loose	180	300
Corrugated cardboard, baled	600–700	1000–1200
Ferrous cans, whole	90	150
Ferrous cans, flattened	500	850
Glass bottles, whole	350–600	600–1000
Glass bottles, crushed	600–1100	1000–1800
Grass clippings	240–900	400–1500
Leaves, uncompacted	150–300	250–500
Leaves, compacted	190–270	320–450
Newsprint, loose	210–470	360–800
Newsprint, baled	425–600	720–1000
Plastic		
HDPE bottles, whole	14	24
HDPE bottles, baled	180–240	300–400
PET bottles, whole	20–24	30–40
PET bottles, baled	180–240	300–400
Wood chips	300	500

Source: Adapted from National Recycling Coalition draft of National Recycling Coalition Measurement Standards and Reporting Guidelines, presented to the NRC membership October 31, 1989. The values have been liberally rounded.

Appendix C

U.S. FEDERAL LAWS RELATED TO WASTE MANAGEMENT

Reprinted from *Reporting on Municipal Solid Waste: A Local Issue* with the permission from the Environmental Health Center of the National Safety Council (Note: P.L. = Public Law)

Clean Air Act of 1970 — P.L. 95-95; Clean Air Act Amendments-P.L. 101-549
Under the Clean Air Act, incinerators must meet performance standards that limit emissions of individual pollutants to the air. Facilities must meet these standards by using the best available technology.

Clean Water Act (l972) — P.L. 95-217
The Clean Water Act applies to waste disposal facilities generating ash-quench water, landfill leachate, and surface water discharges. Disposal of ash-quench water and landfill leachate can present problems for solid waste facilities because many wastewater treatment plants cannot accept these discharges. Facilities generating surface water discharges must use best available technology to control these discharges and must obtain a discharge permit.

The 1987 reauthorization of the Clean Water Act, called the Water Quality Act, mandates site-specific requirements for facilities that discharge to streams where the best available technology still fails to meet water quality standards. It also requires storm water management plans for facilities whose storm runoff volume exceeds specified limits. A facility within a wetlands area needs a Section 404 permit under the Clean Water Act.

Resource Conservation and Recovery Act (RCRA) (1976) — P.L. 94-580
In 1965, the Solid Waste Disposal Act was passed to improve solid waste disposal methods. It was amended in 1970 by the Resource Conservation and Recovery Act (RCRA), which itself was amended in 1980 and 1984.

Subtitle C of RCRA regulates the generation, transportation, treatment, storage, and disposal of hazardous wastes. Wastes designated by RCRA as hazardous are excluded from Subtitle D incinerator and landfill facilities and must be discarded at facilities permitted under the Subtitle C regulations.

Subtitle D of RCRA is for the environmentally safe operation of solid waste management facilities. At a minimum, state waste disposal faciiities must comply with federal standards, although states may adopt more stringent standards. Subtitle D also established a program under which states may develop and implement solid waste management plans. Because this portion of the law is voluntary, USEPA's role has been limited to setting the minimum regulatory requirements that states must follow in designing their plans, and approving plans that comply with these requirements. Responsibility for developing and implementing the plan lies with each state.

Subtitle F of RCRA emphasizes government responsibilities. For example, Section 6002 requires the federal government to participate actively in procurement programs fostering the recovery and use of recycled materials and energy. Specifically, it requires federal agencies and other groups receiving federal funds to procure items composed of the highest percentage of recovered materials practicable and to delete requirements that products be made from virgin materials.

Subtitle I of RCRA regulates underground storage tanks that hold petroleum products or hazardous substances (other than waste). The various sections of Subtitle I pertain to regulations dealing with the prevention, detection, and correction of leaks.

Public Utilities Regulatory and Policy Act (PURPA) (1978)
Developed to encourage cogeneration and small power producers to supplement existing electrical capacity, PURPA requires investor-owned utilities to purchase electrical power from cogenerators or small producers, such as municipal incinerators, at rates developed by state public utilities boards and overseen by the Federal Energy Regulatory Commission. PURPA therefore guarantees a market and a fair price for the energy produced, to control project risk.

Comprehensive Environmental Response, Compensation and Liability Act (CERCLA) (1980) — P.L. 96-510
Also called Superfund. Under Superfund, owners, operators, and users of hazardous waste sites can be held liable for current and past waste disposal practices. Superfund applies to any environmental cleanup, and a substantial number of the sites currently listed as Superfund sites are municipal landfills. Superfund Amendments and Reauthorization Act — P.L. 99-499.

Safe Drinking Water Act (1984) — P.L. 93-523
The protection of water wellhead areas, the sources of springs or streams, as defined in the Safe Drinking Water Act may affect municipal waste disposal facilities. Facilities located in wellhead areas must comply with state and local restrictions on their activities, including design specifications that may add significantly to the cost of the facility.

Other U.S. federal statutes pertaining to waste management

Asbestos Hazard Emergency Response Act — P.L. 99-519
Endangered Species Act — P.L. 93-205
Federal Advisory Committee Act — P.L. 92-463
Federal Insecticide, Fungicide, and Rodenticide Act — P.L. 92-516

Federal Land Policy and Management Act — P.L. 94-579
Federal Oil and Gas Royalty Management Act — P.L. 97-451
Federal Water Pollution Act (1972) — 92-500
Forest and Rangeland Renewable Resources Planning Act — P.L. 93-378
Hazardous and Solid Waste Amendments (1984) — P.L. 98-616
Historic Sites, Buildings and Antiquities Act — Ch. 593, 49 Stat. 666
Intermodal Surface Transportation Efficiency Act of 1991 — P.L. 102-240
Medical Waste Tracking Act — P.L. 100-582
Migratory Bird Treaty Act — P.L. 86-732
Mine Safety and Health Act — P.L. 95-164
Mining Law of 1872
National Energy Conservation Act — P.L. 95-619
National Environmental Policy Act — P.L. 91-190
Occupational Safety and Health Act — P.L. 91-596
Pollution Prevention Act — P.L. 101-508
Resource Recovery Act (1970) — P.L. 91-512
Solid Waste Disposal Act (1965) — P.L. 89-272
Solid Waste Disposal Act Amendments — P.L. 96-482
Superfund Amendments and Reauthorization Act — P.L.99-499
Surface Mining Control and Reclamation Act — P.L. 95-87
Toxic Substances Control Act — P.L. 94-469
Uranium Mine Tailings Radiation Control Act of 1978 — P.L. 95-604

Appendix D

MUNICIPAL SOLID WASTE MANAGEMENT: STATE-BY-STATE

Reproduced from *Reporting on Municipal Solid Waste: A Local Issue*, (July 1993) with permission from the Environmental Health Center of the National Safety Council.

State	Total MSW Generated (t/y)	Percent Recycled	Percent Incinerated	Percent Landfilled
Alabama	4,714,000	12	8	80
Alaska	453,000	6	15	79
Arizona[a]	3,760,000	7	0	93
Arkansas[a]	1,953,000	10	5	85
California[b]	40,376,000	11	2	87
Colorado	3,173,000	26	1	73
Connecticut	2,629,000	19	57	24
Delaware[a]	716,000	16	19	65
D.C.[c]	833,000	30	59	11
Florida[d]	17,588,000	27	23	49
Georgia	5,440,000	12	3	85
Hawaii[a]	1,179,000	4	42	54
Idaho	771,000	10	0	90
Illinois[b]	12,820,000	11	2	87
Indiana[e]	7,616,000	8	17	75
Iowa	1,893,000	23	2	75
Kansas	2,176,000	5	0	95
Kentucky	4,216,000	15	0	85
Louisiana	3,159,000	10[f]	0	90[f]
Maine[c]	1,130,000	30	37	33
Maryland	4,533,000	15	17	68
Massachusetts	5,984,000	30	47	23
Michigan[a]	11,786,000	26	17	57
Minnesota	3,871,000	38	35	27
Mississippi	1,269,000	8	3	89

State	Total MSW Generated (t/y)	Percent Recycled	Percent Incinerated	Percent Landfilled
Missouri	6,800,000	13	0	87
Montana	675,000	5	2	93
Nebraska	1,269,000	10[f]	0	90[f]
Nevada	2,085,000	10	0	90
New Hampshire[a,c]	1,032,000	10	26	64
New Jersey[c]	6,811,000	34	21	45
New Mexico	1,348,000	6	0	94
New York[c,d]	20,671,000	21	17	62
North Carolina[b]	7,061,000	4	1	95
North Dakota	422,000	17	0	83
Ohio[b]	14,869,000	19	6	75
Oklahoma	2,720,000	10	8	82
Oregon[a]	3,037,000	23	6	71
Pennsylvania	8,145,000	11	30	59
Rhode Island	1,088,000	15	0	85
South Carolina[a]	5,258,000	10	5	85
South Dakota[a,f]	725,000	10	0	90
Tennessee[b]	5,528,000	10	8	82
Texas	13,118,000	11	1	88
Utah	1,360,000	13	7	80
Vermont	499,000	25	3	72
Virginia	6,890,000	24	18	58
Washington[c]	5,175,000	33	2	65
West Virginia	997,000	10	0	90
Wisconsin	3,039,000	24	4	72
Wyoming	290,000	4	0	96
Total	291,742,000	17	11	72

Conversion: tonnes/year × 1.102 = tons/year

[a]Includes some industrial waste.

[b]Includes significant industrial waste.

[c]Includes out of state disposal.

[d]Includes construction and demolition waste.

[e]Includes construction and demolition, and sewage sludge.

[f]Data from BioCycle's 1992 "State of Garbage in America" survey.

Source: 1993 Nationwide survey: The state of garbage in America, *BioCycle*, May, 1993.

Appendix E

STATE DISPOSAL BANS AND RECYCLING LAWS

State	Statewide Management Goals	Source Reduction (%)	Recycling[b] (%)	Composting (%)	Vehicle Batteries	Tires	Yard Waste	Motor Oil	White Goods	Other	Mandatory Deposit Laws	Recycling Tax Incentives	Minimum Content Recycling Laws
				Waste Reduction Targets[a]			Disposal Bans						
Alabama	Mandated 1991	↓	25	↓									
Alaska													
Arizona													X
Arkansas	Mandated 2000	↓	40	↓	X	X					X	X	
California	Mandated 2000	↓	50	↓	X	X	X				X	X	
Colorado								X	X			X	X
Connecticut	Mandated 1991	0	25	0						X[c]	X		
Delaware	Not M'd 2000	0	21	0							X	X	X
D.C.	Mandated 1995	0	45										
Florida	Mandated 1995	↓	30	0	X	X	X	X	X				
Georgia	Mandated 1996	↓	25	↓	X	X	X	X	X	X[d]		X	X
Hawaii	Mandated 2000	↓	50	↓	X		X[e]						
Idaho					X								
Illinois	Mandated 2000	0	25	0	X	X	X				X	X	
Indiana	Mandated 2000	↓	50	↓			X		X[f]		X	X	X
Iowa	Mandated 2000	↓	50	↓	X	X	X	X			X	X	
Kansas						X				X[g]	X	X	
Kentucky	Mandated 1997	↓	25	↓								X	
Louisiana	Mandated 1992	↓	25	↓	X	X						X	
Maine	Not M'd 1994	0	50	↓					X	X[h]	X	X	
Maryland	Mandated 1994	0	20	0			X					X	
Massachusetts	Mandated 2000	10	31	15	X	X	X				X	X	
Michigan	Not M'd 2005	8–12	20–30	8–12	X	X	X		X	X[i]	X	X	X
Minnesota	Mandated 1996	0	30–45	0	X		X	X	X	X[j]	X		X
Mississippi	Mandated 1996	0	25	↓	X						X	X	

Waste Reduction Targets[a]

State	Statewide Management Goals	Source Reduction (%)	Recycling[b] (%)	Composting (%)	Disposal Bans: Vehicle Batteries	Tires	Yard Waste	Motor Oil	White Goods	Other	Mandatory Deposit Laws	Recycling Tax Incentives	Minimum Content Recycling Laws
Missouri	Mandated 1998		40		X	X	X	X	X		X		X
Montana	Not M'd 1996		25									X	
Nebraska	Mandated 2002		50		X	X	X	X	X				
Nevada	Mandated 1994		25		X			X	X		X		
New Hampshire	Mandated 2000		40		X					X[c]			
New Jersey	Mandated 1995		60		X		X				X	X	
New Mexico	Mandated 2000		50		X		X[k]	X				X	
New York	Not M'd 2000		60		X						X		
North Carolina	Mandated 1993		25		X	X	X	X	X		X	X	X
North Dakota	Mandated 2000		40		X			X	X		X		
Ohio	Mandated 1994		25			X		X			X		
Oklahoma											X		
Oregon	Mandated 2000		50		X	X	X	X	X	X[l]	X	X	X
Pennsylvania	Mandated 1997	0	25	0	X		X[m]				X		
Rhode Island	Mandated		70		X						X		X
South Carolina	Mandated 1997		30		X	X	X	X	X		X		
South Dakota	Mandated 2001		50		X	X	X	X	X	X[n]		X	
Tennessee	Mandated 1996		25		X	X	X				X		
Texas	Mandated 1994		40		X	X		X	X		X	X	X
Utah					X						X		
Vermont	Not M'd 2000		40	0	X	X		X	X	X[o]	X	X	
Virginia	Mandated 1995	0	25		X							X	
Washington	Mandated 1995		50		X		X	X			X	X	
West Virginia	Mandated 2010		50		X	X	X			X[i]	X	X	X
Wisconsin					X	X	X	X	X			X	X
Wyoming									X				X

[a] The waste reduction target applies to the combination of activities spanned by the arrows.

[b] Includes yard waste composting.

[c] Mercury oxide batteries.

[d] Demolition debris.

[e] Yard trimmings ban passed in 1993 legislative session.

[f] White goods containing CFC gases, mercury switches, and PCBs.

[g] Nondegradable grocery bags, carbonated beverage containers. and liquor bottles with deposits.

[h] Household batteries.

[i] Glass and metal containers. recyclable paper and single polymer plastics.

[j] NiCd batteries, telephone books, and sources of mercury.

[k] Leaves.

[l] Discarded vehicles.

[m] Leaves and brush.

[n] Office and computer paper, newsprint, ccrrugated and paperboard; glass, plastic, steel and aluminum containers.

[o] Various dry cell batteries.

Sources: Miller, C., *Waste Age* 24(3), 26–34. March, 1993; Steuteville, R. and Goldstein, N., The state of garbage in America: Part I, *BioCycle,* 34(5), 42–50, May, 1993; Steuteville, R., Goldstein, N., and Grotz, K., The state of garbage in America: Part II, *BioCycle,* 34(6), 32–37, June, 1993.

Appendix F

OPERATING BIOSOLIDS COMPOSTING FACILITIES IN THE U.S.

State Plant Name	Type	Biosolids Volume (dry tonne/day, unless noted)
Alabama		
Dothan City	In-vessel (Davis)	3.6
Scottsboro	Windrow	450 wet/y
Alaska		
Homer	ASP	18-27/y
Arizona		
Phoenix	Windrow	7.3
Pinetop-Lakeside	Digester/aerated Windrow (BBC)	1.4 (w/11-13 tonnes MSW)
Quartsite	Windrow	33
Arkansas		
Eureka Springs	Windrow	8.2/mo
Hot Springs	Windrow	4.5
Maysville	Windrow	9.1
California		
Las Virgenes	In-vessel (IPS)	3.6
Los Alisos (El Toro)	ASP	1150 m³/y
Los Angeles County	In-vessel (SWS) (for research)	3.2
Lost Hills	Aerated windrow	100 (inc. 230 wet from Los Angeles)
Oakland: East Bay MUD	ASP	5
Riverside County (in Temescal Canyon)	Windrow	455 wet (from 18 communities)
South San Francisco	Windrow	9100/y

State Plant Name	Type	Biosolids Volume (dry tonne/day, unless noted)
Colorado		
Denver Metro	ASP	5.5+
Ft. Collins	Aerated windrow	0.7
Longmont	ASP	1.4
Pitkin County	ASP	0.45
Silverthorne/Dillon	ASP	18/y
Summit County	ASP	1.8
Tri-Lakes	ASP	1-1.3
Upper Eagle Valley (Vail area)	ASP	0.9 (2.5 days/wk)
Connecticut		
Bristol	ASP	8.1
Fairfield	In-vessel (IPS)	3
Farmington	In-vessel (IPS)	4.5
Greenwich	ASP	1900 m^3/y
Delaware		
Seaford	ASP	5.4/wk
Florida		
Cooper City Utilities	Windrow	0.45-0.9
Meadowood Utility	Windrow	0.18
Miami-Dade Water Sewer: Central Plant	ASP (without amendment)	9
Palm Beach County	In-vessel (IPS)	20
Reedy Creek	ASP	7.2
Sarasota	In-vessel (Purac)	3.6
Georgia		
Brunswick	In-vessel (ISP)	7.3
Clayton Co.	In-vessel (Davis)	2
Preston	ASP	90 wet
Hawaii		
Maui County	Windrow	9-14
Idaho		
Coeur d'Alene	ASP	1.5
Lewiston	ASP	720/y (serv. 4 communities)
Illinois		
DuPage County Woodridge-Greene Valley	ASP	1.8-2.7
Indiana		
Bloomington:Blucher Poole WWTP	Windrow	2.3-2.7
Pike County	ASP	23

State Plant Name	Type	Biosolids Volume (dry tonne/day, unless noted)
Kansas		
Topeka: Oakland WWTP	Windrow	11
Maine		
Bar Harbor	ASP	1100 m^3/y
Ft. Fairfield	Windrow	380 m^3/y
Kennebunkport	ASP	620 m^3/y
Lewiston/Auburn Water	In-vessel	9
Pollution Control auth.	(Longwood Mfg.)	
Lincoln	Windrow	480 m^3/y
Old Orchard Beach	ASP	1600 m^3/y
& Saco		
Old Town & Orono	ASP	4100 m^3/y
Rockland	In-vessel	n/a (septage)
Rumford & Mexico	ASP	0.56
Scarborough	ASP	1030 m^3/y
Unity	In-vessel	18
(Hawk Ridge)	(Gicon tunnels)	
Wilton	ASP	250 m^3/y
Yarmouth	Windrow	940 m^3/y
Maryland		
Aberdeen	ASP	1.1
Baltimore:Back River	In-vessel	44
	(CSC Paygro)	
Elkton	ASP	1.8
Havre de Grace	ASP	0.6
Montgomery County	ASP	63
Perryville	ASP	2.3
Queen Anne's County	ASP	46 m^3/mo
		(total generated)
Massachusetts		
Barre	ASP	3.8 m^3/wk
Billerica WWTP	ASP	1.4
Bridgewater	ASP	1.4
Dartmouth	In-vessel (IPS)	36 wet
Holyoke	In-vessel (IPS)	16-17
Hopedale	Aerated windrow	n/a
Ipswich	Aerated windrow	0.18 wet/y
Leicester WWTP	ASP	0.7/wk
Mansfield	ASP	23m^3/wk
Marlborough East Plant	ASP	9
Nantucket	ASP	0.4-0.9
Pepperell	ASP	9/mo
Somerset	ASP	1.8
Southbridge	ASP	3.6
Springfield	In-vessel (Davis)	27
Williamstown/Hoosac	ASP	4.5
Yarmouth	In-vessel (Royer)	16 m^3 (septage)

State Plant Name	Type	Biosolids Volume (dry tonne/day, unless noted)
Michigan		
Mackinac Island	ASP	n/a (w/MSW)
Montana		
Kalispell	ASP	27 wet/wk
Missoula	ASP	2.7
Nebraska		
Beatrice	Windrow	1.4
Grand Island	ASP	6.3-9
Kearney	Windrow	5.4/wk
Omaha	Windrow	180/y
New Hampshire		
Bristol	ASP	214 m^3/y
Claremont	ASP	2.7-4.5
Dover	ASP	1.8
Durham	ASP	1.8
Lebanon	ASP	1
Merrimack	ASP	9-11
Milford	ASP	0.8
Plymouth	In-vessel (IPS)	0.54
Rochester	In-vessel (IPS)	15 (30 full scale)
Wolfboro	Windrow	1400 m^3/y
New Jersey		
Buena Borough	ASP	0.45
Camden MUA WPCF #1	In-vessel (WS)	45
Cape May Co. MUA	In-vessel (Purac)	18
Middletown Twp.	ASP	4.5
Pennsville	Windrow	0.9
Stanhope: (Musconetcong Sewerage Authority)	In-vessel (Davis)	3.6
Sussex Co. MUA	ASP	6
New Mexico		
Albuquerque	Aerated windrow	1.7
Artesia	Turned pile	n/a
Rosswell	Windrow	137/y
New York		
Binghamton/Johnson	In-vessel (Davis)	9
Cedarhurst	ASP	0.22 m^3/d
Churchville	ASP	32/y
Clinton Co. (Plattsburgh)	In-vessel (Fairfield)	11
Cobleskill	Windrow	146/y

State Plant Name	Type	Biosolids Volume (dry tonne/day, unless noted)
Endicott	In-vessel (Davis)	1.8
Geneva	In-vessel (Davis)	1.8
Gowanda	ASP	1 m^3
Greene	Windrow	137/y
Groveland/Livingston Correctional Facility	n/a	n/a
Guilderland	In-vessel (IPS)	0.9
Lawrence STP	ASP	0.45
LeRoy	In-vessel (IPS)	5.4
Lockport	In-vessel (IPS)	9
Manchester/Shortsville	ASP	0.9
Medina	ASP	90/y
Minoa	Windrow	0.18
Rouses Point	ASP	0.45
S & B Sanitation	ASP	n/a
Schenectady	In-vessel (ABT)	9.5
Thompson	ASP	1.4
Tompkins County	ASP	4.5
Tri-Municipal Sewage	ASP	1.8
Verona	ASP	600/y
Webster	ASP/Windrow	1.35
Yorktown Heights	Windrow	0.9
North Carolina		
Beech Mountain	ASP	0.009
Burnsville	ASP	0.045
Jackson County	ASP	1.4
Lexington	In-vessel (Royer)	1.8
Morganton: Catawba River Plant	ASP	3
Pilot Mountain	ASP	1.4
Rockingham	ASP	1.4
Statesville	Windrow	4.5
Valdese	ASP	1.8
Ohio		
Akron	In-vessel (CSC Paygro)	59
Columbus	ASP	23
Hamilton WWTP	In-vessel (SWS)	14
Lake County	Aerated Window	8
Oregon		
Douglas County: Winston Green WWTF	ASP	0.9
Douglas County: Glide-Idleyld	Aerated windrow	7.7/mo (septage)
Klamath Falls	Aerated windrow	2.3/wk
Newburg	In-vessel (WS)	0.9-1.4
Portland	In-vessel (Davis)	9

State Plant Name	Type	Biosolids Volume (dry tonne/day, unless noted)
Redwood Sewer Dist.	Aerated windrow	0.45
Washington County (Unified Sewerage Ag.)	Static Pile	9/d (for 6 mo)
Pennsylvania		
Butler	ASP	4.5-6.3
Centre County: Univ. Area Jt. Auth.	In-vessel (IPS)	7
Lancaster	In-vessel (Davis)	7.2
Lancaster	ASP (construction upgrade)	
Mansfield	ASP	n/a
Philadelphia	ASP	90
Scranton	ASP (in winter)	16
Springettsbury Twp.	ASP	9-10 (not daily)
Rhode Island		
Bristol	In-vessel (IPS)	2.3
Jamestown	Windrow	214 m³/y
West Warwick	ASP	4.5
South Carolina		
East Richland Co. PSD:Gills Creek	In-vessel (Davis)	4.5
Hilton Head: Broad Creek Plant	In-vessel (CSC Dynatherm)	4.5
Kiawah Island	ASP	13/y
Myrtle Beach	ASP	0.9
South Dakota		
Pierre	Windrow	0.27
Tennessee		
Bristol	In-vessel (Davis)	13
Elizabethton	ASP	1
Sevierville (Sevier SWA)	Digester (BBC) with aerated windrows	68 wet (with 135 t/d MSW)
Texas		
Austin	Windrow	13
Belton: Brazos River Auth.	Windrow	2.3
Bryan	Windrow	5.4
Luling	Windrow	15 m³/y
Pampa	Windrow	n/a
San Antonio: Leon Creek & Dos River WWTPs	Windrow	4.5
Texarkana	Windrow	44 m³/d

State Plant Name	Type	Biosolids Volume (dry tonne/day, unless noted)
Utah		
Provo	Windrow	300 m^3/mo
Moroni	Windrow	n/a
Timpanagos	Windrow	n/a (startup)
Vermont		
Bennington	In-vessel (IPS)	1.8
Springfield	ASP	1.1
Virginia		
Charlottesville (Moores Creek WWTP)	ASP	7.2
Hampton Roads (Peninsula Compost Facility)	ASP	11
Upper Occoquan	Aerated windrow	6.8-9
Washington		
Cowlitz	Windrow	310 m^3/mo
Granite Falls	Turned piles in bays	180 m^3/y
King County Metro	Static pile	54
Monroe	Static pile	3.6
Olympus Terrace Sewer District	Static pile	110 wet/mo
Port Angeles	Static pile	1380 m^3/y
Port Townsend	ASP	15/mo
Puyallup	Windrow	18
Southwest Suburban Sewer District (Miller Creek, Seattle)	Static pile	1.45
Vashon Sewer District	Static Pile	14/y
West Virginia		
Brooke County	ASP/Windrow	90
Wetzel County	ASP/Windrow	72
White Sulphur Springs	Windrow	1.8-2.7
Wisconsin		
Columbia County Portage	Drum with windrows	(with 18 t/d MSW)

ASP = Aerated Static Pile.

ABT = American Biotech Systems.

BBC = Bedminster Bioconversion.

CSC = Compost Systems Company.

Davis = Davis Composting & Residuals Management (formerly Taulman).

IPS = International Process Systems.

RCCI = Resource Control Composting Inc.

Wheelabrator CWS = Clean Water Systems.

WS = Waste Solutions.

Source: Adapted with permission from Goldstein, N.., Riggle, D., and Steuteville, R., Biosolids composting strengthens its base, *Biocycle*, 35(12), 48–57, Dec., 1994.

Appendix G

OPERATING MUNICIPAL WASTE-TO-ENERGY FACILITIES IN THE U.S.

Facility	Startup Year	Type[a]	Design Capacity (t/d)	Energy Generation[b]	Air Emission Control Equipment[c]	Owner	Operator
Alabama							
Huntsville	1990	MB	630	STM	SDA,FF	Public	Ogden Martin
Tuscaloosa	1984	MOD	270	STM	ESP	Public	Consumat
Alaska							
Sitka	1985	MOD	23	STM	Cyclone, ESP	Public	Public
Arkansas							
Batesville	1986	MOD	90	STM	None	Public	Public
N. Little Rock	1977	MOD	90	STM	None	Public	Public
Osceola	1980	MOD	45	STM	None	Public	Public
California							
Commerce	1987	MB	342	ELE	SDA,FF		
Long Beach	1990	MB	1240	ELE	SDA,FF		
Stanislaus	1989	MB	720	ELE	SDA,FF	Ogden Martin	Ogden Martin
Colorado							
Denver (Airport)	N/A	RDF	90	STM	N/A	Public	N/A
Connecticut							
Bridgeport	1988	MB	2025	ELE	SDA,FF	Public	Wheelabrator
Bristol	1988	MB	585	ELE	SDA,FF	Ogden Martin	Ogden Martin
Hartford	1988	RDF	1800	ELE	SDA,FF	Public	ABB RRS &MDC

Location	Year	Technology	Capacity	Energy	APC	Ownership	Vendor
Preston	1992	MB	540	ELE	SDA/FF	Ogden Martin	Ogden Martin
Walingford	1990	MB	378	COG	SDA,FF	Public	Public
Windham	1981	MOD	97	COG	FF,scrub		
Delaware							
Wilmington	1984	RDF-P	900	RDF	FF,cyclone	Public	Raytheon
Wilmington	1987	RDF-B	540	COG	ESP	Private	Private
Florida							
Panama City	1987	MB	460	ELE	ESP	Private	Private
Ft. Lauderdale	1991	MB	2025	ELE	SDA,FF	Wheelabrator	Wheelabrator
Dade Co. (Miami)	1989	RDF	2700	ELE	ESP	Public	Montenay
Hillsborough Co.	1987	MB	1080	ELE	ESP	Public	Ogden Martin
Key West	1986	MB	135	ELE	ESP	Private	Montenay
Lake Co.	1991	MB	475	ELE	SDA,FF	Ogden Martin	Ogden Martin
Lakeland	1983	RDF	270	ELE	Wscrub/ESP	Public	Public
McKay Bay	1985	MB	900	ELE	ESP	Wheelabrator	Wheelabrator
Miami Airport	1984	MOD	540	STM	None	N/A	N/A
Palm Beach Co.	1989	RDF	1800	ELE	SDA,ESP	Public	National Ecology
Pasco Co.	1991	MB	945	ELE	SDA,FF	Public	Ogden Martin
Pinellas Co.	1983	MB	2700	ELE	ESP	Public	Wheelabrator
Pompano Beach	1992	MB	2025	ELE	ESP	Wheelabrator	Wheelabrator
Georgia							
Savannah	1987	MB	450	COG	N/A	Private	Private
Hawaii							
Honolulu	1990	RDF	1944	ELE	SDA,FF	Private	ABB RRS
Idaho							
Cassia Co.	1981	MOD	45	STM	None	Public	Public

Facility	Startup Year	Type[a]	Design Capacity (t/d)	Energy Generation[b]	Air Emission Control Equipment[c]	Owner	Operator
Illinois							
Chicago NW	1970	MB	1440	COG	ESP	Public	Public
Indiana							
Indianapolis	1988	MB	2126	STM	SDA,FF	Ogden Martin	Ogden Martin
Iowa							
Ames	1975	RDF	180	ELE	ESP,FF	Ames Utility	Ames Utility
Maine							
Auburn	1992	MB	180	ELE	SDA/FF	Public	American Energy
Biddeford/Saco	1987	RDF	540	ELE	SDA,FF	Public	Private
Orrington	1988	RDF	900	ELE	SDA,FF	Public	Private
Portland	1988	MB	450	ELE	SDA,FF	Public	Private
Maryland							
Baltimore	1985	MB	2025	COG	ESP	Public	Wheelabrator
Hartford Co. (U.S. Army)	1988	MOD	324	STM	ESP	Ensco	Ensco
Massachusetts							
Agawan	1988	MOD	324	COG	Dscrub,FF	Private	Private
Haverhill/ Haverhill/	1989	MB	1485	ELE	SDA,ESP	Ogden Martin	Ogden Martin
Lawrence	1984	RDF	855	COG	ESP	SBR Assoc.	Ogden Martin
Millbury	1987	MB	1350	ELE	SDA,ESP	Public	Wheelabrator
North Andover	1985	MB	1350	ELE	ESP	Wheelabrator	Wheelabrator
Pittsfield	1981	MOD	216	STM	ESP	Vicon	Vicon

Location	Year	Type	Capacity	Combustion	APC	Owner	Operator
Rochester	1988	RDF	1620	ELE	SDA,ESP	Private	Bechtel
Saugus	1975	MB	1350	ELE	SDA,FF	Wheelabrator	Wheelabrator
Springfield	1988	MOD	324	COG	DSCRUB/FF	Private	Private
Michigan							
Detroit	1990	RDF	2970	COG	SDA,ESP	Public	ABB RRS
Detroit (Fisher)	1986	MOD	90	STM	Cyclone	Private	Private
Jackson	1987	MB	180	COG	SDA,FF	Public	N/A
Kent Co.	1990	MB	563	ELE	SDA,FF	Public	Ogden Martin
Minnesota							
Duluth	1985	RDF	360	STM	Wscrub	Public	Public
Eden Prairie	1987	RDF	504	ELE	N/A	Reuter	Reuter & NSP
Elk River	1989	RDF	1350	ELE	Dscrub,FF	Northern States Power & UPA	Northern States Power & UPA
Fergus Falls	1988	MOD	56	STM	Wscrub	Public	Public
Hennepin Co.	1989	MB	1080	ELE	SDA,FF	Private	Ogden Martin
Olmstead Co.	1987	MB	180	COG	ESP	Public	Public
Perham	1986	MOD	90	STM	ESP	Private	Private
Polk Co.	1988	MOD	72	STM	ESP	Public	Public
Pope/Douglas Co.	1987	MOD	65	STM	ESP	Public	Public
Ramsey/Washington Co.	1987	RDF	900	ELE	ESP	Northern States Power	Northern States Power
Red Wing	1988	MOD	65	STM	ESP	Public	Public
Savage	1982	MOD	54	STM	ESP	Private	Private
Thief River Falls	1985	RDF	80	STM	N/A	N/A	N/A
Mississippi							
Pascagoula	1985	MOD	135	STM	N/A	Public	Sigoure Freres

Facility	Startup Year	Type[a]	Design Capacity (t/d)	Energy Generation[b]	Air Emission Control Equipment[c]	Owner	Operator
Missouri							
Ft. Leonard Wood	1982	MOD	68	STM	None	U.S. Army	Private
Montana							
Livingston	1982	MOD	65	STM	None	Public	Public
New Hampshire							
Claremont	1987	MB	180	ELE	Dscrub,FF	Wheelabrator	Wheelabrator
Concord	1989	MB	450	ELE	Dscrub,FF	Wheelabrator	Wheelabrator
Durham	1980	MOD	97	STM	Cyclone	Public	Public
Groveton	1975	MOD	22	STM	None	Private	Private
New Jersey							
Atlantic Co. Jail	1990	MOD	13	STM	N/A	Public	Public
Camden Co.	1991	MB	945	ELE	SDA,ESP	Foster Wheeler	Foster Wheeler
Essex Co.	1990	MB	2050	ELE	SDA,ESP	Private	Private
Fort Dix	1986	MOD	72	STM	Wscrub,FF	U.S. Army	Private
Gloucester Co.	1990	MB	518	COG	SDA,FF	Wheelabrator	Wheelabrator
Warren Co.	1988	MB	360	ELE	SDA,FF	Ogden Martin	Ogden Martin
New York							
Albany	1981	RDF	675	STM	ESP	Public	EAC Operations
Babylon	1989	MB	675	ELE	SDA,FF	Ogden Martin	Ogden Martin
Betts Ave., Queens	1964	MB	900	STM	ESP	Public	Public
Cattaraugas Co.	1983	MOD	100	STM	None	Public	Public
Dutchess Co.	1989	MB	360	COG	Dscrub,FF	Public	Private
Glen Cove	1983	MB	225	ELE	ESP	Public	Montenay

Location							
Hempstead	1989	MB	2090	ELE	SDA,FF	Am. Ref-Fuel	Am. Ref-Fuel
Hudson Falls	1992	MB	405	ELE	SDA/ESP	Industrial Dev. Agency	Foster Wheeler
Huntington	1992	MB	675	ELE	SDA,FF	Ogden Martin	Ogden Martin
Islip	1990	MB	470	ELE	Dscrub,FF	Public	Montenay
Long Beach	1988	MB	180	ELE	ESP	Catalyst Energy	Montenay
Niagara Falls	1981	RDF	1800	COG	ESP	Occidental Chemical	Occidental Chemical
Oneida Co.	1985	MOD	180	COG	ESP	Public	Public
Oswego Co.	1986	MOD	180	COG	ESP	Public	Public
Rochester (Kodak)	1970	RDF	135	COG	ESP	Kodak	Kodak
Westchester Co.	1985	MB	2025	N/A	N/A	N/A	N/A
North Carolina							
Mecklenburg Co.	1989	MB	212	COG	ESP	Public	Private
New Hanover Co.	1984	MB	405	COG	ESP	Public	Public
Ohio							
Akron	1979	RDF	900	COG	ESP	Public	WTE Corp.
Columbus	1982	RDF	1800	ELE	ESP	Public	Public
Montgomery Co.	1970	MB	810	ELE	ESP	Public	Public
Oklahoma							
Miami	1982	MOD	97	STM	None	Public	Consumat
Tulsa	1986	MB	1013	COG	ESP	Private	Ogden Martin
Oregon							
Marion Co.	1987	MB	495	ELE	SDA,FF	Ogden Martin	Ogden Martin
Pennsylvania							
Delaware County	1991	MB	2420	ELE	SDA/FF	Public	Private
Harrisburg	1972	MB	650	ELE	SDA,FF	Public	Ogden Martin

Facility	Startup Year	Type[a]	Design Capacity (t/d)	Energy Generation[b]	Air Emission Control Equipment[c]	Owner	Operator
Lancaster Co.	1991	MB	1080	ELE	SDA,FF	Public	Ogden Martin
Westmoreland Co.	1986	MOD	45	COG	ESP	Public	Public
South Carolina							
Charleston	1989	MB	540	COG	SDA,ESP	AT&T	Foster Wheeler
Hampton	1985	MOD	243	STM	Scrub,ESP	Public	Public
Tennessee							
Dyersburg	1980	MOD	90	STM	None	Public	Public
Humboldt	1990	RDF	135	N/A	N/A	N/A	N/A
Lewisburg	1980	MOD	54	STM	Wscrub	Public	Public
Nashville	1974	MB	1000	COG	ESP	Public	Nashville Thermal Transfer
Sumner Co.	1981	MB	180	COG	ESP	Public	Public
Texas							
Carthage City	1986	MOD	36	STM	Wscrub	Public	Public
Center	1986	MOD	36	STM	Wscrub	Public	Public
Cleburne	1986	MOD	104	ELE	ESP	Public	Public
Gatesville	1981	MOD	18	MOD	None	Public	Public
Palestine	1980	MOD	23	MOD	None	Public	Public
Waxahachie	1982	MOD	45	MOD	None	Public	Public
Utah							
Davis Co.	1988	MB	360	STM	Dscrub,FF	Public	Private

			tons/d				
Virginia							
Alexandria	1988	MB	878	ELE	ESP	Ogden Martin	Ogden Martin
Fairfax Co.	1990	MB	2700	ELE	SDA,FF	Ogden Martin	Ogden Martin
Fort Eustis	1981	MOD	36	STM	None	U.S. Army	U.S. Army
Galax	1986	MOD	50	STM	FF	Public	N/A
Hampton	1980	MB	180	STM	ESP	NASA/Public	Public
Harrisonburg	1982	MOD	90	STM	ESP	Public	Public
Norfolk	1967	MB	324	STM	ESP	U.S. Navy	U.S. Navy
Portsmouth	1988	RDF	1800	COG	ESP,FF	Public	Public
Salem	1978	MOD	90	STM	None	Public	Public
Washington							
Bellingham	1985	MOD	90	ELE	Wscrub,FF	Private	Private
Skagit Co.	1988	MB	160	ELE	SDA,FF	Public	Private
Spokane County	1991	MB	720	ELE	SDA/FF	Public	Wheelabrator
Tacoma	1979	RDF	450	ELE	Dscrub,FF	Public &Utility	Public & Utility
Wisconsin							
Barron Co.	1986	MOD	72	COG	ESP	Public	Consumat
LaCrosse Co.	1988	RDF	360	ELE	Gravel bed	Public & Northern States Power	Public & Northern States Power
St. Croix Co.	1988	MOD	104	COG	Scrub,FF	Private	Private

tons/d = 1.102 × tonnes/d

[a] MB = mass burn; MOD = modular combustor; RDF = refuse-derived fuel.

[b] COG = cogeneration of steam and electricity; ELE = electricity generated from steam ; STM = steam generation only.

[c] Dscrub = dry scrubber; ESP = electrostatic precipitator; FF = fabric filter (baghouse); Scrub = scrubber; SDA = spray dry absorber; Wscrub = wet scrubber.

Source: Kiser, J.V.L., The 1992 municipal waste combustion guide, *Waste Age*, 23(11), 99–117, Nov., 1992. Copyright 1992 by National Solid Waste Management Association. Reprinted with permission.

Appendix H

PROXIMATE AND ELEMENTAL (ULTIMATE) ANALYSES AND HEATING VALUES OF REFUSE COMPONENTS

Refuse Component	Proximate Analysis (as-received) weight %				Elemental Analysis (dry) weight %						Higher Heating Value (kJ/kg)		
	Moisture	Volatile Matter	Fixed Carbon	Non-Comb.	C	H	O	N	S	Non-Comb.	As-Received	Dry	Moisture and Ash Free
Paper and paper products													
Paper, mixed	10.24	75.94	8.44	5.38	43.41	5.82	44.32	0.25	0.20	6.00	15800	17610	18730
Newsprint	5.97	81.12	11.48	1.43	49.14	6.10	43.03	0.05	0.16	1.52	18540	19720	20000
Brown paper	5.83	83.92	9.24	1.01	44.90	6.08	47.84	0.00	0.11	1.07	16870	17920	18140
Trade magazine	4.11	66.39	7.03	22.47	32.91	4.95	38.55	0.07	0.09	23.43	12220	12740	16630
Corrugated boxes	5.20	77.47	12.27	5.06	43.73	5.70	44.93	0.09	0.21	5.34	16380	17280	18260
Plastic-coated paper	4.71	84.20	8.45	2.64	45.30	6.17	45.50	0.18	0.08	2.77	17070	17910	18470
Waxed milk cartons	3.45	90.92	4.46	1.17	59.18	9.25	30.13	0.12	0.10	1.22	26340	27280	27650
Paper food cartons	6.11	75.59	11.80	6.50	44.74	6.10	41.92	0.15	0.16	6.93	16880	17980	19190
Junk mail	4.56	73.32	9.03	13.09	37.87	5.41	42.74	0.17	0.09	13.72	14160	14830	17210
Food and food waste													
Vegetable food waste	78.29	17.10	3.55	1.06	49.06	6.62	37.55	1.68	0.20	4.89	4174	19230	20230
Citrus rinds and seeds	78.70	16.55	4.01	0.74	47.96	5.68	41.67	1.11	0.12	3.46	3970	18640	19300
Meat scraps (cooked)	38.74	56.34	1.81	3.11	59.59	9.47	24.65	1.02	0.19	5.08	17730	28940	30490
Fried fats	0.00	97.64	2.36	0.00	73.14	11.54	14.82	0.43	0.07	0.00	38290	38290	38290
Mixed garbage I	72.00	20.26	3.26	4.48	44.99	6.43	28.76	3.30	0.52	16.00	5512	19730	23490
Mixed garbage II	—	—	—	—	41.72	5.75	27.62	2.97	0.25	21.87	—	16850	21540

Trees, wood, brush, plants

Green logs	50.00	42.25	7.25	0.50	50.12	6.40	42.26	0.14	0.08	1.00	4890	9780	9880
Rotten timbers	26.80	55.01	16.13	2.06	52.3	5.5	39.0	0.2	1.2	2.8	10950	14810	15260
Demolition softwood	7.70	77.62	13.93	0.75	51.0	6.2	41.8	0.1	<0.1	0.8	16980	1810	18590
Waste hardwood	12.00	75.05	12.41	0.54	49.4	6.1	43.7	0.1	<0.1	0.6	14950	16980	17070
Furniture wood	6.00	80.92	11.74	1.34	49.7	6.1	42.6	0.1	<0.1	1.4	17090	18170	18470
Evergreen shrubs	69.00	25.18	5.01	0.81	48.51	6.54	40.44	1.71	0.19	2.61	6298	20310	20840
Balsam spruce	74.35	20.70	4.13	0.82	53.30	6.66	35.17	1.49	0.20	3.18	5691	22190	22910
Flowering plants	53.94	35.64	8.08	2.34	46.65	6.61	40.18	1.21	0.26	5.09	8598	18670	19670
Lawn grass I	75.24	18.64	4.50	1.62	46.18	5.96	36.43	4.46	0.42	6.55	4786	19330	20700
Lawn grass II	65.00	—	—	2.37	43.33	6.04	41.68	2.15	0.05	6.75	6256	17890	19190
Ripe leaves I	9.97	66.92	19.29	3.82	52.15	6.11	30.34	6.99	0.16	4.25	18570	20630	21560
Ripe leaves II	50.00	—	—	4.10	40.50	5.95	45.10	0.20	0.05	8.20	8221	16440	17910
Wood and bark	20.00	67.89	11.31	0.80	50.46	5.97	42.37	0.15	0.05	1.00	16050	20030	20230
Brush	40.00	—	—	5.00	42.52	5.90	41.20	2.00	0.05	8.33	11040	18370	20000
Mixed greens	62.00	26.74	6.32	4.94	40.31	5.64	39.00	2.00	0.05	13.00	6256	16460	18920
Grass, dirt, leaves	21-62	—	—	—	36.20	4.75	26.61	2.10	0.26	30.08	—	14610	20910

Domestic wastes

Upholstery	6.9	75.96	14.52	2.62	47.1	6.1	43.6	0.3	<0.1	2.8	16190	17390	17880
Tires	1.02	64.92	27.51	6.55	79.1	6.8	5.9	0.1	1.5	6.6	32090	32340	34650
Leather	10.00	68.46	12.49	9.10	60.00	8.00	11.50	10.00	0.40	10.10	18510	20580	22910
Leather shoe	7.46	57.12	14.26	21.16	42.01	5.32	22.83	5.98	1.00	22.86	16840	18200	23610
Shoe heel and sole	1.15	67.03	2.08	29.74	53.22	7.09	7.76	0.50	1.34	30.09	25350	25640	36720
Rubber	1.20	83.98	4.94	9.88	77.65	10.35	—	—	2.00	10.00	26050	26350	29300
Mixed plastics	2.0	—	—	10.00	60.00	7.20	22.60	—	—	10.20	32790	33410	37210
Plastic film	3-20	—	—	—	67.21	9.72	15.82	0.46	0.07	6.72	—	32200	34580
Polyethylene	0.20	98.54	0.07	1.19	84.54	14.18	0.00	0.06	0.03	1.19	43460	45880	46510
Polystyrene	0.20	98.67	0.68	0.45	87.10	8.45	3.96	0.21	0.02	0.45	38180	38260	38400

Refuse Component	Proximate Analysis (as-received) weight %				Elemental Analysis (dry) weight %						Higher Heating Value (kJ/kg)		
	Moisture	Volatile Matter	Fixed Carbon	Non-Comb.	C	H	O	N	S	Non-Comb.	As-Received	Dry	Moisture and Ash Free
Polyurethane (a)	0.20	87.12	8.30	4.38	63.27	6.26	17.65	5.99	0.02	4.38	26050	26110	27280
Polyvinyl chloride (b)	0.20	86.89	10.85	2.06	45.14	5.61	1.56	0.08	0.14	2.06	22680	22730	23260
Linoleum	2.10	64.50	6.60	26.80	48.06	5.34	18.70	0.10	0.40	27.40	18950	19330	26630
Rags	10.00	84.34	3.46	2.20	55.00	6.60	31.20	4.62	0.13	2.45	16050	17800	18240
Textiles	15–31	—	—	—	46.19	6.41	41.85	2.18	0.20	3.17	—	18690	19300
Oils, paints	0	—	—	16.30	66.85	9.65	5.20	2.00	—	16.30	31160	31160	37210
Vacuum cleaner dirt	5.47	55.68	8.51	30.34	35.69	4.73	20.08	6.26	1.15	32.09	14850	15710	23160
Household dirt	3.20	20.54	6.26	70.00	20.62	2.87	4.00	0.50	0.01	72.30	8535	8810	31740
Other municipal wastes													
Street sweepings	20.00	54.00	6.00	20.00	34.70	4.76	35.20	0.14	0.20	25.00	11160	13950	18510
Ashes	10.00	2.68	24.12	63.2	28.0	0.5	0.8	—	0.5	70.2	8740	9700	32560

Conversions: kJ/kg × 0.4300 = BTU/lb
kJ/kg × 0.2389 = kcal/kg

[a] Remaining 2.42% is chlorine.
[b] Remaining 45.41% is chlorine.

Source: Niessen, W.R., in *Handbook of Solid Waste Management*, Wilson, D.G., ed., Van Nostrand Reinhold Co., New York, NY, 1977. Reprinted with permission.

Appendix I

USEPA ANALYTICAL METHODS
FOR CHEMICAL PARAMETERS AT LANDFILLS

Source: U.S. Environmental Protection Agency, Test Methods for Evaluating Solid Waste, Report SW-846, 1986.

METALLIC ANALYTES

Sample Preparation Methods

Method 3005: Acid Digestion of Waters for Total Recoverable or Dissolved Metals for Analysis by Flame Atomic Absorption Spectroscopy or Inductively Coupled Plasma Spectroscopy

Method 3010: Acid Digestion of Aqueous Samples and Extracts for Total Metals for Analysis by Flame Atomic Absorption Spectroscopy or Inductively Coupled Plasma Spectroscopy

Method 3020: Acid Digestion of Aqueous Samples and Extracts for Total Metals for Analysis by Furnace Atomic Absorption Spectroscopy

Method 3040: Dissolution Procedure for Oils, Greases, or Waxes

Method 3050: Acid Digestion of Sediments, Sludges, and Soils

Methods for Determination of Metals

Method 6010: Inductively Coupled Plasma Atomic Emission Spectroscopy

Method 7000: Atomic Absorption Methods

Method 7020: Aluminum (AA, Direct Aspiration)

Method 7040: Antimony (AA, Direct Aspiration)

Method 7041: Antimony (AA, Furnace Technique)

Method 7060:	Arsenic (AA, Furnace Technique)
Method 7061:	Arsenic (AA, Gaseous Hydride)
Method 7080:	Barium (AA, Direct Aspiration)
Method 7081:	Barium (AA, Furnace Technique)
Method 7090:	Beryllium (AA, Direct Aspiration)
Method 7091:	Beryllium (AA, Furnace Technique)
Method 7130:	Cadmium (AA, Direct Aspiration)
Method 7131:	Cadmium (AA, Furnace Technique)
Method 7140:	Calcium (AA, Direct Aspiration)
Method 7190:	Chromium (AA, Direct Aspiration)
Method 7191:	Chromium (AA, Furnace Technique)
Method 7195:	Chromium, Hexavalent (Coprecipitation)
Method 7196:	Chromium, Hexavalent (Colorimetric)
Method 7197:	Chromium, Hexavalent (Chelation/Extraction)
Method 7198:	Chromium, Hexavalent (Differential Pulse Polarography)
Method 7200:	Cobalt (AA, Direct Aspiration)
Method 7201:	Cobalt (AA, Furnace Technique)
Method 7210:	Copper (AA, Direct Aspiration)
Method 7211:	Copper (AA, Furnace Technique)
Method 7380:	Iron (AA, Direct Aspiration)
Method 7381:	Iron (AA, Furnace Technique)
Method 7420:	Lead (AA, Direct Aspiration)
Method 7421:	Lead (AA, Furnace Technique)
Method 7430:	Lithium (AA, Direct Aspiration)
Method 7450:	Magnesium (AA, Direct Aspiration)
Method 7460:	Manganese (AA, Direct Aspiration)
Method 7461:	Manganese (AA, Furnace Technique)
Method 7470:	Mercury in Liquid Waste (Manual Cold-Vapor Technique
Method 7411:	Mercury in Solid or Semisolid Waste (Manual Cold-Vapor Technique)
Method 7480:	Molybdenum (AA, Direct Aspiration)
Method 7481:	Molybdenum (AA, Furnace Technique)
Method 7520:	Nickel (AA, Direct Aspiration)
Method 7550:	Osmium (AA, Direct Aspiration)
Method 7610:	Potassium (AA, Direct Aspiration)
Method 7740:	Selenium (AA, Furnace Technique)
Method 7741:	Selenium (Gaseous Hydride Method)

Method 7760:   Silver (AA, Direct Aspiration)

Method 7761:   Silver (AA, Furnace Technique)

Method 7770:   Sodium (AA, Direct Aspiration)

Method 7180:   Strontium (AA, Direct Aspiration)

Method 7840:   Thallium (AA, Direct Aspiration)

Method 7841:   Thallium (AA, Furnace Technique)

Method 7870:   Tin (AA, Direct Aspiration)

Method 7910:   Vanadium (AA, Direct Aspiration)

Method 7911:   Vanadium (AA, Furnace Technique)

Method 7950:   Zinc (AA, Direct Aspiration)

Method 7951:   Zinc (AA, Furnace Technique)

ORGANIC ANALYTES

Sample Preparation Methods

Extractions and Preparations

Method 3500:   Organic Extraction and Sample Preparation

Method 3510:   Separatory Funnel Liquid-Liquid Extraction

Method 3520:   Continuous Liquid-Liquid Extraction

Method 3540:   Soxhlet Extraction

Method 3550:   Sonication Extraction

Method 3580:   Waste Dilution

Method 5030:   Purge-and-Trap

Method 5040:   Protocol for Analysis of Sorbent Cartridges from Volatile Organic Sampling Train

Cleanup

Method 3600:   Cleanup

Method 3610:   Alumina Column Cleanup

Method 3611:   Alumina Column Cleanup and Separation of Petroleum Wastes

Method 3620:   Florisil Column Cleanup

Method 3630:   Silica Gel Cleanup

Method 3640:   Gel-Permeation Cleanup

Method 3650:   Acid-Base Partition Cleanup

Method 3660:   Sulfur Cleanup

Determination of Organic Analytes

Gas Chromatographic Methods

Method 8000: Gas Chromatography

Method 8010: Halogenated Volatile Organics

Method 8011: 1,2-Dibromoethane and 1,2-Dibromo-3-chloropropane in Water by Microextraction and Gas Chromatography

Method 8015: Nonhalogenated Volatile Organics

Method 8020: Aromatic Volatile Organics

Method 8021: Volatile Organic Compounds in Water by Purge-and-Trap Capillary Column Gas Chromatography with Photoionization and Electrolytic Conductivity Detectors in Series

Method 8030: Acrolein, Acrylonitrile, Acetonitrile

Method 8040: Phenols

Method 8060: Phthalate Esters

Method 8070: Nitrosamines

Method 8080: Organochlorine Pesticides and PCBs

Method 8090: Nitroaromatics and Cyclic Ketones

Method 8100: Polynuclear Aromatic Hydrocarbons

Method 8110: Haloethers

Method 8120: Chlorinated Hydrocarbons

Method 8140: Organophosphorus Pesticides

Method 8141: Organophosphorus Pesticides: Capillary Column

Method 8150: Chlorinated Herbicides

Gas Chromatographic/Mass Spectroscopic Methods

Method 8240: Gas Chromatography/Mass Spectrometry for Volatile Organics

Method 8250: Gas Chromatography/Mass Spectrometry for Semivolatile Organics: Packed Column Technique

Method 8260: Gas Chromatography/Mass Spectrometry for Volatile Organics: Capillary Column Technique

Method 8270: Gas Chromatography/Mass Spectrometry for Semivolatile Organics: Capillary Column Technique

Method 8280: The Analysis of Polychlorinated Dibenzo-p-Dioxins and Polychlorinated Dibenzofurans

High Performance Liquid Chromatographic Methods

Method 8310: Polynuclear Aromatic Hydrocarbons

MISCELLANEOUS SCREENING METHODS

Method 3820:   Headspace

Method 3820:   Hexadecane Extraction and Screening of Purgeable Organics

MISCELLANEOUS TEST METHODS

Method 9010:   Total and Amenable Cyanide (Colorimetric, Manual)

Method 9012:   Total and Amenable Cyanide (Colorimetric, Automated UV)

Method 9020:   Total Organic Halides (TOX)

Method 9021:   Purgeable Organic Halides (POX)

Method 9022:   Total Organic Halides (TOX) by Neutron Activation Analysis

Method 9030:   Acid-Soluble and Acid-Insoluble Sulfides

Method 9031:   Extractable Sulfides

Method 9035:   Sulfate (Colorimetric, Automated, Chloranilate)

Method 9036:   Sulfate (Colorimetric, Automated, Methylthymol Blue, AA II)

Method 9038:   Sulfate (Turbidimetric)

Method 9060:   Total Organic Carbon

Method 9065:   Phenolics (Spectrophotometric, Manual 4-AAP with Distillation)

Method 9066:   Phenolics (Colorimetric, Automated 4-AAP with Distillation)

Method 9067:   Phenolics (Spectrophotometric, MBTH with Distillation)

Method 9070:   Total Recoverable Oil & Grease (Gravimetric, Separatory Funnel Extraction)

Method 9071:   Oil & Grease Extraction Method for Sludge Samples

Method 9131:   Total Coliform: Multiple Tube Fermentation Technique

Method 9132:   Total Coliform: Membrane Filter Technique

Method 9200:   Nitrate

Method 9250:   Chloride (Colorimetric, Automated Ferricyanide AAI)

Method 9251:   Chloride (Colorimetric, Automated Ferricyanide AAII)

Method 9252:   Chloride (Titrimetric, Mercuric Nitrate)

Method 9320:   Radium-228

PROPERTIES

Method 1320:   Multiple Extraction Procedure

Method 1330:   Extraction Procedure for Oily Wastes

Method 9040:   pH Electrometric Measurement

Method 9041:   pH Paper Method

Method 9045:   Soil pH

Method 9050:   Specific Conductance

Method 9080:   Cation-Exchange Capacity of Soils (Ammonium Acetate)

Method 9081:   Cation-Exchange Capacity of Soils (Sodium Acetate)

Method 9090:   Compatibility Test for Wastes and Membrane Liners

Method 9095:   Paint Filter Liquids Test

Method 9100:   Saturated Hydraulic Conductivity, Saturated Leachate Conductivity, and Intrinsic Permeability

Method 9310:   Gross Alpha & Gross Beta

Method 9315:   Alpha-Emitting Radium Isotopes

HAZARDOUS CHARACTERISTICS

Ignitability

Method 1010:   Pensky-Martens Closed-Cup Method for Determining Ignitability

Method 1020:   Setaflash Closed-Cup Method for Determining Ignitability

Corrosivity

Method 1110:   Corrosivity Toward Steel

Toxicity

Method 1310:   Extraction Procedure (EP) Toxicity Test Method and Structural Integrity Test

Method 1311:   Toxicity Characteristic Leaching Procedure (TCLP)

Appendix J

INFORMATION SOURCES

INDUSTRY AND TRADE ORGANIZATIONS

Aluminum Association
900 19th Street, N.W.
Washington, DC 20006
(202) 862–5100

Aluminum Recycling Association
1000 16th Street, N.W., Suite 603
Washington, DC 20036
(202) 785–0951

American Forest and Paper Association
1111 19th Street, N.W., Suite 800
Washington, DC 20036
(800) 878–8878

American Iron and Steel Institute
1101 17th Street, N.W., 13th Floor
Washington, DC 20036
(202) 452–7100

American Newspaper Publishers
    Association
11600 Sunrise Valley Drive
Reston, VA 22091
(703) 648–1000

American Plastics Council
1275 K Street, N.W., Suite 400
Washington, DC 20005
(800) 243–5790
(202) 371–5320

American Pulpwood Association, Inc.
1025 Vermont Avenue, N.W., Suite 1020
Washington, DC 20005
(202) 347–2900

American Retreaders Association, Inc.
P.O. Box 17203
Louisville, KY 40217
(800) 426–8835
(502) 367–9133

Aseptic Packaging Council
1000 Potomac Street, N.W., Suite 401
Washington, DC 20007
(202) 785–4020

Association of Petroleum Re–Refiners
P.O. Box 427
Ellicott Station
Buffalo, NY 14205
(716) 855–2212

Automotive Dismantlers & Recyclers
    Association
10400 Eaton Place, Suite 203
Fairfax, VA 22030
(703) 385–1001

Battery Council International
111 East Wacker Drive
Chicago, IL 60601
(312) 644–6610

Brass and Bronze Ingot Manufacturers
300 W. Washington, Suite 1500
Chicago, IL 60606
(312) 236–2715

Can Manufacturers Institute
1625 Massachusetts Avenue, N.W.
Washington, DC 20036
(202) 232–4677

Canadian Association of Recycling
     Industries
415 Rue Yonge Street, Suite 1620
Toronto, Ontario M5B 2E7, Canada
(416) 595–5552

Composting Council
114 S. Pitt Street
Alexandria, VA 22314
(703) 739–2401

Council on Packaging in the
     Environment
1001 Connecticut Avenue, N.W.,
     Suite 401
Washington, DC 20036
(202) 331–0099

Environmental Industry Associations
4301 Connecticut Avenue, N.W.,
     Suite 300
Washington, DC 20008
(202) 244–4700

Foodservice and Packaging Institute
1025 Connecticut Avenue, N.W.,
     Suite 513
Washington, DC 20036
(202) 822–6420

Glass Packaging Institute
1801 K Street, N.W., Suite 1105L
Washington, DC 20006
(202) 887–4850

Institute of Resource Recovery
1730 Rhode Island Avenue, N.W.,
     Suite 1000
Washington, DC 20036
(202) 659–4613

Institute of Scrap Recycling Industries,
     Inc.
1627 K Street, N.W., Suite 700
Washington, DC 20006
(202) 466–4050
FAX: (202) 775–9109

International Association of Wiping Cloth
     Manufacturers
7910 Woodmont, Suite 1212
Bethesda, MD 20814
(301) 656–1077

National Association of Chemical
     Recyclers
1333 New Hampshire Avenue, N.W.,
     Suite 1100
Washington, DC 20036
(202) 463–6956

National Association for Plastic
     Container Recovery
3770 Nations Bank Corporate Center
100 N. Tryon Street
Charlotte, NC 28202
(704) 358–8882
(800) 7NAPCOR

National Oil Recyclers Association
2777 Broadway Avenue
Cleveland, OH 44106
(216) 623–8383
FAX: (216) 623–8393

National Paper Trade Association, Inc.
111 Great Neck Road
Great Neck, NY 11021
(516) 829–3070

National Polystyrene Recycling Council
P.O. Box 66495
Washington. 20035
(202) 296–1954

National Recycling Coalition
1101 30th Street, N.W., Suite 305
Washington, DC 20007
(202) 625–6406
(able to provide state recycling
    organization contacts)

National Resource Recovery Association
1620 I Street, N.W.
Washington, DC 20006
(202) 293–7330

National Soft Drink Association
Solid Waste Management Dept.
1101 16th Street, N.W.
Washington, DC 20036
(202) 463–6700

National Solid Waste Institute
10928 North 56th Street
Tampa, FL 33617
(813) 995–3208

National Tire Dealers & Retreaders
    Association
1250 I Street, N.W., Suite 400
Washington, DC 20005
(202) 789–2300

Non–Ferrous Founders' Society, Inc.
455 State Street, Suite 100
Des Plaines, IL 60016
(312) 299–0950

The Paper Bag Institute Inc.
505 White Plains Road
Tarrytown, NY 10594
(914) 631–0696

Plastics Recycling Foundation
1275 K Street, N.W., Suite 400
Washington, DC 20005
(202) 371–5200

Polystyrene Packaging Council Inc.
1025 Connecticut Avenue, N.W., Suite 515
Washington, DC 20036
(202) 822–6424

Pulp & Paper Manufacturers Association
2000 S. Memorial Drive
Appleton, WI 54915
(414) 734–5778

Rubber Manufacturers Association
1400 K Street, N.W.
Washington, DC 20005
(202) 682–4800

Scrap Tire Management Council
1400 K Street, N.W.
Washington, DC 20005
(201) 408–7781
FAX: (202) 682–4854

The Silver Institute
1026 16th Street, N.W., Suite 101
Washington, DC 20036
(202) 783–0500

Society of the Plastics Industry, Inc.
1275 K Street, N.W., Suite 400
Washington, DC 20005
(202) 371–5200

Steel Manufacturers Association
815 Connecticut Avenue, N.W., Suite
    304
Washington, DC 20006
(202) 331–7027
FAX: (202) 331–7675

Steel Recycling Institute
Foster Plaza 10
680 Andersen Drive
Pittsburgh, PA 15220
(800) 876–7274

TAPPI (Technical Association of the Pulp
    and Paper Industry)
Technology Park
P.O. Box 105113
Atlanta, GA 30348
(404) 446–1400

Textile Fibers & By–Products
    Association
4108 Park Road, Suite 202
P.O. Box 11065
Charlotte, NC 28220
(704) 527–5593

The Vinyl Institute
155 Route 46 West
Wayne, NJ 07470
(201) 890–9299

PUBLIC INTEREST GROUPS

Center for the Biology of Natural Systems
Queens College
Flushing, NY 11367
(718) 670–4180

Citizen's Clearinghouse for Hazardous
 Waste
P.O. Box 926
Arlington, VA 22216
(703) 276–7070

Concern, Inc.
1794 Columbia Road, N.W.
Washington, DC 20009
(202) 328–8160

Environmental Action
1525 New Hampshire Avenue, N.W.
Washington, DC 20036
(202) 745–4870

Environmental Defense Fund
257 Park Avenue, South
New York, NY 10010
(212) 505–2100

Environmental Institute for International
 Research
331 Madison Avenue, 6th Floor
New York, NY 10017
(212) 883–1770

Environmental Policy Institute
218 D Street, S.E.
Washington, DC 20003
(202)544–2600

Greenpeace
1436 U Street, N.W.
Washington, DC 20009
(202) 462–1177

INFORM
381 Park Avenue, South
New York, NY 10016
(212) 689–4040

Institute for Local Self–Reliance
2425 18th Street, N.W.
Washington, DC 20009
(202) 232–4108

Institute of Resource Recovery
1730 Rhode Island Avenue, N.W., Suite
 1000
Washington, DC 20036
(202) 659–4613

Keep America Beautiful, Inc.
Mill River Plaza
9 West Broad Street
Stamford, CT 06902
(203) 323–8987

League of Women Voters
1730 M Street, N.W.
Washington, DC 20036
(202) 429–1965

National Wildlife Federation
1400 16th Street, N.W.
Washington, DC 20036–6800
(202) 797–680

Natural Resources Defense Council
40 W. 20th Street
New York, NY 10011
(212) 727–2700

Renew America
1400 16th Street, N.W., Suite 710
Washington, DC 20036
(202) 232–2252

Sierra Club
730 Polk Street
San Francisco, CA 94109
(415) 776–2211

U.S. Public Interest Research Group
215 Pennsylvania Avenue, S.E.
Washington, DC 20003
(202) 546–9707
(able to provide PIRG state affiliate
 organizations)

Worldwatch Institute
1776 Massachusetts Avenue, N.W.
Washington, DC 20036
(202) 452–1999

GOVERNMENT/PUBLIC ORGANIZATIONS

Government–Affiliated Groups

Association of State and Territorial Solid
    Waste Management Officials
444 N. Capitol Street, Suite 388
Washington, DC 20001
(202) 624–5828

Council of State Governments
Iron Works Pike
P.O. Box 11910
Lexington, KY 40578–1910
(606) 231–1866

International City Managers Association
777 N. Capital Street, N.E., Suite 500
Washington, DC 20002
(202) 289–4262

National Appropriate Technology
    Assistance Service (NATAS)
U.S. Department of Energy
P.O. Box 2525
Butte, MT 59702–2525
(800) 428–2525

National Association of Counties
440 First Street, N.W., 8th Floor
Washington, DC 20001
(202) 393–6226

National Conference of State
    Legislatures
1050 Seventeenth Street, Suite 2100
Denver, CO 80265
(303) 623–7800

National League of Cities
1301 Pennsylvania Avenue, N.W.
Washington, DC 20004
(202) 626–3000

RCRA/Superfund Hotline
(800) 424–9346
(703) 920–9810

Solid Waste Association of North
    America (SWANA)
P.O. Box 6126
Silver Spring, MD 20916
(301) 585–2898

Solid Waste Information Clearinghouse
P.O. Box 7219
8750 Georgia Avenue, Suite 140
Silver Spring, MD 20910
(301) 67–SWICH

U.S. Conference of Mayors
Institute for Resource Recovery
1620 I Street, N.W., 4th Floor
Washington, DC 20006
(202) 293–7330

U.S. Department of Agriculture
Agricultural Research Service
Soil Microbial Systems Lab
Building 318, Barc–E
Beltsville, MD 200705
(301) 344–3327

Federal Agencies

U.S. Environmental Protection Agency
Municipal and Industrial Solid Waste
    Division
401 M Street, S.W.
Washington, DC 20460
(202) 382–6261

U.S. EPA, Region 1
(CT, ME, MA, NH, RI, VT)
Research Library for Solid Waste
JFK Federal Building
Boston, MA 02203
(617) 573–9687

U.S. EPA, Region 2
(NJ, NY)
Air and Waste Management Division
26 Federal Plaza
New York, NY 10278
(212) 264–3384

U.S. EPA, Region 3
(DE, DC, MD, PA, VA, WV)
Waste Management Division
841 Chestnut Street
Philadelphia, PA 19107
(215) 597–0982

U.S. EPA, Region 4
(AL, FL, GA, KY, MS, NC, SC, TN)
RCRA & Federal Facilities Branch
345 Courtland Street, N.E.
Atlanta, GA 30365
(404) 347–3016

U.S. EPA, Region 5
(IL, IN, MI, MN, OH, WI)
RCRA Permitting Branch
77 Jackson Blvd. (HRP–8J)
Chicago, IL 60604
(312) 886–7452

U.S. EPA, Region 6
(AR, LA, NM, OK, TX)
RCRA Programs Branch
1445 Ross Avenue (6H–HW)
Dallas, TX 75202–2733
(214) 655–6760

U.S. EPA, Region 7
(IA, KS, MO, NE)
State Programs Section
726 Minnesota Avenue (STPG)
Kansas City, KS 66101
(913) 551–7055

U.S. EPA, Region 8
(CO, MT, ND, SD, UT, WY)
Waste Management Branch
999 18th Street, Suite 500 (8HWM–WM)
Denver, CO 80202–2466
(303) 293–1845

U.S. EPA, Region 9
(AZ, CA, HI, NV, AS, GU)
Hazardous Waste Management Division
75 Hawthorne (H–3–1)
San Francisco, CA 94105
(415) 744–1500

U.S. EPA, Region 10
(AK, ID, OR, WA)
Hazardous Waste Division
1200 Sixth Avenue
Seattle, WA 98101
(206) 553–2857

State Agencies

Alabama Department of Environmental
   Management
Division of Solid & Hazardous Waste
1751 Congressman Dickinson Drive
Montgomery, AL 36130
(217) 271–7700

Alaska Department of Environmental
   Conservation
P.O. Box 0
Juneau, AK 99801–1795
(907) 465–5060

Arizona Department of Commerce
Energy Office
3800 N. Central Avenue, Suite 1200
Phoenix, AZ 85102
(602) 280–1300

Arkansas Department of Pollution
   Control & Ecology
State Marketing Board for Recyclables
P.O. Box 8913
Little Rock, AR 72219–8913
(501) 562–6533

California Integrated Waste
   Management Board
8800 Cal Center Drive
Sacramento, CA 95826–3268
(916) 255–2200

Colorado Department of Health
Hazardous Material & Waste
   Management Division
4300 Cherry Creek Drive, S.
Denver, CO 80222–1530
(303) 692–2000

Connecticut Department of
   Environmental Protection
Division of Planning & Standards
165 Capitol Avenue
Hartford, CT 06106
(203) 566–8722

Delaware Department of Natural
   Resources & Environmental Control
Solid Waste Management
P.O. Box 1401
Dover, DE 19903
(302) 739–3822

District of Columbia Department of
    Public Works
2000 14th Street, N.W., 6th Floor
Washington, DC 20009
(202) 939–8000

Florida Department of Environmental
    Regulation
Division of Waste Management/Bureau
    of Solid & Hazardous Waste
2600 Blairstone Road
Tallahassee, FL 32399–2400
(904) 488–0300

Georgia Department of Natural
    Resources
Environmental Protection Division
4244 International Parkway, #100
Atlanta, GA 30354
(404) 362–2692

Hawaii Department of Health
Department of Solid Waste Management
P.O. Box 3378
Honolulu, HI 96801
(808) 543–8227

Idaho Department of Health and Welfare
Environmental Quality/Solid Waste
1410 N. Hilton Street
Boise, ID 83706
(208) 334–5880

Illinois Environmental Protection Agency
Disposal Alternatives Unit
P.O. Box 19276
Springfield, IL 62794–9276
(217) 782–6762

Illinois Office of Waste Recycling &
    Waste Reduction
325 W. Adams Street
Springfield, IL 62704
(217) 524–5454

Indiana Department of Environmental
    Management
P.O. Box 6015
Indianapolis, IN 46206–6015
(317) 232–8172

Iowa Department of Natural Resources
Waste Management Authority Division
900 E. Grand Avenue, Wallace Building
Des Moines, IA 50319
(515) 281–8975

Kansas Department of Health &
    Environment
Solid Waste Section
Building 740, Forbes Field
Topeka, KS 66620
(913) 296–1590

Kentucky Division of Waste
    Management
Resource Conservation & Local
    Assistance
14 Reilly Road
Frankfort, KY 40601
(502) 564–6716

Louisiana Department of Environmental
    Quality
P.O. Box 82263
Baton Rouge, LA 70884–2263
(504) 765–0741

Louisiana Department of Solid &
    Hazardous Waste
P.O. Box 82178
Baton Rouge, LA 70884–2178
(504) 765–0249

Maine Waste Management Agency
State House, Station #154
Augusta, ME 04333
(207) 289–5300

Maryland Department of the
    Environment
Office of Waste Minimization
2500 Broening Highway
Baltimore, MD 21224
(301) 631–3315

Massachusetts Department of
    Environmental Protection
Division of Solid Waste Management
1 Winter Street, 4th Floor
Boston, MA 02108
(617) 292–5960

Michigan Department of Natural
   Resources
Waste Management Division
P.O. Box 30038
Lansing, MI 48909
(517) 373–2730

Minnesota Office of Waste Management
1350 Energy Lane, Suite 201
St. Paul, MN 55108
(612) 649–5750

Mississippi Department of
   Environmental Quality
Bureau of Pollution Control/Solid Waste
P.O. Box 10385
Jackson, MS 39289–0385
(601) 961–5171

Missouri Department of Natural
   Resources
Solid Waste Management Program
P.O. Box 176
Jefferson City, MO 65102–0176
(314) 751–5401

Montana Department of Health &
   Environmental Sciences
Solid & Hazardous Waste Bureau
836 Front Street
Helena, MT 59601
(406) 444–2821

Nebraska Department of Environmental
   Quality
Litter Reduction & Recycling Program
P.O. Box 98922, State House Station
Lincoln, NE 68509–8922
(402) 471–2186

Nevada Department of Environmental
   Protection
Bureau of Waste Management
333 W. Nye, Capital Complex
Carson City, NV 89710
(702) 687–5872

New Hampshire Department of
   Environmental Services
Waste Management Division
6 Hazen Drive
Concord, NH 03301–2925
(603) 271–2925

New Jersey Bureau of Source Reduction
   & Market Development
CN414, 840 Bear Tavern Road
Trenton, NJ 08625–0414
(609) 530–8000

New Mexico Environmental Department
P.O. Box 26110
Santa Fe, NM 87502
(505) 827–2850

New York Department of Environmental
   Conservation
Bureau of Municipal Waste
50 Wolf Road, Room 230
Albany, NY 12233–4013
(518) 457–2051

New York Department of Environmental
   Conservation
Bureau of Waste Reduction & Recycling
50 Wolf Road, Room 200
Albany, NY 12233–4015
(518) 457–7337

North Carolina Department of
   Environmental Health & Natural
   Resources
Division of Solid Waste Management
P.O. Box 27687
Raleigh, NC 27611–7687
(919) 733–2178

North Dakota Department of Health &
   Consolidated Laboratories
Division of Waste Management
P.O. Box 5520
Bismarck, ND 58502–5520
(701) 221–5166

Ohio Department of Natural Resources
Division of Litter Prevention & Recycling
1889 Fountain Square, Building F–2
Columbus, OH 43224
(614) 265–6353

Oklahoma Department of Health
Solid Waste Management Services
   (0206)
1000 – NE 10th Street
Oklahoma City, OK 73117–1299
(405) 271–7169

Oregon Department of Environmental
   Quality
Waste Reduction Section
811 – SW Sixth Avenue
Portland, OR 97204
(503) 229–5913

Pennsylvania Department of
   Environmental Resources
Division of Waste Minimization & Planning
P.O. Box 2063
Harrisburg, PA 17105–2063
(717) 787–7382

Rhode Island Department of
   Environmental Management
83 Park Street, 5th Floor
Providence, RI 02903
(401) 277–3434

South Carolina Department of Health &
   Environmental Control
Bureau of Solid & Hazardous Waste
2600 Bull Street
Columbia, SC 29201
(803) 734–5200

South Dakota Department of
   Environment & Natural Resources
Office of Waste Management
319 S. Coteau, c/o 500 E. Capital
Pierre, SD 57501
(605) 773–3153

Tennessee Department of
   Environmental & Conservation
Division of Solid Waste Assistance
401 Church Street, L&C Tower,
   14th Floor
Nashville, TN 37243–0455
(615) 532–0091

Texas Department of Health
Division of Solid Waste Management
1100 – W. 49th Street
Austin, TX 78756–3199
(512) 458–7271

Utah Department of Environmental Quality
Division of Solid & Hazardous Waste
P.O. Box 144880
Salt Lake City, UT 84114–4880
(801) 538–6170

Vermont Department of Environmental
   Conservation
Division of Solid Waste
103 S. Main Street, Laundry Building
Waterbury, VT 05671–0407
(802) 244–3831

Vermont Department of Environmental
   Conservation
Division of Hazardous Materials
103 S. Main Street
Waterbury, VT 05671–0404
(802) 244–8702

Virginia Department of Environmental
   Quality
101 – N. 14th Street, 11th Floor
Richmond, VA 23219
(804) 225–2667

Washington Department of Ecology
Waste Reduction/Recycling & Litter
P.O. Box 47600
Olympia, WA 98504–7600
(206) 438–7541

West Virginia Division of Environmental
   Protection
Solid Waste Section
1356 Hansford Street
Charleston, WV 25301
(304) 558–5993

Wisconsin Department of Natural
   Resources
Bureau of Solid & Hazardous Waste
   Management
P.O. Box 7921
Madison, WI 53707–7921
(608) 267–7566

Wyoming Department of Environmental
   Quality
Solid & Hazardous Waste Division
122 – W. 25th Street
Cheyenne, WY 82002
(307) 777–7752

WASTEWATER AND RELATED ORGANIZATIONS

American Water Resources Association
5410 Grosvenor Lane
Suite 220
Bethesda, MD 20814–2192
(301) 493–8600

Association of Metropolitan Sewerage
    Agencies
1000 Connecticut Avenue, N.W.
Suite 410
Washington, DC 20036
(202) 833–2672

Water Environment Federation
601 Wythe Street
Alexandria, VA 22314–1994
(202) 337–2500

MAGAZINES AND PERIODICALS

General Waste Management

BioCycle Magazine
The JG Press, Inc. (formerly Rodale
    Press)
419 State Avenue
Emmaus, PA 18049
(717) 967–4135

Bottle/Can Recycling Update
Resource Recycling, Inc.
P.O. Box 10540
Portland, OR 97210
(503) 227–1319

Chemical Engineering Progress
345 E. 47th Street
New York, NY 10017
(212) 705–7496

Critical Reviews in Environmental
    Control
CRC Press
2000 Corporate Blvd, N.W.
Boca Raton, FL 33431
(800) 272–7737

Environmental Science and Technology
American Chemical Society
155 16th Street, N.W.
Washington, DC 20036
(800) 333–9511

Journal of Environmental Engineering
ASCE
345 E. 47th Street
New York, NY 10017–2398
(212) 705–7496

Journal of Hazardous Materials
Elsevier Science B.V.
Molenwerg 1, P.O. Box 211
1000AE Amsterdam, Netherlands
(212) 633–3750

Materials Recycling Markets
P.O. Box 577
Ogdensburg, NY 13669
(800) 267–0707

MSW Management
Forester Communications
216 East Gutierrez
Santa Barbara, CA 93101
(805) 899–3355

Organic Gardening Magazine
The JG Press, Inc. (formerly Rodale
    Press)
Emmaus, PA 18098
(610) 967–5171

Recycling Times
Environmental Industry Associates
4301 Connecticut Avenue, Suite 300
Washington, DC 20008
(800) 424–2869

Recycling Today
GIE, Inc., Publishers
4012 Bridge Avenue
Cleveland, OH 44113–3320
(216) 961–4130

Resource Recovery and Conservation
Elsevier Science
655 Avenue of the Americas
New York, NY 10010
(212) 633–3750

Resource Recovery Report
5313 38th Street, N.W.
Washington, DC 20015
(202) 362–6034

Resource Recycling
P.O. Box 10540
Portland, OR 97210–0540
(503) 227–1319

Scrap Processing and Recycling
Institute of Scrap Recycling Industries
1325 G Street, N.W., Suite 1000
Washington, DC 20005–3104
(202) 466–4050

Solid Waste Report
Business Publishers, Inc.
951 Pershing Drive
Silver Spring, MD 20910
(301) 587–6300

Solid Waste Technologies
HCI Publications
410 Archibald Street
Kansas City, MO 64111–3046
(816) 931–1311

Waste Age
4301 Connecticut Avenue, N.W.,
    Suite 300
Washington, DC 20008
(202) 244–4700

Waste Management
Elsevier Science Inc.
660 White Plains Road
Tarrytown, NY 10591–5153
(914) 333–2400

Waste Management and Research
Academic Press
24–28 Oval Road
London NW1 7DX
United Kingdom

Water, Air and Soil Pollution Journal
Kluwer Academic Publishers Group
P.O. Box 358
Accord Station
Hingham, MA 02018–0358
(617) 871–6600

World Wastes
Argus Communication
6255 Barfield Road
Atlanta, GA 30328
(404) 955–2500

Glass–Related

Glass Industry
310 Madison Avenue
New York, NY 10017–6098
(212) 682–7681

Metal–Related

American Metal Market
Chilton Publishing
825 7th Avenue
New York, NY 10019
(212) 887–8493

Iron Age
Chilton Publishing Co.
191 S. Gary Avenue
Carol Stream, IL 60188
(708) 665–1000

Modern Metals
Delta Communications, Inc.
400 N. Michigan Avenue
Chicago, IL 60611
(312) 222–2000

Paper–Related

Fibre Market News
GIE Publishing Company
4012 Bridge Avenue
Cleveland, OH 44113–3399

Official Board Markets
131 W. First Street
Duluth, MN 55802
(218) 723–9477

Paper Stock Report
McEntee Media Corporation
13727 Holland Road
Cleveland, OH 44142–3920
(216) 362–7979

Pulp & Paper
Miller Freeman, Inc.
600 Harrison Street
San Francisco, CA 94107
(415) 905–2200

Pulp & Paper Week
Miller Freeman Inc.
600 Harrison Street
San Francisco, CA 94107
(415) 905–2200

Plastic–Related

Modern Plastics
McGraw–Hill Publications Company
1221 Avenue of the Americas
New York, NY 10020
(800) 257–9402

Plastics Engineering
Society of Plastics Engineers, Inc.
14 Fairfield Drive
Brookfield Center, CT 06805
(203) 775–0471

Plastic News
Crain Communications, Inc.
965 E. Jefferson Avenue
Detroit, MI 48207
(313) 446–6000

Plastics Recycling Update
P.O. Box 10540
Portland, OR 97210
(503) 227–1319

Plastics Technology
Bill Communications, Inc.
633 Third Avenue
New York, NY 10017
(212) 973–4901

Plastics World
Cahners Publishing Company
44 Cook Street
Denver, CO 80206
(303) 388–4511

Rubber–Related

Scrap Tire News
133 Mountain Road
Suffield, CT 06078
(203) 668–5422

Construction and Demolition Related

C+D Debris Recycling
29 North Wacker Drive
Chicago, IL 60606–3298
(312) 726–2802

Wastewater or Sludge Related

Clean Water Report
CJE Associates
237 Gretna Green Court
Alexandria VA 22304
(703) 823–0662

Pollution Engineering
P. O. Box 173377
Denver CO 80217–3377

Sludge Newsletter
951 Pershing Drive
Silver Springs, MD 20910

US Water News
230 Main Street
Halstead KA 67056
(316) 835–2222

Water Engineering and Management
380 E. Northwest Highway
Des Plaines, IL 60016–2282
(708) 298–6622

Water Environment Research
601 Wythe Street
Alexandria, VA 22314–1994
(202) 337–2500

Water Resources Bulletin
5410 Grosvenor Lane
Suite 220
Bethesda, MD 20814–2192
(301) 493–8600

Water Research
Elsevier Science, Ltd.
Bampfylde Street
Exeter, EXI
2AH ENGLAND
Tel:(0392)–51558

SOURCES

Gills, L., Who's who among environmental organizations, *Waste Age*, 20(1), 124–142, January, 1989.

1994 Directory of U.S. and Canadian Scrap Plastics Processors and Buyers, Resource Recycling, Inc., P.O. Box 10540, Portland, OR 97210, April, 1994.

Recycled Products Guide, Recycling Data Management Corp., P.O. Box 577, Ogdensburg, NY 13669, October, 1993.

Recycling Contact List, National Recycling Coalition, 1101 30th Street, N.W., Suite 305, Washington, DC 20007, 1993.

Reporting on Municipal Solid Waste: A Local Issue, The Environmental Health Center of the National Safety Council, 1019 19th Street, N.W. #401, Washington, DC 20036.

Powelson, D.R. and Powelson, M.A., *The Recycler's Manual for Business, Government, and the Environmental Community*, Van Nostrand Reinhold, New York, NY, 1992.

Index